ALSO BY ROBERT BUDERI

The Invention That Changed the World: How a Small
Group of Radar Pioneers Won the Second World War
and Launched a Technological Revolution

ENGINES OF

■

*How the World's Best Companies
Are Using Their Research Labs
to Win the Future*

TOMORROW

■

ROBERT BUDERI

A TOUCHSTONE BOOK
PUBLISHED BY SIMON & SCHUSTER
NEW YORK   LONDON   TORONTO   SYDNEY   SINGAPORE

 TOUCHSTONE
Rockefeller Center
1230 Avenue of the Americas
New York, NY 10020

10   9   8   7   6   5   4   3   2   1

The Library of Congress has cataloged the
Simon & Schuster edition as follows:

Buderi, Robert.
  Engines of tomorrow : how the world's best companies
are using their research labs to win the future / Robert
Buderi.
    p. cm.
  Includes bibliographical references and index.
  1. Research, Industrial.   2. Technological
innovations.   I. Title.
T175 .B83 2000
607'.2—dc21   99-059910

ISBN 0-684-83900-8
     0-7432-0022-5 (Pbk)

ACKNOWLEDGMENTS

John Armstrong patiently provided his perspective and sound and gracious advice from the planning stages through the final copy.

I am (again) greatly indebted to the Alfred P. Sloan Foundation. Art Singer proved critical in getting things going, and Doron Weber helped carry them to fruition. Frank Mayadas talked candidly about his IBM days. And Ralph Gomory, giving generously of his time and rich experience, really carried the day.

On the other end of the spectrum, thanks for administering the Sloan grant goes to Herb Bernstein and ISIS, The Institute for Science and Interdisciplinary Studies. This is an outstanding (and thankfully not greedy) program devoted to public understanding of science based at Hampshire College in Amherst, Massachusetts.

It is not easy to live as a writer. Special thanks to John Benditt, editor-in-chief of *Technology Review*, and Richard Brandt and Galen Gruman at *Upside*, all of whom afforded me the opportunity to develop and adapt aspects of this book in feature articles for their publications. In the case of *Upside*, portions also appeared in my monthly "Lab Watch" column. Thanks as well to my "hands-on" (and even hands-off) editors at these publications: Rebecca Zacks and Herb Brody at *TR* and Chuck Lenatti and Deborah Branscum at *Upside*. They all did much to make things better. Similar, much-appreciated assistance came from Mike Wolff at *Research Technology Management* and Jack Beatty and colleagues at *The Atlantic Monthly*. Last but not least on this front, John Carey gave willingly of his time to provide valuable additional advice about key sections of the manuscript.

All through the long course of researching and writing this book, a host of people shared their experiences, provided information, and helped me understand it. Warm thanks to all who gave interviews—and to the almost-always underappreciated and incredible support staffs around the world who made it all happen. Several researchers went above and beyond the call. Tony Tyson not only helped me comprehend his work, he led me to a better understanding of what fires innovation. Peter Price helped organize a strong and thoughtful contingent from the early days of IBM Research, including Alan Fowler and the late Rolf Landauer. In England, Jeremy Gunawardena arranged a memorable dinner discussion.

Friendly librarians, archivists, and researchers proved indispensable to combing through a wealth of data, especially about the history of corporate research. Special thanks go out to Lynn Labun at the Mellon Institute Library; Dr. Margarete Busch at Bayer archives; George Wise, formerly of GE; Joan A'Hearn from the Schenectady Museum Archives; Helmuth Trischler at the Deutsches Museum; and Diane Currens D'Almeida at M.I.T. Alex de Ravel helped with the early research. At different times, Vivien Marx, Manuela Ventu, Pino Oppo, and Ruth Buderi pitched in with German translations.

Nor can the social front be ignored. In Germany, Hartmut Runge introduced me to the Bavarian Alps. Aston Bridgeman and Bob Neff showed me the real Japan. In England, Alun and Rie Anderson proved especially generous hosts—from gardening to margarita-making. In California, it was Ottavia Bassetti; Charlie Fager and Sandy Geller; and Trip and Lisa Hawkins, who got more than they bargained for in a guest with the flu.

Thanks as well to my agent, Rafe Sagalyn, editor Alice Mayhew, and associate editor Roger Labrie.

And, as always, to my family: Kacey, Robbie, and Nancy.

To Dick Thompson

The author acknowledges with great gratitude the support of the Alfred P. Sloan Foundation.

CONTENTS

ENGINES
OF TOMORROW

CHANGE

"The value and even the mark of true science consists, in my opinion, in the useful inventions which can be derived from it."

—G. W. LEIBNIZ

"Science today is everybody's business."

—I. BERNARD COHEN

THE PROLIFIC rags-to-riches inventor Charles F. Kettering, backbone of the mighty General Motors research machine that gave the world four-wheel brakes and the refrigerant Freon, liked to call a company's research house its "change-making department." In a 1929 speech to members of the United States Chamber of Commerce, he put it this way: "I am not pleading with you to make changes. I am telling you you have got to make them—not because I say so, but because old Father Time will take care of you if you don't change. Advancing waves of other people's progress sweep over the unchanging man and wash him out. Consequently, you need to organize a department of systematic change-making."

Corporate change-making, in the form of a company's central research operation, is the focus of this book. As a pivotal force behind a plethora of key industries, research can be crucial not only to winning in the marketplace but to national economic viability. Yet serious discussion of the subject—examining the nature and evolution of successful research, and who still manages to pull it off—has been missing from virtually every popular management treatise in vogue today. This failure has become all the more glaring amidst the sweeping cutbacks in research spending that marked the early and mid-1990s. The upheaval has led to the widespread perception—among Congress, the President, leading policymakers, academics, and the press—that companies have almost unilaterally, and shortsightedly, pulled back from far-ranging and potentially revolutionary studies to concentrate on incremental improvements to existing products.

This book challenges that perception by throwing open the spotlight on corporate research—laying out its history, identifying the real issues, and delving inside the labs of some of the most noble practitioners to illuminate the various approaches to managing innovation in today's quicksilver global economy. The focus is on the rapidly blurring electronics-computer-telecommunications industries. It can be argued that especially with the rise of the Internet, which is bringing together voice, data, telephony, computers, and a variety of office technologies, these fields impact our daily lives more profoundly, or at least far more visibly, than other industries. But more than that, these sectors have undergone the most upheaval in recent years, in the process becoming the focal point of the vigorous public debate about research and development policy and the future of competitiveness. The hard choices companies have made—about spending, targeting resources, and managing innovation across geographic and cultural boundaries—hold crucial lessons for what lies ahead in a fast-changing world.

Industrial research first appeared in the 1870s and quickly took its place as a hallmark of the industrial age. Companies learned that, when successful, systematic research could provide a huge edge. It helped fight fires, protect core business lines in a myriad of small and often hidden ways, and even create powerful new industries—like the radio, wireless communications, and television empires that grew out of early twentieth-century electrical investigations at General Electric and telephony studies at American Telephone & Telegraph.

Taken as a whole, these innovations helped forge a strong sense, not just from Charles Kettering, but many other scions of industry and government, that corporate research could play a vital role in competitiveness. But despite a widely accepted view that research was important to corporate well-being and vigor, it wasn't until the 1950s and early 1960s that things really took off. That was when young Massachusetts Institute of Technology assistant professor Robert Solow showed mathematically—the basis for his 1987 Nobel Prize—that in the long run growth in gross national product per worker is not due so much to capital investment, the traditional pillar of classic economic analyses. Instead, such growth is in very large part the result of technological progress. In a nutshell, advanced economies grow largely on the strength of what one can call New Knowledge—in most cases the high-technology fruits of research and development.

Although it was not high in the minds of research leaders, Solow's work debuted amidst the Cold War and the resulting surge in government and industrial R&D spending that helped shift inventive activity away from straightforward mechanical engineering to an affair deeply rooted in science—primarily physics and chemistry. With that transformation came a dramatic increase in the growth of corporate labs, to the point that the Washington-based Industrial Research Institute now counts close to 15,000 corporate labs in the United States alone. These organizations employ an estimated 750,000 scientists and engineers—some 70 percent of the total number of these professionals—and spend roughly $150 billion annually on research and development. All told, industry funds about 65 percent of the R&D conducted in the country, and performs closer to three-quarters. And this is only part of the story. In rough terms, the U.S. spending is matched by the rest of the world combined—meaning the saga of corporate research today is truly a worldwide tale.

The research story goes far beyond corporations, of course. The drive for successful innovation encompasses government labs, universities, and even still the basement inventor—with the outcome dependent on a staggering array of factors from grade school education to state and federal tax credits, national R&D spending policies, and patent laws. However, companies reside on the leading edge of this ongoing, global struggle. They are the prime venue where New Knowledge is converted into Useful Products, and where success and failure can be most plainly gauged in terms of patents, market share, sales, stock prices, and the like. By focusing on this front, I hope to provide deep insights into invention and innovation today—from the trials and tribulations of management to what's coming down the pike, and who has the edge in key technology-driven industries that will shape the way we live in the twenty-first century. To do this, I've gone inside the labs of some of the biggest names in innovation on three continents: IBM, Siemens, NEC, General Electric, Lucent Technologies, Xerox, Hewlett-Packard, Intel, and Microsoft.

Of course, there is absolutely no assurance that the companies, or even the nations, whose labs generate key inventions are the ones who will reap the economic rewards. On the back of key discoveries, Britain served as the initial center of the chemical industry; but it was the Germans who rose to control global markets for the half century leading up to the First World War. Similarly, the United States became a leading economic power long before it achieved its present position of research and scientific dominance. The annals of industrial research history are rife with examples of usurpers riding to glory on the waves of other people's innovations.

Marketing, sales, service, packaging, distribution, manufacturing, and development, the close sibling of research, all remain vital cogs in the wheel of corporate fortune. Which is most important? Well, it's like the case of the drunken juggler, notes John A. Armstrong, a former IBM vice president for science and technology. "Let's say the juggler has two tennis balls and a flaming sword, and a club," he supposes. "Now you say, which one of those things is most important? Well the thing that's most important is the next one he's got to catch. And success is keeping them all up in the air."

The point is that a host of factors can outstrip the benefits of the strongest research program, as indicated by recent woes at Microsoft, Xerox, and Lucent, whose labs are profiled here. Yet if wielded effectively, research can sharpen the vision of an otherwise all-too-hazy future, raising the chances for long-term success. This book focuses on the management philosophies, funding paradigms, incentives, and all the rest employed by top labs to do just that. There is no single formula; there exist many paths to maintaining vital research organizations. Which one is most successful depends on such factors as the economy, the industry, and the firm's role in it—as well as corporate culture.

Indeed, fitting into that culture—not just of the research organization but of the entire company—and helping scientists and engineers see themselves as part of a larger whole, turns out to be one of the keys to successful industrial research. Some company research arms, like Hewlett-Packard's, possess a strong sense of history and a deep connection to the rest of the organization—an enviable situation made easier because the corporation itself is only fifty years old, and by the fact that the founders and all the chief executives until Carly Fiorina took over in mid-1999 have been engineers. An outfit such as Bell Labs, by contrast, nurtures an equally strong research culture but became increasingly disconnected from AT&T, partly as a result of the sweeping success of its scientific investigations that transformed a corporate laboratory into a university-like situation. Combined with years of uncertainty over AT&T's future, the result was a kind of slow death that was only averted by research managers taking drastic action. Even then, it wasn't until the Bell Labs core spun off into a separate company, Lucent Technologies, that it rekindled the old spirit of innovation and reclaimed its place as one of the world's great industrial laboratories, a position it maintains despite Lucent's recent troubles.

In many ways, the book is an extension of my work as *Business-Week*'s technology editor, where I oversaw coverage of corporate labs worldwide and planned and wrote major stories on innovation and research strategy, including those involving the now-defunct R&D Scoreboard, which tracked the research and development expenditures of individual firms in a variety of industries. But it is also greatly influenced by my research for a previous book, *The Invention That Changed the World*, which told the story of World War II radar development at M.I.T.'s top-secret Radiation Laboratory.

In a sense, the Rad Lab was the first Manhattan project. The nation's top physicists were recruited to the enterprise, mixing with engineers, mathematicians, and even biologists and astronomers, to develop radar systems that had a dramatic effect on the course of the war. Even more than that, though, the radar work played a critical role in the evolution of postwar science and technology. Besides having a direct impact on specific landmark creations and discoveries—including the transistor, nuclear magnetic resonance, wireless communications, microwave ovens, radio astronomy, and the maser and laser—the radar effort helped spark a whole new attitude about the management of research as a fast-paced, collaborative, and cross-disciplinary affair. Seeing this crucial aspect of the evolution of corporate research greatly increased my insight into what went on in corporate labs throughout the 1990s. Indeed, a lack of historical and evolutionary focus has far too often caused corporate research to be viewed simplistically. Many of the common perceptions are old, or wrong—and myths and easy stereotypes abound about the innovation process, as well as the very role and importance of research in modern corporations.

One of the most fundamental of these misconceptions is the treatment of research and development as if they were one thing—namely, an enterprise called R&D. This is an easy trap to fall into, because spending on these two variables is always tracked together, and the goal of nearly every corporation is to fuse them better, so that products flow more quickly out of the lab into the marketplace. Nevertheless, the two are vastly different—in both substance and management approach.

Development is the stage where ideas or prototypes are taken and engineered into real-life products, ready to be affordably mass-produced and able to meet reliability standards and fit into the existing infrastructure. It is by far and away the bulk of the R&D equation—typically somewhere around 90 percent in the industries examined here. It is a sprawling, sweeping mass of a subject almost impossible to get one's arms around in a comprehensible way.

Research, by contrast, is where ideas are investigated, refined, and shaped into the beginnings of a new product, system, or process. Though it is an extremely small part of R&D, I've homed in on it because in the companies with a long-standing reputation for innovation—those that continually blaze trails and create new industries—invariably it is the research side that lights the way into the future. Although they often work on small-scale improvements to specific products, central research arms also concern themselves with the farther-out problems that often apply to different businesses and product lines. They garner a plurality of patents—usually the more important ones—and win the occasional Nobel Prize. Even more than that, central research is the focal point of most debates surrounding R&D policy. So, perhaps, there is more about it that needs to be illuminated.

An additional widespread misperception lies in the idea that we progress linearly from science to technology, from research breakthrough to developing products and then the market. If this were true, then managing research might be far more straightforward than it really is. In actuality, science and technology constantly feed into each other on all sorts of levels. The semiconductors at the heart of modern computers trace their roots to advances in solid-state physics that gave birth to the transistor. Yet these breakthroughs owed a large debt to attempts to analyze and control the silicon and germanium crystals used as radar detectors during World War II. More recently, semiconductors have evolved technically—becoming ever smaller, faster, and more powerful. But now these devices are becoming so small that researchers envision building chips on the atomic scale—spurring places like Hewlett-Packard, IBM, and NEC to devote more resources to studies of basic quantum physics. Staying atop this interplay between science and technology, finding novel ways to channel the fruits of one into the other, and motivating people to do it, form key parts of the research challenge.

Still another common pitfall is that in considering and debating research issues we tend to tackle the subject in sweeping terms, as if the same rules and conditions apply to every industry. But the uses of science and technology actually vary widely. In chemicals, one basic patent—say for nylon or Kevlar—can spawn an industry. But more usual is the situation found in the telecommunications industry. Breakthroughs do occur, but advances typically hinge on evolutionary change, so that it takes a multitude of piecemeal improvements to add up to a revolution. From the research and development standpoint, then, invention must be laid on top of invention—oftentimes other people's inventions. And with companies today operating research labs on several continents, managing this give-and-take across geographic and cultural chasms can be a daunting challenge.

Not to beat a dead horse, but a final dangerous practice comes in reading too much into the numbers. The general feeling seems to be that more research is better research. Therefore, when corporate R&D budgets declined sharply in the early 1990s (they have risen dramatically since 1994), it was perceived as a crisis. However, in some regards this was simply a spending readjustment after the boom times of the 1960s and 1970s—when it was easier to increase funding across the board—and therefore better reflective of reality. It also illustrates a refocusing of some resources. In the computer industry, for example, as hardware lines mature software has emerged as the driving force in sales and profits. AT&T recognized the trend back in 1990, when it slashed research and development spending 8 percent. Part of those cuts stemmed from shifting its focus from the infrastructure-heavy physical sciences to more streamlined software research. But while the move made sense, the pundits didn't much care. The cuts were viewed almost as a national tragedy, even though in AT&T's view it was doing what it was supposed to do: adapting to reflect the real world.

Even if Bell Labs slashed its budget, so what? Merely spending money is no guarantee for success—and can even be a sign of poor R&D planning. IBM's research and development outlays were once greater than the sum of the R&D budgets of its next dozen or so largest competitors—including DEC, Hewlett-Packard, Hitachi, and Fujitsu—yet Big Blue was far from a sure bet in the marketplace. Big budgets can mean the research pie is spread too thin, or locked on the wrong target, so slashing them actually can be a positive sign that a company is tightening its focus.

To provide a more accurate portrait of the research endeavor, and to unearth the secrets of standout management, I have profiled labs in Europe, Japan, and the United States. Along the way, I have visited with everyone from top policymakers to individual scientists, from long-time veterans to fresh hires. And I have looked at projects running the gamut from creating a better electric range to fashioning transistors from individual atoms. In this way, I hope to illustrate a range of actions and strategies relevant to the debate, while also bringing readers inside the labs and showcasing the often exciting technologies making their way, ever faster, toward the market, the office, and the living room.

The nine companies profiled have been selected on the basis of extensive background reading, site visits, and scores of interviews as being among the world's best in the computer-telecommunications-electronics sectors. Although Siemens and NEC mark the only corporations examined that are based outside the United States, nearly all the companies operate labs on multiple continents. All told, I visited more than two dozen facilities in Asia, Europe, and the U.S., including several run by firms that were not profiled but still contributed greatly to my overall understanding of how research operates.

Often, I try to illustrate management philosophies through descriptions of individual projects. I look mainly at success stories. However, the research organization that does not have a significant portion of failed projects is not pushing the envelope enough. Therefore, I have also included examples of some seeming failures—including IBM's well-over-$100-million effort to develop a revolutionary class of computers based on Josephson junction principles and Microsoft's Talisman technology for rendering high-quality PC graphics—as well as a fresh look at the lessons learned from Xerox's legendary "fumble" of the pioneering personal computing advances from its famous Palo Alto Research Center (PARC), several of which provide insight into the firm's current problems.

Because central research arms are so big—often more than a thousand people—I have focused on projects and organizational issues that stand out as unique or especially evocative of the management philosophy. General Electric's research arm, for example, is unusual for leading one aspect of the company's Six Sigma quality initiative: a program called Design for Six Sigma. The idea is that achieving the highest quality products cannot be guaranteed only through traditional manufacturing initiatives, but must be built into products from the get-go—and that often means starting in the research lab. At Lucent Technologies, the focus is on Bell Labs' famous Physical Research Laboratory, home to its farthest-out projects—everything from new kinds of lasers to mapping the dark matter of the universe. The Xerox profile concentrates largely on the role of anthropology and the social sciences at PARC.

The book is divided into three main sections. The first sets the table, surveying the global situation and then laying out the basically optimistic theme—that industrial innovation continues with more vigor than is typically suspected. To more fully develop this picture, I have tried to place modern corporate research in its rarely understood historical context—a vital framework for understanding change and distinguishing fundamental shifts from periodic swings of the pendulum. This task involves tracing the rise of industrial research, from its origins in the German dye industry of the late 1800s through the establishment of GE's pioneering facility in 1900 and the Bell Telephone Laboratories on New Year's Day 1925. It chronicles the tremendous influence of World War II on today's labs and management style, and shows how the first two postwar decades—a time of unprecedented American hegemony—warped the common view of what corporate research should be, a view that ended abruptly with the resurgence of European firms and the rise of the Japanese. The section concludes with a detailed examination of the reasons behind what former HP Labs director Joel Birnbaum once termed the "research bloodbath" that hit labs in the late 1980s and early 1990s, with an eye on newly evolving strategies for harnessing and encouraging research in the twenty-first century.

The second part consists of detailed looks inside the research operations of three electronics and computer giants. One is IBM, arguably the world's most powerful corporate research house when it comes to the physical sciences. Of Big Blue's eight research arms, I have visited four—the main Thomas J. Watson Research Center in Yorktown Heights, New York, and three satellites—in Zurich, Tokyo, and San Jose. And I have discussed the work at the other four, in Delhi, Tel Aviv, Beijing, and Austin, Texas.

The remaining two firms are staunch, longtime competitors from different continents—Siemens in Germany, the largest research spender outside the United States, and NEC in Japan, which has shown the most dramatic gains of all Japanese firms in its patent portfolio during the 1990s while also swimming against the tide by pursuing an exciting array of basic studies—both at home and at its unique lab in Princeton, New Jersey. Like IBM, both Siemens and NEC maintain facilities and laboratories around the world—and I have visited the key sites and spoken to all the top research leaders.

The book's final section consists of three additional chapters, each containing shorter profiles of two companies paired together either for their historical connection, intense competition, or some combination of these factors.

The Pioneers chapter looks at General Electric and Bell Labs. Their venerable laboratories helped define the institution of corporate research early in the century and competed heavily for decades in the infant days of radio and telephony, but have since drifted onto separate paths as their companies evolved apart. Both have endured major trauma and makeovers in recent years, but have emerged as leading centers of innovation. The next chapter, Children of the Sixties, examines the central research arms of Xerox and Hewlett-Packard, two much newer organizations that arose during the heyday of science in the 1960s. These two enterprises followed a path diametrically opposed to that taken by GE and Bell Labs: they did not compete for more than two decades but have recently become fierce antagonists in copiers, facsimile machines, laser printers, and general office technology. Finally, The New Pioneers offers the first detailed looks inside the upstart research labs of Intel and Microsoft. These modern-day giants followed the path of the original Pioneers by waiting until achieving dominance in their industries before creating long-range research ventures: both opened labs in the 1990s that are seeking to shape the future of personal computing.

I am not, nor have I ever been while researching and writing this book, an investor in any of the firms I have profiled. I have not invested in or been associated with any of their direct competitors—nor have I ever served as an employee or consultant to any of them. In July 1999, I participated in a Bell Labs panel for its Global Science Scholars Summit, for which I received an honorarium. Otherwise, the extent of my financial indebtedness to them runs to a few meals and one tee-shirt. However, I am greatly beholden to all these companies for their open cooperation, as well as their willingness to discuss complicated issues and patiently answer all my follow-up questions in an effort to cultivate a deeper understanding of managing innovation in modern times.

Although I have covered business issues and technology for the better part of two decades, I have never been a particular fan of corporate management or practices. I have always approached the companies I cover with a somewhat wary, even adversarial, eye—one that recognizes they are striving to put their best foot forward and are not about to tell an outsider complete details of their future projects and plans. All that aside, however, I learned a lot. Writing a book is far different from banging out a magazine or newspaper article. Because of my project's long time frame—nearly three years all told—I was able to visit top managers time and time again and really listen to and test what they were saying, luxuries virtually impossible with normal news deadlines and a job that requires constant flitting between subjects. The same held true when dealing with the researchers. In some cases, I followed their work for two years, checking things periodically to stay on top of developments.

The overall picture is one of optimism—not in the sense that research operations are wonderful entities that don't make mistakes, or even that those profiled here will stay atop their fields. Rather, I'm optimistic in the broader sense—that no matter the state of any individual firm, the enterprise as a whole is advancing. Progress is not always pretty, or even steady. But as a rule, the best corporations learn from their mistakes and strive continuously to improve the innovation machine, whipping inventions into shape faster and tailoring them to better meet customer needs—while still maintaining the balance between so-called incremental improvements in existing products and more pathbreaking work that can create new lines of business, even entire industries.

This last issue, concerning the balance between short-term and longer-range studies, basic or strategic research and applied projects, lies at the heart of much of the debate over the direction of industrial research today. It comes up repeatedly throughout the book. Basic research has always been a small part of corporate labs, despite all the hullabaloo. Nevertheless, my finding of health flies in the face of conventional perception.

This is not to deny that corporate labs have undergone a major transformation in recent years. Arno Penzias, who led Bell Labs through its darkest days in the early 1990s, says the heyday of basic industrial research—at least in the physical sciences—actually came in the 1960s and 1970s, around the time "U.S. cars lost their fins." Since then the field has undergone a series of ups and downs that include the dramatic cutbacks of the late 1980s and early 1990s, followed by a slow and cautious rebirth in recent years. It's hardly like the old days, though. Research arms have become a lot more focused on corporate needs and goals. As with all other aspects of business today, they are judged increasingly by productivity and quality measures—and must be accountable for their actions.

All this is necessary and good. But while facing this reality, the best companies—and all those profiled in this book—remain keenly aware that without a program of bold, longer-term research they run the risk of falling dramatically behind and never catching up. As Charles Kettering said when talking about his change-making department: "The Lord has given a fellow the right to choose the kind of troubles he will have. He can have either those that go with being a pioneer or those that go with being a trailer."

This book is mainly about the pioneers—old and new—who chose the troubles that go with managing innovation on a variety of time frames, including those far beyond the horizon. They provide the framework and engines of tomorrow.

CHAPTER ONE

A MATTER OF DEATH AND LIFE

"Research is a high-hat word that scares a lot of people. It needn't. It is rather simple. Essentially, it is nothing but a state of mind—a friendly, welcoming attitude toward change. . . ."

—CHARLES KETTERING

THE RESURRECTION of Bell Labs research was conceived on a warm August night, in a flooding of Arno Penzias's soul almost as swift and total as an epiphany. The year was 1989. The lab's Nobel Prize–winning research vice president was vacationing at Tanglewood, home to a series of summer music concerts in the Berkshires town of Lenox, Massachusetts. But he wasn't relaxing. A few months earlier, AT&T president Robert Allen had outlined plans for breaking the company into a series of business units designed to help it compete better in fast-moving markets—and Penzias had spent much of the day on the phone, preparing for his upcoming annual budget presentation and fretting over how to keep his scientists comfortable while also proving to senior management that research would pull its weight under the new structure. That evening, Penzias tried to relax by attending a Boston Symphony Orchestra performance. Only instead of enjoying the music, he couldn't shake the feeling that something had gone terribly wrong with his great research organization.

"My gut was starting to grab," Penzias remembers. "I was listening to Beethoven's Third and my stomach was churning and I didn't know what to do." Suddenly, it hit him: the core research side of Bell Labs—about a tenth of the overall operation—had in many ways become outdated. Since its 1925 founding, the great enterprise had enjoyed a long and glorious ride highlighted by thousands of patents and seven Nobel Prize winners, including himself. But AT&T was no longer a regulated monopoly, able to run research at a leisurely, academic-like pace—and instead of worrying about writing scientific papers or building the world's smallest laser, it needed to focus far better on business objectives.

Penzias had resisted accepting this reality for several years. But everything changed for him that summer evening, in light of Allen's recently announced decentralization plans and surrounded by Beethoven's powerful music. "I just couldn't do it anymore," Penzias relates. "Somewhere along the line the business units were going to need help, and we were going to have to do something about it. . . . And I realized I had more power to make that happen than I had thought."

Within a few weeks, Penzias had begun plotting a dramatic overhaul of AT&T's research activities. It took months of careful and exhaustive study. Seeking to identify redundant programs and inefficient practices, he launched a massive review of all endeavors—calling in a management consultant for strategic guidance. Then, on July 20, 1990, he presented his make-over in Bell Lab's big auditorium, stunning those assembled by vowing to "reverse the trend toward university-like research" that had come to characterize the place. Penzias declared that he was eliminating entire laboratories and that the organization would shift many resources to software studies—curtailing much of the physical sciences explorations that had made Bell Labs famous. "AT&T's senior management continues to invest more than one million dollars every working day in our work, and they give us the freedom to use it as we think best," he explained to his researchers. "That gives me a deep sense of responsibility to use our resources and our freedom wisely."

The changes put in effect that day shook Bell Labs to its soul. Soon, Penzias's policies were attracting headlines, inciting cries of outrage from national policy leaders that the legendary enterprise was abandoning the kind of fundamental scientific investigations that had given rise to the transistor and other great advances.

Even as the Bell Labs research head carried out his plans, a similar story was unfolding at IBM's Thomas J. Watson Research Center in Yorktown Heights, New York. There, at the helm of another of the world's great corporate research organizations, James McGroddy was spearheading his own brutal self-examination with that same bad feeling in his gut: a growing awareness that despite its own sparkling history of five Nobel Prize winners, research was not making much difference to the company.

Ever since 1990, much like his counterpart Penzias, McGroddy had been working to couple activities more tightly to IBM's various businesses—concentrating more on providing customers with "solutions" rather than just technology. By late 1992, with the coming year's budget under fire, he had also begun to streamline support services, eliminate redundant or seemingly dead-end programs, and find outside sources of funding by bringing in government contracts or launching spinoff companies. Prophetically, with the company in dire straits, McGroddy felt certain chairman John Akers would soon resign. Since any replacement likely would know little of the research operation, he prepared a complete documentation of the value the division created for and provided to IBM.

The sixteen-page document was only a few months old when new chairman and chief executive officer Louis Gerstner, Jr., arrived on the scene. The former head of American Express and RJR Nabisco liked to move fast. On April 1, 1993, the very day his appointment was announced, he called senior executives, including McGroddy, to the big board room at IBM's Armonk headquarters. The new boss asked his execs for a ten- or twelve-page description of their segment of the company to bring him up to speed on Big Blue's place in the world—a first step in determining the hard measures needed to put the lethargic giant back on the fast track. In particular, he wanted to know the details of each business, the economic model it followed, its customers and competition, strengths and weaknesses. The reports were to be submitted within a few weeks—and after Gerstner had time to study the documents, he would schedule a meeting with each executive to discuss things in more depth.

In essence, Gerstner sought the very kind of report McGroddy had recently wrapped up on his own—a well-written statement of organizational practices and goals, coupled with a frank assessment of strengths and weaknesses, and an action plan for the future. McGroddy went back to the Watson center and slashed four pages from the report—turning the document over to Gerstner a few days later. Research was handily the first group to complete its assignment—and the Watson lab was Gerstner's first IBM visit outside headquarters. The chairman liked what he saw. At the end of the five-hour visit, he told McGroddy that he had long served on the AT&T board, where he had witnessed the troubles associated with a great research lab that made little impact on customers or the bottom line. "I want to thank you for not putting that problem on my plate here at IBM."

Over the next two years, driven partly by McGroddy's vision, Gerstner eliminated one-third of IBM's total R&D budget, moving from about $5.1 billion a year to a tad under $3.4 billion. It didn't matter that inside IBM the battle had been won—and that the new chief executive began featuring research as a key to the company's revitalization. Just as with AT&T before it, many leading academics and policy wonks concluded Big Blue was abandoning science, and with it the future. You could hear the screams from Harvard to Washington, D.C.

Jump forward almost a decade, to the last years of the twentieth century. Arno Penzias and Jim McGroddy are both retired, though active in a multitude of research and development issues. Picture Penzias, still a Bell Labs consultant, just arrived at lab headquarters in Murray Hill, New Jersey, from his home in San Francisco. The cathedral-like lobby harbors a firsthand compendium of the electronics age. Display cases contain the original transistor and a model of Alexander Graham Bell's pioneering telephone. A Telstar 1 satellite hangs from the ceiling, while a series of plaques describe the pioneering work of the lab's now eleven Nobel laureates—including Penzias's 1965 discovery, with colleague Robert Wilson, of the cosmic background radiation presumably left over from the Big Bang.

But the former research director pays them no heed. Still fit in his late sixties from swimming and running, a laptop computer slung over one shoulder, he strides purposefully past the guard counter into the facility itself—where it's a whole new day. Whereas many once considered Bell Labs a national asset, today's operation belongs solidly to its stockholders. Fundamental investigations have been scaled back, especially in Penzias's beloved physics. Arun Netravali, his chosen successor, later named head of all Bell Labs, emphasizes connection to business goals and customers. Projects lean to the shorter term, with many managers attuned far more to products than scientific accomplishments. And it's working. Under the tutelage of rising star Lucent Technologies—one of three companies formed by AT&T's 1996 self-inflicted breakup—the lab has unleashed a cavalcade of innovations that Penzias asserts "by my count have added tens of billions of dollars to Lucent." He gets good feelings from Beethoven these days.

Jim McGroddy seems just as content. While serving on a variety of small company boards and government or professional committees, he devotes much of his time to a nonprofit business—Advanced Networking and Services—that promotes high school science education. At the research chief's retirement party in late 1996, chairman Lou Gerstner told how he had considered breaking up the Research division and parceling its resources and personnel out to individual business units so that those assets could be brought directly to bear on Big Blue's product needs. But McGroddy's tremendous first impression convinced him to stay the course. Within a few years, the research ship had turned around almost completely. Three new labs opened. Meanwhile, despite what was widely perceived as savaging its R&D budget in 1993 and 1994, IBM began winning more U.S. patents than any group in the world—a streak that in 1999 reached its seventh consecutive year. "People bitch about input measures," states McGroddy. "But where's the missing output?"

Corporate research is dead. Long live corporate research. From the hallowed, warren-like corridors of Bell Labs to the graceful stone-and-glass crescent of the Thomas J. Watson Research Center, from Xerox's famed Palo Alto Research Center (PARC) cascading down the sunny California hills to the glimmering high-tech sheen of NEC's basic research lab in Tsukuba or the fish-stocked ponds of Siemens's sprawling Erlangen facility, come stories of trauma and renewal, death and rebirth.

The dark days lingered for years—as the research bloodbath in many ways spurred by the big early-1990s upheavals at Bell Labs and IBM rolled around the world. In 1992, after decades of growth that generally far outshot inflation, industrial research spending dropped in real terms. It continued to spiral down the next two years, an unprecedented period of decline that affected nearly every major research lab in computers, telecommunications, and . . . name-your-industry.

Those sweeping cutbacks have drawn impassioned cries of lament—and outrage. The biggest concern is nothing less than the fate of national economies. With research costs soaring, and global competition intensifying, various experts have warned repeatedly that corporations are putting the brakes on science. The gravest danger, the argument goes, is that companies have focused so much on the "D" side of R&D that they are forsaking the more fundamental "R" work that creates the breakthroughs needed to spawn new industries. The *New York Times* summed up these sentiments with a front page article late in 1996: "Basic Research is Losing Out as Companies Stress Results." The *Times* warned the resulting shortfall of new technology could one day "shackle the economy."

But the various forces at work have been widely misunderstood. The axed laboratories and slashed budgets that characterized much of the 1990s obscured a vitally needed realignment—and the revival of corporate research has been almost universally missed. Rid of many of their vestigial ways and bad habits, the best labs today have moved research into another dimension—shifting their orientation beyond the old standards of merely inventing things to the more ambitious problem of *innovation*, artfully described by Xerox PARC director John Seely Brown as "invention implemented."

This drive for innovation takes a different face at every company and evolved at varying times and rates for each. But everywhere the aim is to vanquish the old linear model that says ideas progress from research to development to manufacturing to market in favor of a much more dynamic enterprise that includes constant interactions along this entire chain. Innovation requires researchers to simultaneously seize responsibility for escorting creations through corporate and marketplace barriers, listen to advice from inside and outside the company, retool their work based on that feedback, and do anything else necessary to get products to the customer.

It's true that in the face of new realities, including higher costs and stiffer competition, company research arms have had to scale back some longer-range projects. A hard line was especially needed in the ferociously competitive and tumultuous computer, telecommunications, and electronics industries, where technologies have converged at lightspeed to bring about what economist Raymond Vernon calls the "spectacular shrinkage of space."

But while the pressures of rapid change have forced research to be far more tightly coupled to the here and now—or at least the sooner rather than the later—the change is actually positive. In Europe, Japan, and around the United States, companies are evaluating projects with greater care, finding more effective ways of conducting research, and bringing innovations to market faster, spurring economic growth in the process. Tom Anthony, who with more than 160 patents ranks as the third-most-prolific inventor in General Electric's glorious research history, puts it this way: "In the old era, if you did something the chance of a business using it was zilch, so you couldn't get that satisfaction. Now, if something works, there's a really good chance it will be used."

Even more, the best labs today have regained their equilibrium and begun to see potential opportunity and wealth in all the mayhem. As colleagues complain about brutal and often unfair competition from the far corners of the globe, these leaders point to untapped resources, collaborators, and ultimately new markets. And while some pundits scream about a shortsighted focus on incremental improvements to existing products that shortchanges the kind of path-breaking investigations that gave the world the transistor, the new guard stresses the payoffs from making hard choices that fit the times.

But basic research—a better term in the case of corporations is "pioneering" or "strategic" studies—is far from dead. In fact, as companies recover from the financially brutal early and mid-1990s—industrial research spending in the United States has actually climbed almost 5 percent annually since 1996—one can even find clear signs of a resurgence in more fundamental pursuits. After big cuts early in the decade, IBM is once again cautiously increasing its long-term research. Hewlett-Packard has nearly tripled research spending in recent years while ramping up basic or pioneering investigations into atomic structure and chaotic systems. Microsoft plans to boldly expand its fledgling research organization to six hundred staffers by mid-2000 as it explores such far-out software and computer science issues as advanced interactivity and decision systems. Meanwhile, in 1995 Intel launched a research arm devoted to long-term studies in such areas as computer architectures and user interfaces.

What's different today is that such far-out prospecting exists in an era of increased attention to corporate relevance in which projects are chosen more carefully to bear on areas likely to benefit the firm. Xerox's John Seely Brown likes to talk of a "bold but grounded" approach to research. "You get steeped in the real problems of a corporation and then go to the root and reframe when necessary," he explains. To pull off such a feat, researchers must combine fundamental studies with routine interaction with customers and counterparts from all around the company in order to develop products that not only work, but are needed in the marketplace. The way things used to be, Brown notes, "We were these elite scientists sitting in this building inventing the future. Already, talking about 'inventing the future' smacks exactly of the ontological problem. We don't invent the future. We can help enact the future—but we must work with others in making that happen."

And what a future is in the works. The world unfolding today is at once fully wired—and wireless. It's an era of super-smart cars that anticipate when a driver's about to turn or change lanes and make sure it's safe—and of "calm computing," when network computers and web servers will be quietly and invisibly embedded in appliances and walls, allowing clocks to reset themselves after a power failure, or paint to detect intruders. It's a world of atomic scale transistors, spurred to life by that universal high-tech mantra: smaller, better, faster, cheaper. But it's also a time where the transistor has taken another path: bigger, worse, slower— only *way* cheaper. Under this vision, low-tech plastic transistors that can be mass printed without clean rooms, etching, lithography, and all the rest are used to smarten up new generations of toys, luggage tags, and appliances—or even teamed with sensors to flash up-to-the-minute freshness dates for drinks and vitamins that take into account storage conditions.

Above all, the world being created today is friendlier. Computers turn on as users sit down, talk to their owners, even recognize basic moods and do things like hold calls after detecting a look of annoyance when the phone rings. It's a world of easier scrolling and easier searching through the Internet, based on intuitive combinations of words and pictures. Web surfers have been freed from the computer terminal by systems that enable people to *listen* to the Internet and check e-mail over their cell phones or car radios. Then, too, there's the vision of personal area networks, in which people carry a specialized smart card that transmits a digital aura and securely conveys their bank account, driver's license, access codes, and so forth without the need to swipe cards through "readers." Imagine: you walk onto a train or plane and are billed automatically, or rent cars and check into hotels without ever stopping in a line, slowing down only to scan a computer screen to find your parking space or room number.

By all indications that's just a start of the fun that may be in store, for change is coming fast—and, like some benevolent rust, good research today never sleeps. Those at the best labs not only accept the fact, they like it that way. At General Electric's research center overlooking the Mohawk River outside Schenectady, materials scientist Minyoung Lee says tough competitors drive him to greater accomplishments. "It's no fun when you compete with a dummy," the Korea native asserts. "Then you got no work to do."

The days of one-shift isolation, where scientists ponder problems all day—and maybe into the night—in a university-like lab and worry mainly about beating scientific competitors to publication, are fast dying out. In select areas like programming the new era is moving closer to three-shift innovation. Computer whizzes attack a problem throughout the day, hand it off to counterparts a few time zones away as their stint in the lab begins, only to have it relayed eight hours later to a third shift in yet another part of the globe before returning to the roost to begin the next morning. "You have the program developed as the Earth turns," marvels Bell Labs president Arun Netravali, "and that's a very exciting concept." Exults Robert Spinrad, vice president of technology strategy for Xerox, "I visualize an image of the globe turning in space, and we begin at the interface between morning and day as people dash off to work on the same problem. It's fabulous what's going on."

If this portrait of surging vitality and renewed spirit flies in the face of headlines bemoaning the effects of all the early and mid-1990s funding cutbacks and reconfigurations kicked off by research leaders like Arno Penzias and Jim McGroddy, it's because too many people don't understand the evolution that has taken place and continue to mourn the old regime.

Corporate labs evolved not to produce scientific breakthroughs, but to bridge the gap between science and technology and create useful products. The first industrial research houses arose in the German dye industry of the 1870s, when manufacturers grew wary of their dependence on buying patent rights from independent chemists and realized the advantages of establishing their own laboratories. The practice soon spread beyond Europe. In late 1900, with Thomas Edison no longer a presence in company affairs and his key patents expiring, General Electric decided that its own lab could be critical to maintaining leadership in the electric lighting arena.

On the heels of the GE move, DuPont, Eastman Kodak, and others opened research arms. World War I saw a host of firms going full-out to fulfill military contracts and dramatically increase the scope of corporate research—largely to reduce U.S. dependence on imported dyes, chemicals, drugs, metals, and other products. The push to better control the future through research continued during the interwar years. New Year's Day 1925 saw the creation of the Bell Telephone Laboratories on the western border of Greenwich Village. By then more than five hundred American corporations ran their own research shops.

Occasionally these labs engendered fundamental scientific breakthroughs. In 1932, for instance, GE's Irving Langmuir won the Nobel Prize for his contributions to surface state chemistry. But science was never the real aim of industrial labs. Rather, writes historian of technology George Basalla in *The Evolution of Technology*, the point was to employ "research scientists to advance industrial goals."

World War II, though, fueled another era in research. The success of the atomic bomb effort at Los Alamos and the even bigger project to develop microwave radar at the M.I.T. Radiation Laboratory set the tone for an upbeat style of research: fast-paced and involving large interdisciplinary teams of engineers, biologists, mathematicians, and physicists—both experimentalists and theorists. The war was all about applying their skills to turn out new and better weapons and technologies—and in the process scientists became national heroes, holders of the keys to the future. As the Cold War heated up, warming with the first Soviet nuclear explosion in August 1949 and then sizzling with the 1957 Sputnik launch, the United States witnessed an unprecedented federal openness toward funding basic science in universities, government labs, and industry, especially when it came to farther-out projects with possible military applications.

Such factors helped spur phenomenal growth in the numbers of industrial labs—and with the great expansion came a far more science-friendly attitude toward corporate research. When General Motors and Ford opened major labs in the 1950s, they saw fit to probe subjects in basic chemistry and physics far beyond anything likely to impact automotive technology. In both cases, President Dwight D. Eisenhower spoke at the dedication ceremony via a special closed-circuit television broadcast, depicting the centers as nothing less than defenders of democracy. Pegged to the 1956 opening of the $125 million GM center, the *New York Times Magazine* ran an article entitled, "Key Men of Business— Scientists." "The welcome mat is out for eggheads in industry," it began. "With atomic energy, jet engines, automation and other scientific revolutions erasing the boundary between university and factory, scientists are becoming as fundamental as salesmen in industry's scheme of things."

As part of this new Ivory Tower climate, lounges were stocked with tea and cookies to encourage interaction between staffers. Especially at the biggest facilities, scientists found a unique freedom to pursue ideas. Researchers basked in this freedom, often tossing inventions over the wall to development without regard to making them work in the real world.

The era of science worship had reached full swing when AT&T scientists Arno Penzias and Robert Wilson measured the cosmic background radiation in 1965, and Sony physicist Leo Esaki and GE's Ivar Giaever made pioneering investigations in so-called tunneling phenomena, for which they shared the 1973 Nobel Prize in physics with Brian Josephson of Cambridge University in England. The basic climate lasted into the late 1980s, when four members of IBM's Zurich Research Laboratory won Nobel Prizes: Gerd Binnig and Heinrich Rohrer in 1986 for their invention of the scanning tunneling microscope, followed the next year by J. Georg Bednorz and K. Alex Müller for their discovery of new high-temperature superconducting materials.

But corporations need far more than good science to thrive in the marketplace—and already by probably the mid-1960s a few companies were beginning to realize that something was wrong with the innovation machine. Over the next two decades, U.S. firms found themselves clearly outmaneuvered and out-innovated by a host of Japanese companies in everything from cars to televisions to Walkmans. The message came through loud and clear.

Much of the attention focused—and rightly—on the research and development pipeline. The very hugeness and scale of R&D in the United States, and a widespread lack of communication between a company's research, development, manufacturing, and marketing arms, had created vast cultures of inefficiency. Even when researchers weren't busy worrying about their scientific careers and concentrated on creating new or better technologies, the often chasm-like divides between the labs and the corporate mainstream made it extremely difficult to consistently translate good research into products.

By the late 1980s, with many companies moribund and widespread layoffs the rule, corporate research seemed poised for outright disaster. The 1990s opened with AT&T slashing R&D spending 8 percent— before adjusting for inflation. Aerospace giants Lockheed, General Dynamics, and Northrop also showed double-digit dips in their research and development outlays. The next year saw big cuts at General Electric. Then in 1992, embodied by Jim McGroddy's axe-wielding over at IBM, came the full-blown "bloodbath" that extended to companies around the world, and in nearly every industry.

In all these venues the particulars may have been different but the central aim was the same: bring down barriers separating research and development, speed up technology cycle times, and work more closely with business units to hasten products to market. These same goals continue to dominate research agendas as industry charges into the millennium. That's because in these heady times of cell phones and wireless Internet links, Dells and Gateways, virtually all the factors that contributed to the rise of industrial Ivory Towers have evaporated. Consider:

- Corporate research blossomed in companies like GE, IBM, AT&T, and Kodak, whose dominant market positions or outright monopolies cushioned budgets from narrow margins and allowed ample room and time to fully profit from new technologies. Today, with deregulation and global competition the norm, companies essentially have no alternative but to hold research operations much more accountable. As Arno Penzias puts it: "The genie of international competition has escaped from its bottle and shows no signs of returning. . . . With billion-dollar contracts riding on who can supply the best technology at the lowest price, today's scientists can hardly afford to behave as they did when we were the only game in town."

- A dramatic increase in the volume of scientific and technical research also makes it difficult to get a clear edge from fundamental inquiries. Andrew M. Odlyzko, head of AT&T Laboratories' mathematics and cryptography research department, notes that when IBM researchers Bednorz and Müller announced high-temperature superconductivity in 1987, rival groups at the University of Alabama, the University of Houston, and Bell Labs pushed the field forward with further discoveries in only a few weeks. In the past, research planners could count on a reasonable shot of turning a scientific breakthrough into a dominant market position. However, Odlyzko asserts, "These assumptions are no longer believed by industrial R&D managers, and are being questioned by national policymakers."

• Many firms these days succeed despite doing little research. In the case of personal computer makers, market leaders such as Compaq, Dell, and Gateway have comparatively little technology of their own. Instead of research, their innovations lie in more business-oriented realms such as manufacturing, marketing, and distribution—adding to the pressure to compete in real time and not far down the road. As erstwhile Xerox research head and Sematech chief executive William Spencer has joined Harvard Business School professor (now professor emeritus) Richard Rosenbloom in noting: "The success of these free riders deters further investment by traditional pioneers in industrial research."

Coupled with rising costs of doing science and the end of steamrolling growth in federal support of research (in the United States, at least), these pressures have spurred the need for a more applications-oriented strategy—a reality that is here for the foreseeable future. Although industrial scientists may want to be valued solely for their knowledge, Lucent Technologies theoretical physicist J. C. Phillips told a European Physical Society meeting, "Specific industries . . . cannot achieve prosperity in the highly competitive markets of an overpopulated world simply by consulting physicists who understand general principles. The physicists must be eager to take their understanding and use it to develop specific products."

In light of such realities, research managers have had little choice but to question basic assumptions about how they went about their jobs—even what constitutes good research. One of the first things to go as budgets tightened was the old notion that more is better—both in terms of overall research expenditures and the number of projects supported. That meant focusing activities on areas of strategic importance, oftentimes core technologies in which the firm excelled and that applied to a wide range of products or business lines. It has also dictated making research far more accountable to the bottom line. "Basically, we are not a profit center, we are a cost center," notes Thomas Grandke, president of Siemens Corporate Research Inc. in Princeton, one of the German giant's three main research houses. "It requires a continual justification that we are worth our money."

It was primarily the need to justify the existence of Bell Labs' research that prompted Arno Penzias's Tanglewood epiphany. His right-hand man, Bill Brinkman, was speaking specifically about Bell Labs—but he might just as well have been summing up the state of research at scores of other corporations when he said: "What was very clear . . . was that we were in deep trouble. The business suits in AT&T were really coming at us and saying, 'Hey, what in the hell are you doing for us?' The answer was we were just too researchy."

Rather than marking the end of corporate research, the transformation that has rumbled through labs around the world heralds a new era— maybe the most exciting yet. The watchword for the millennium is "relevance," and what the historian of technology David Hounshell calls "a crazy quilt of avenues, approaches, and opportunities" has emerged to augment the innovation process and ensure that what's conjured up in labs matters to the parent company and its customers.

Research organizations today act more quickly to jump-start promising projects—often affording them special fast-track status. But they're also faster to pull the plug if things seem to veer off target. At the same time, technology is at once so diversified and specialized that companies can no longer do everything in-house. So they band together to form consortia and R&D joint partnerships, contract for work from private institutes or government labs, license technology, buy start-ups, bankroll start-ups of their own, and fund university studies. Tying everything together, teleconferences and virtual meetings via the Internet link scientists and engineers across nations and continents—and the day is coming fast when every computer will have its own video camera to allow this interaction to occur more easily and spontaneously.

Some of these measures or tools have been around for years, others are brand new—and still others represent mutations of old ideas. And while many steps are small and commonsensical, a growing number flip conventional views on their heads. Lee Davenport spent fifteen years as research director for General Telephone and Electronics in the 1960s and 1970s, taking part in numerous government committees addressing research and development issues and establishing himself as a leading voice on corporate innovation. The best organizations, he maintains, cultivate innovation by creating a climate that connects staffers to the real world, but also encourages out-of-the-box thinking. "Research is never just a gamble," he says. "You can definitely shape the odds in your favor." In a series of lectures at the University of Virginia's Darden Graduate School of Business Administration, he spelled out some basic steps to successful research that remain essential guidelines for corporate labs (see box, p. 39).

The vigor and vim of corporate research stands out clearly on the broader global canvas. In recent years, competitors have do-si-doed into each others' beckoning markets, greatly leveling the playing field in Asia, Europe, and the United States. At the same time, more equal science and technology education—combined with a growing tendency of first-

---

Shaping the Odds:
Lee Davenport's Seven Rules for Innovation

It's not easy bringing good things to life. Finding standout researchers is only a first step. Far more difficult is turning exciting ideas into things that matter. A world of strategies tries to bring this about—from touchy-feely teamwork sessions to hard-wired number-crunching. These can work, or fail, depending on lab style, culture, and implementation. The best companies, though, seem to share a few simple measures that forge a framework for innovation.

Octogenarian Lee Davenport caravans in road rallies and restores vintage cars. But as a physicist and industrial research director he has enjoyed a ringside seat on the electronics age—from tubes to chips, analog to digital. In war and peace, he's seen ideas come and go—and come again. That gives him an all-too-rare commodity: perspective.

During World War II, at the M.I.T. Radiation Laboratory, Davenport helped create a revolutionary fire-control radar that was instrumental in shooting down buzz bombs over England. After the war, he served many years as research director for General Telephone and Electronics, now GTE Corp.—a major local phone service provider that in those days owned Sylvania. Based on his lecture notes from the Darden School of Business Administration, he culled out seven commonsense rules of corporate research proven to set the stage for innovation:

1. Success is based on schedules and results—not effort, job difficulty, or loyalty. "You must expect your R&D people to produce results and reward them accordingly."

2. Since most projects last several years, managers must break them into shorter segments, with measurable goals at each phase.

3. Never allow general goals. Avoid such words as: approve, advance, increase, investigate, study, explore. All are false goals—immeasurable.

4. Look for idea people. Only a few individuals have truly unique, even harebrained ideas. Encourage them.

5. Find product champions—internal entrepreneurs who understand technology, explain it clearly, and can push ideas through corporate barriers. These traits typically elude top researchers.

6. Keep a little something on the side. A bootleg research budget is sometimes the only way to pursue ideas that break the mold.

7. Hire young blood. A research staff's average age must not increase even one year per annum. In a high-tech lab, a nice average is under thirty-five.

---

rate scientists to stay in their home countries—has distributed the talent pool far beyond American borders, where it had been largely contained in the first three postwar decades.

Top research-oriented companies have tried to tap these far-flung resources by opening labs around the world. Operating these outposts can add geographic, language, and cultural barriers to the innovation problem. American high-definition television expert Jack Fuhrer was hired by Hitachi in 1991 as senior director of its new Digital Media Systems Laboratory in Princeton. It took Fuhrer two years to realize that his Japanese managers didn't like to say "No" outright, but instead relied on subtle clues such as a lack of enthusiasm or silence to get the message across. One day, though, Fuhrer was stunned when a senior manager flatly declined a suggestion. "I said, 'Matsuura, this is weird, you don't usually say no.' He said, 'I'm trying to help. I've learned when I think "no" it helps you if I say no.' "

Despite such pitfalls, though, foreign labs are better attuned to the needs of local customers and markets than the mother ship back home—and can more easily leverage developments in nearby companies and universities. They also offer a welcome source of diversity by bringing into the innovation stream people with varying cultural views and perspectives that add rich dimensions to attacking problems. That's why the tendency to go abroad has picked up steam in recent years. Beyond home shore labs in New York, California, and Texas, IBM runs research operations in Israel, Japan, Switzerland, China, and India: the last two have opened since 1995. Underscoring Microsoft's newfound commitment to research was its 1997 decision to invest $80 million in a new facility in Cambridge, England, a move followed late the next year by the creation of a Beijing lab. "The market environment is different among the United States, Japan, and Europe—and each region has its own strengths, its own characteristics," notes Tatsuo Ishiguro, associate senior vice president in charge of research and development for NEC, which has opened labs on all three continents in the past decade. "So by mixing these different strings we want to get higher, or stronger."

But starting labs is just one facet of what's happening. More telling, perhaps, is where companies put their researchers. For some forty years beginning in the 1950s, the trend for major corporations was to place labs near academic institutions or in secluded, pastoral havens. But when Microsoft launched its central research effort in 1991, the facility went up on the company's main campus in Redmond, Washington, where it was easier to build ties to the product development groups next door. Intel chose a similar path when creating its far-looking Microprocessor Research Laboratory four years later. Other companies worry about researchers throwing inventions over the transom to developers. In Intel's case, there was no transom. Research set up shop on the sixth floor of company headquarters, with nary a sign to distinguish it from the development and business groups all around it.

The assault on time-honored practices extends to all levels of research—down to the workbench. A prevailing theme these days is to hand scientists more responsibility for getting their creations to market. Before licensing its technology to Norton, IBM launched major updates of its AntiVirus application directly from research. Theorists, physicists, programmers, and interface designers all worked together on the program, which includes a neural network to detect previously unknown viruses. In 1997, this ad hoc collaboration brought the latest version of the package from conception to market in five months, the second time in the product's five-year history that researchers had performed a similar feat. Jeffrey O. Kephart, former manager of IBM's Anti-virus Science and Technology group, called the process "almost instant gratification. You can have a seemingly wild, improbable idea, and a few months later millions of people are benefiting from it."

Xerox PARC has added yet another twist to the idea of linking researchers with markets. As in most labs today, scientists and engineers routinely bring clients to the lab, and in turn visit businesses in order to better understand what companies want. But a long time ago PARC managers realized that while computer jockeys and other technologists are usually adept at seeing how their products and expertise can help fill customer needs, even more value can be created by also letting customer needs shape their inventions. Today, the center's three-hundred-odd researchers include a half-dozen anthropologists who have gone into a law firm, an airline operations room, and government engineering offices to study the hidden and intuitive ways people do their jobs—and then work with computer scientists to create novel technologies designed to facilitate those often overlooked workstyles.

These strategies and initiatives mark only the tip of the innovation iceberg. To help spur creativity, Siemens sponsored a company-wide ideas contest. Through thick and thin, Hewlett-Packard has maintained its long-standing practice of allowing researchers to devote 10 percent of their time to unofficial, bootleg-type projects. Each year IBM grants some products or developments special status by deeming them an Accomplishment, Outstanding, or, at the highest level, Extraordinary. Ratings depend either on the innovation's general impact on science and technology, or its effect on IBM's business. The latest Extraordinary rating came in 1998, when the SP-1 scalable programmable parallel computer known as Deep Blue—the machine that demoralized world chess champion Garry Kasparov in 1997—was elevated in rank from Outstanding.

Wielding all the weapons in the contemporary research arsenal certainly adds excitement to the process. What's especially different these days is the sheer number of these tools and strategies, many more than available to the previous generation of managers. Raising the stakes even higher, all must be interwoven and operated in parallel—for everything takes place in the context of an intensely competitive world where alliances, products, fashions, prices, and most other aspects of business change colors faster than a chameleon.

Heading into the twenty-first century, Bell Labs president Arun Netravali cites three fundamental tenets of successful corporate research: speed, complexity, and cannibalization. Largely due to competition, the need for speed is paramount—in evaluating projects, pursuing advances, adopting outside technologies, and creating novel products. Complexity is another way of saying that nothing is straightforward. For example, Lucent's business hinges on data switching and transmission. Now that PCs are often connected to local area networks, data packets can be transmitted at incredibly varying rates—from a snail-like one an hour to a blurry million per second. The seemingly incalculable degree of randomness—combined with the fact that modern systems utilize equipment from a series of vendors, all of which must be maintained with incredible reliability—has created data management headaches orders of magnitude greater than just a few years ago.

Overcoming all that, though, is almost the easy part. The next step is to obsolete everything: cannibalization. This goes far beyond simply coming up with subsequent generations of products to eating your own lunch. For instance, modems have evolved from boards to chips. Yet at AT&T, before trivestiture, the organizations stamping out modems and chips were entirely separate. So in some senses chip-makers had to think about rendering their counterparts obsolete. Says Netravali, "Let us become better at doing this than some outside company, because it's going to happen anyway."

But is there a price to pay? The metamorphosis in corporate research aims to cement ties between scientists and the world outside the lab—an action everyone associated with the field calls warranted. The big worry in academic and policy circles, though, is that as companies move toward such product-centric work they will abandon the risky fundamental studies that often ignite future growth. In the 1980s and 1990s, for example, General Electric, Siemens, and a number of other major corporations retooled budget formulas so that research labs now depend on contracts with their own firms' business units for much more of their funding than in the past. The contracted work almost never involves anything more than two or three years out.

"The fact is we are no longer getting the major contribution of long-term research from the corporate labs that we had done for decades," asserts Allan Bromley, science adviser to President George Bush. Of investigations such as those leading to the transistor, laser, and fiber optics, he states, "That has simply been cut out altogether." Former Sematech head Spencer, joined by Harvard's Rosenbloom, sees the recent trends as transforming corporate research centers from prime sources of innovation to the pipeline through which corporations filter innovations from the outside. "Without a concerted effort by government, industry, and universities, the engines of innovation that have worked so well for the United States during the past five decades could be significantly impaired," the two conclude. Going a step further, Edward E. David, Jr., another former presidential science adviser who also directed Exxon's research, once predicted that if the drive toward incrementalism and applied R&D were continued to its logical conclusion, central labs would be eliminated and research enslaved to technical service.

These views—or fears—are shared by many others. But while central labs certainly aren't for every company, the institution is not going away—and in some ways is becoming *more* important. In November 1998, out to strengthen its handle on the future, Motorola boosted research spending and combined previously separated investigations in semiconductors, wireless communications, and other fields into a new organization called Motorola Labs. Innovative thinkers like Xerox PARC director Brown go so far as to assert that this type of centralized research holds the power to "reinvent the corporation" and provide the "genetic variance" that ensures a company's long-term survival.

It's not that big companies expect their central labs to be first to market with new products: that rarely happens. Instead, argue research managers, the purpose of places like Bell Labs and GE's historic Schenectady complex is to keep their parents in the game for the long haul, by providing a full range of integrated services and technologies that customers can count on with no significant loss in lead time. This is possible because central labs not only harbor a well of expertise to evaluate and assimilate the rising tide of outside inventions, they possess the ability to supplement those developments with a stream of innovations and discoveries of their own. Venture companies may be more nimble than large organizations because they can focus on a single project, notes NEC research head Ishiguro, "but the merit of the big organization is that we have a wide range of technologies. If we can get synergy of this wide range of technologies, we can have some merit or some strengths."

Xerox vice president of research and technology Mark Myers, who oversees all the company's worldwide research and advanced technology operations, points out that central labs have also emerged as critical players in research alliances or consortia. That's partly due to their being well situated to absorb the fruits of external research. But it's also because firms entering into such increasingly important partnerships must typically bring something to the table: often that is proprietary technology developed in the central lab that is broadly applicable to the entire industry. Furthermore, says Myers, unlike research operations tied to individual business units, a central organization offers the unique capability to look across product lines, identify the critical forces of change, and shape the firm's strategy to meet the likely consequences. Finally, being a step removed from market pressures, its managers are often more ready, willing, and able to commit strategic resources to longer-term issues.

The extended time horizon of central labs is why many directors also insist that basic research is alive and well—if not thriving. The 1990s *have* seen corporate labs cut back significantly on fundamental studies. A first step came in reining in "pure" science—that is, pursuits characteristic of a university aimed solely at advancing the understanding of how nature works. In truth, only the biggest and most dominant labs, such as IBM and Bell Labs, ever engaged in this kind of basic research—and it was never more than perhaps 1 percent of the total research effort. Far more typical, but still only a tenth of the budget at major labs, were quasi-fundamental investigations such as the solid-state studies leading to the transistor. Although they can advance science, these "strategic" or "pioneering" efforts are a step removed from "pure" studies because they are well focused on areas likely to benefit the company.

Many of these pioneering investigations have also been phased out. But those painting a dark picture of the future of fundamental research misread the situation. Theirs is an overly impatient view that does not allow time for firms to respond to the acute need to cut costs, speed up research and development cycle times, and come to a new equilibrium. John Armstrong, former IBM vice president for science and technology, likes to say the critics suffer from "binary syndrome." That is, they think basic or strategic research is either on or off—and that "if it isn't the same or greater, it's zero." Moreover, these critics fail to realize that by refining areas of focus, managers can often improve their odds of hitting a home run.

In today's research world, with the merits of different types of research—basic, strategic, pioneering, applied, long-term, short-term—being debated up, down, and sideways, and in several cases blurred almost beyond recognition, some favor doing away with these terms entirely. "I do not like this differentiation between basic research and applied and fundamental, et cetera," asserts Claus Weyrich, who as head of Corporate Technology for Siemens oversees its three central labs worldwide. "I'd like only to differentiate between good research and bad research." Bad research helps nobody, Weyrich says. Good research is what helps the company, no matter its nature or time horizons.

That being said, even though sustaining vigorous programs of fundamental research is typically viewed as risky, another way of looking at the issue is that not pursuing such studies means rolling the dice in another way: by forsaking the future. That explains why the best companies still walk the fine line between short-term research and the need for farther-out, path-breaking investigations.

While successfully implementing Arno Penzias's moves to shed much of its academic style, Bell Labs has not cut the overall size of its Physical Research Laboratory, where some 140 scientists pursue investigations not likely to bear fruit for two decades or longer. Physical Research is where Alan Gelperin experiments with live slug brains to uncover basic principles for constructing biological computers, astrophysicist Tony Tyson maps the universe's invisible dark matter by studying the ways its gravitational forces bend light, and Federico Capasso pursues his invention of the quantum cascade laser, a novel and highly sensitive device with applications in everything from detecting illegal drugs to analyzing auto pollution.

It's the same at other top labs around the world. In the Japanese high-tech enclave of Tsukuba—dubbed Science City—NEC researcher Sumio Iijima strives to advance his discovery of carbon nanotubes, strings of super-strong carbon atoms whose unique electrical properties might be perfect for creating nanometer-size transistors. On the other side of the world, at IBM's famous research laboratory in the hills above Zurich, James Gimzewski plots to fabricate molecular machines too small to see with the human eye, showing off pictures of history's smallest computer—an abacus he fashioned out of single molecules.

Indeed, some firms which never engaged in basic or strategic investigations are now taking the plunge—though with the twenty-first century twist of asking researchers to think more about potential commercial payoffs than in the past, when corporate scientists were judged largely on the volume and importance of their published papers. Take Microsoft natural language processing expert Lucy Vanderwende, who pursues fundamental linguistics studies in an effort to enable computers to understand everyday human speech and respond in kind. Vanderwende and her colleagues have constructed a Product Rainbow, which consists of two arcs laying out milestones they hope to reach at various stages. The "rainbow's" inner arc describes the basic scientific goals, such as morphology and logical form. Just outside it, another arc details some products the group expects to create as it achieves the scientific goals—including a grammar and style checker that debuted in late 1998.

Hewlett-Packard, longtime king of the applied labs, also has dramatically expanded pioneering studies in recent years. The establishment of the Basic Research Institute in the Mathematical Sciences at its lab in Bristol, England, marks just one example. There, theoretical physicist Jeremy Gunawardena supervises work that includes probing the possibilities of building quantum computers vastly more powerful than today's supercomputers. To achieve that goal, he relates, scientists will have to resolve some basic questions debated by Albert Einstein and Niels Bohr. "That's really when the physics changes. It changes from being semiconductor physics which people have known and studied for forty years and really brings you up against some foundational questions in quantum physics."

That's just the kind of basic research Joel Birnbaum, HP's retired senior vice president of R&D, might have encountered years ago when he worked for IBM. The difference today, he relates: "We don't say, 'Think great thoughts for the rest of your life.' We say, 'We are working on topics in early stages of fundamental understanding. We hope you'll be very disappointed if people don't apply what you've learned in ten to fifteen years.'"

Ralph Gomory, a mathematician who led IBM's esteemed research organization for seventeen years during the heyday of science, tells the story of growing up in Brooklyn Heights in the 1930s. Neighborhood kids liked to congregate on a one-block-long street called Garden Place, where they played stoop ball, bunch ball, and roller-skate hockey. Parents never worried about their safety. No cars were parked on the street, and few even drove down it. Over time, though, traffic gradually increased. It was hard to notice each change as it happened, but now Garden Place is lined with cars, and somewhere along the way it became impossible for kids to play those same games. "It was unplanned evolutionary change," notes Gomory. "Yet the whole structure of life for children changed."

A lot of that same kind of unplanned evolution has taken place in corporate research. A few decades ago, Gomory notes, corporate labs needed to offer some researchers a great deal of freedom and latitude to pursue fundamental studies: otherwise, they wouldn't have come. But the world has changed, dictating pursuits more directly relevant to company aims and goals. Just like in his old neighborhood, Gomory relates, people have to face that reality. "There's no choice."

Maybe it all could have been done more peacefully, smoothly, elegantly, caringly, sparingly—a kinder, gentler end of an era. After all, notes Harvard Business School's Richard Rosenbloom, "The proper way to run a research organization is on a steady course." If labs undergo cycles, swings, and traumatic change, he asserts, "I think that's a confession of major error." And indeed, Rosenbloom warns that the true consequences of the shift and cutbacks in longer-term research might not be visible until early in the twenty-first century.

There's a lot to be said for that point of view. Places like IBM and Bell Labs felt the pain deep in their souls, whereas Hewlett-Packard hardly experienced any discomfort at all—so it's possible to argue that an HP-like, steady-as-she-goes management style could have minimized the trauma at other labs. Still, some things are outside anyone's control—and the changes in corporate research have proven so universal that it's hard to simply blame management. Even if companies made serious mistakes—and many did—it doesn't alter the fact that corporations must adapt to face a rapidly evolving world. So while naysayers in government, academe, and even corporate labs themselves often lament those changes—especially where it relates to fundamental research—the same truths that reshaped Ralph Gomory's old neighborhood have worked their will on corporate research.

Even more than that, the need for reinvention has always been part of life in the research world—as it is in any dynamic enterprise. Or, put another way, every healthy organism invests periodically in self-renewal. Just as Charles Kettering warned many decades ago, the alternative is to be washed away in "advancing waves of other people's progress." Of all the household names in consumer electronics from the vacuum tube days, only Motorola remains as a leading maker of electronic components. It had to shed traditional businesses and reinvent itself several times, moving from mobile radios to portable radios to pagers and cell phones—all activities in which research was heavily engaged.

That's what the best are doing today—using their research operations as the linchpins for transformation. It's even possible that to the extent labs get more connected to business needs, and shed themselves of the old academic style while still maintaining some far-ranging studies, the chances of making path-breaking discoveries actually increase. The reason: science alone was never what made places like Bell Labs and General Electric great. The truly special aspect of these places was that more than almost any university, the labs brought together world-class scientists with experts in such areas as electronics and antenna design, theoreticians with experimentalists, chemists with physicists and engineers—and from that mix rose a tremendous spark of discovery.

Bell Labs astrophysicist Tony Tyson says the dynamic for such a spark may actually be better now than at any time since the 1950s—at least where he works. An increased focus on relevance has put short-term pressures on researchers and made it harder to pursue "pure" science. However, he states, "I think it's healthy to have this tension. Otherwise you're just sitting in the Ivory Tower doing nothing for anybody. It really does help to be immersed in the needs of the corporation at the same time you're trying to make some new discovery. If you're immersed in other cross streams of technology, of ideas, of demands . . . that's a very rich environment for completely new ideas to spring forward."

At the dawn of the twenty-first century, the essence of the inherent vitality of industrial research is often lost in a confusing eddy of forces. A first step in making sense of the current times is to resurrect and examine the past with a contemporary eye. Indeed, when tracing the evolution of industrial research from its German roots to the present day, it becomes clear that in many ways corporate research is actually more connected to its original purpose of advancing company goals than at any time in the past half century.

CHAPTER TWO

THE INVENTION OF INVENTION

"The greatest invention of the nineteenth century was the invention of the method of invention."
— ALFRED NORTH WHITEHEAD

"We can't be like the old German professor who as long as he can get his black bread and beer is content to spend his whole life studying the fuzz on a bee!"
— THOMAS EDISON

AN EIGHTEEN-YEAR-OLD college student on Easter vacation stumbled onto the discovery that created the industry that gave birth to corporate research. William Henry Perkin was the son of a London contractor and builder. But already as a teenager he had emerged as a prodigy at Britain's Royal College of Chemistry. Pursuing his studies during the 1856 spring break, he kicked off a saga that soon crossed the English Channel to the fashion-conscious court of Napoleon III, only to jump borders again to the heartlands of Germany and sweep over the Atlantic to American shores.

Like many young men, Perkin thought big. He was trying to synthesize quinine, an antimalarial vital to the expanding British empire but in limited supply because the cinchona tree from whose bark it was extracted could not be acclimated to lands beyond the Far East Dutch Colonies. Curious after an experiment with an impure coal-tar derivative called aniline produced a strange black substance, he kept at it and finally extracted a compound that he found stained silk a surprisingly fast and rich shade of purple.

The discovery's market potential hit Perkin immediately. While textile machinery had undergone great advances, the same monotonous colors had adorned fabrics for some two centuries—and a fortune awaited the originators of lush new colors. The energized youth patented the creation that August and quit the Royal College over the strenuous objections of his mentor, the famed German-born chemist August Wilhelm Hofmann, to begin commercialization. Perkin's father, G. F., risked nearly all of the family's savings to back the venture. Elder brother Thomas quit the building trade to manage the business. Meanwhile, William labored to improve dyeing properties, line up raw materials suppliers, and devise manufacturing methods. A factory went up hastily outside London; and by December 1957 the first aniline dye, Tyrian purple, was being sold to commercial silk dyers. From there, the operation quickly extended its line to dyes for cotton, wool, and print cloth—launching the organic dye industry.

The debut of Perkin's purple set off a European color frenzy. Rival chemists—English, French, Swiss, Austrian, German—joined the hunt for dye-giving organic compounds. It was all rough and tumble. Over in France, Perkin's patent was ruled invalid, "owing to a mistake as to date," establishing the country as a magnet for copycat firms. Still, the French gave Perkin's aniline purple the name "mauve" and did much to make it a hit. Then, in 1859, Lyon schoolteacher Emanuel Verguin discovered a red aniline dye he christened fuchsin, since its shade called to mind the fuchsia flower. Featured in the court of Napoleon III—the center of high fashion—fuchsin quickly surpassed mauve in popularity.

On the heels of these successes, the small core of pioneers soon faced a host of competitors. Twenty-nine firms showed up at the London International Exhibition of 1862, nine from the host country and another dozen from France, with the rest made up of Austrian and German concerns, and a lone start-up from Basel, Switzerland. By then synthetic dye-makers had put out five hit colors: the original mauve, Imperial Purple, a yellow, a blue, and the wildly popular fuchsin, dubbed magenta in England because it appeared around the time the French and Sardinians defeated Austria at the northern Italian town of Magenta.

Conjuring up organic dyes, though, remained largely a game of luck and high-stakes nerves. In the absence of exact knowledge of coal-tar chemistry, the invention of colors remained almost totally trial and error. More important, while exhibiting great beauty the anilines ran or faded easily and could not supplant natural dyes. As a result, colors enjoyed only fleeting popularity, and could be sold for high profits for just two or three years before prices plunged.

It all augured a far more systematic approach—rooted in science—that could deliver a steady stream of new and improved products. Ultimately, German companies moved most forcefully down this path. However, even their conversion didn't come quickly. Over the first decade of the industry's existence about a dozen firms jumped into the fray. Barely profitable outfits that relied on copying French and British manufacturing processes, these enterprises hardly stood out as innovators. Though they employed staff chemists from early on, scientists served essentially as chemical engineers—fretting over production methods and quality control. "Research" was often led by powerful foremen known as *Meisteren*, who typically possessed a few reagent bottles and some glassware with which they could alter the various ingredients in a batch of dyes to try to enhance yield or quality. Like alchemists or witch doctors, *Meisteren* guarded their secrets closely, using the mystery of it all to entrench their standing. At Bayer, for example, a recipe for aniline blue required forty-eight egg whites in each batch. Workers made pancakes from the leftovers, but when management contracted to sell the yolks to a local baker a sudden process "improvement" vanquished the need for eggs.

Though often extremely talented and adept at improving yields or processes, foremen rarely came up with novel shades. Instead, companies seeking to expand or revitalize their product lines had to rely on independent inventors, outside academics, and even rival firms—learning the secrets of new colors by hook or by crook. Twice in its infancy the struggling firm of Höchst—named after the village near Frankfurt where it was based—narrowly escaped disaster. In 1863, barely staying afloat producing aniline blue and fuchsin dyes, co-founder Eugen Lucius heard of a French chemist's discovery of a green fuchsin-based dye and began experimenting with it. Only after a nearby manufacturer asked about applying the green to leather did Lucius stumble on the path to reprieve. A chemist himself, Lucius dipped a sample of tanned hide into his dye bath and found that the green coagulated and settled out. He traced the behavior to tannic acid in the leather, and learned that the acid could be used to precipitate the color out of solution, enabling the dye to be sold as an easily packaged and applied paste. The discovery saw the tiny concern through its first dark days. Dyers Renard & Villet bought up a monopoly on exports and delivered several yards of fine green cloth to Empress Eugenie's dressmaker "as a present from the city of Lyon." The cloth debuted as a stunning evening gown worn by the Empress to the opera a few days later; and the tint soon became all the rage.

But even this provided only a fleeting redemption. In 1868, two academics at the Berliner Gewerbe Institut lab of noted chemist Adolf Baeyer synthesized alizarin, a much more color-fast dye than aniline, second only to indigo in importance and derived naturally from the madder plant cultivated widely in Provence. Their process was quickly commercialized—first by Badische Anilin- und Soda-Fabrik (BASF), and then by Bayer. By the early 1870s, Höchst again faced bankruptcy. Disaster was avoided when chemist Adolf Brüning advertised in Bayer's hometown Elberfeld paper for the secret to his rival's alizarin process—and got it.

Luck, bribery, or even great skills applied on-the-fly could hardly be counted on for the long haul, however. A series of events or milestones in the 1870s galvanized the leading German dye firms to lay the foundations for in-house research as a means of smoothing the road to the future. First came political stability. Following the end of the Franco-Prussian War in 1871, the Germanic states were unified as the Deutsches Reich. Of equal or greater significance: the sweeping success of alizarin. The laboratory product quickly replaced madder, forcing huge tracts of land in Provence to be turned over to vineyards or left fallow. Largely behind alizarin, revenues in the emerging dyestuffs industry climbed from 7.54 million marks in 1862 to 92 million marks in 1883, as synthetic dyes came to surpass natural colors in cost, ease of application, clarity, variety, and even fastness. By then, the once barely significant German concerns accounted for roughly two-thirds of total sales. On the strength of its head start, BASF led the way. Höchst ran a close second, Bayer a distant third.

Alizarin rescued German dyemakers from the constant battle just to stay alive. BASF, Höchst, and Bayer all enjoyed strong management that exploited economies of scale to churn out huge quantities of dyes. Each erected vast works along the Rhine or its tributaries, which provided the never-ending stream of water needed in production processes and the access to boats and railroads that ferried in fuel and raw materials such as alkalis, acids, coal oil, and pyrite ores; the low-bulk, high-value finished dyes could then be transported out with little impact on cost. By the late 1870s and early 1880s, the leaders had also invested heavily in marketing and distribution. The legal system augmented these moves by allowing for cartels and interfirm agreements; and German corporations entered into sales agreements, set prices, and pooled profits in a system some economic historians have called organized capitalism.

Alizarin also drove home the power of harnessing good science. As the 1870s got underway and their huge investments in marketing and distribution paid off in expanded sales, dye houses began raising the ante on research and development. Their first moves often involved hiring a few additional chemists who spent at least part of their time hunting for new colors—but the real effort was plowed into enhancing and formalizing university ties. Germany's unparalleled system of scientific and technical education made this relatively easy. The first of Germany's famous *technische Hochschulen*, the polytechnic institutions modeled after France's

École Polytechnique, opened in 1825 at Karlsruhe in the Grand Duchy of Baden. The model spread rapidly and by century's end permeated the country. Moreover, these schools existed in addition to the great universities such as Giessen, where Justus Liebig—who had trained Hofmann, who in turn had taught Perkin—had pioneered organic chemistry.

Many of these labs were held as personal fiefdoms by the famous professors at their helms. Companies wooed these scientists by doling out generous consulting fees, providing the legal support necessary to patent and defend an invention, and paying royalties on any inventions. They supplied rare chemicals, undertook tedious analytical tests, and even sent staff chemists to assist in university labs. In this way, since professors typically assigned students to specific research projects, the dyemakers ensured a steady supply of chemists familiar with industry problems. At the same time, they locked up rights to any discoveries from key individuals or entire departments: BASF and Höchst cultivated Adolf Baeyer, rival Bayer maintained close ties to universities at Würtzburg and Göttingen.

In the face of stiffening competition, however, even the most fruitful part-time efforts and partnerships fell short of the mark. For one thing, through an outpouring of experiments from French, English, and German chemists the science behind dye-making was becoming better elucidated and far more complex, dictating an increasingly scientific approach to the creation of colors. Chemists were also starting to realize that the same raw materials, intermediate products, and processes used to manufacture dyestuffs could also be applied to churning out heavy chemicals, pharmaceuticals, and photographic film. This fact, especially, compelled firms to commit to in-house research, since the same core investigations could be applied to many lines of business. A final galvanizing force was Germany's unified patent law, which took effect in 1877. By ending the scattershot protection of intellectual property that had existed under the various Germanic states, the law encouraged the creation of intellectual property as the basis for continual growth by giving companies the safeguards needed to negotiate agreements and entrench positions once new dyes and processes had been invented. Records the chemical industry historian John J. Beer, "Almost overnight the copying practices, used heretofore with telling success by the German dye firms, came to an end."

Almost as if a starter's gun had sounded, companies raced to hire chemists—wooing university graduates and raiding rival firms—whose full-time jobs consisted solely of creating new dyes. At first these recruits did not belong to any formal research structure. But as staffs swelled the benefits of assembling small teams to tackle problems cooperatively became apparent, creating the need for direction and structure. That entailed allocating budgets, calculating risks, setting up training programs, and, eventually, placing scientific operations on the same financial and administrative footing required of other major operations. It wasn't long before research broke off as a separate division.

Elaborate facilities were erected to support the growing demands of research. Organic chemists were joined by inorganic chemists, biologists, biochemists, bacteriologists, medical doctors, veterinarians, botanists, and entomologists. The wide array of disciplines made it easier to attack problems from several angles, and also to expand into areas beyond dyes. Backing up the researchers came scores of technical helpers and assistants: chemical analysts, instrument makers, clerks, storeroom keepers, glass blowers, bottle washers. Company libraries bought up scientific books and subscribed to key technical journals, while growing patent staffs waited in the wings to protect the fruits of any investigations. These state-of-the-art operations soon were able to focus on specific industry challenges with resources and skills far beyond those found in universities.

On virtually all the fronts encouraging the expansion and globalization of business—from exploiting economies of scale and scope to framing trade and patent laws and investing heavily in research and development—the upstart Germans outshone their British and French competitors. William Perkin sold his company in 1874 to Messrs. Brooke, Simpson, and Spiller, realizing a considerable fortune that allowed him to retire at age thirty-five to a private lab, where for more than three decades he added much to organic synthesis techniques. Even as Perkin eased into the life of the gentleman scientist, however, the decline of the British dyemakers became evident.

Initial strong demand had kept the money flowing so fast that companies neglected to create worldwide marketing and distribution channels or bring in professional management to bolster what were typically family-run enterprises. Specifically when it came to research, although dye houses often recruited top chemists, they failed to invest in the expanded laboratories and staffs necessary to keep the pipeline full. A stream of German scientists soon left Britain for more hospitable environs in their native land. They included BASF's Heinrich Caro, who had provided the key to alizarin commercialization. Carl Alexander Martius, Caro's colleague in the Manchester firm of Roberts, Dale and Co., also left England to join Paul Mendelsohn-Bartholdy in forming Aktiengesellschaft für Anilin-Fabrikation. AGFA quickly gained fourth place in the German dye industry, securing a virtual monopoly on the findings of August Wilhelm Hofmann, who had joined the exodus from England after the Prussian state dangled funds to build a lab that made his Royal College digs look like a shack.

A like tale unfolded in France. Dyemakers prospered briefly based on aggressive leadership and good luck finding colors. But their largely empirical methods could not keep pace with the times. Once bulked up in manufacturing, marketing, distribution, and management, the Reich's dye houses further distanced themselves from the pack through their increasingly powerful research arms, unleashing a blitzkrieg of innovation that overwhelmed the old piecemeal ways.

In 1880 Adolf Baeyer synthesized indigo—setting off massive campaigns to render it commercially viable, especially at BASF and Höchst. The BASF effort, probably the world's first large-scale industrial research project, eventually spanned seventeen years and yielded 152 patents—but was not a direct success. It wasn't until a series of separate advancements—building on yet another pathway to indigo synthesis developed by Karl Heumann at the Eidgenössische Technische Hochschule—that BASF finally got indigo production off the ground in 1897. Still, the trials and tribulations seem to have paid off. The company was first out of the blocks with synthetic indigo, and within three years was churning out enough of its dye to equal the yield from 250,000 acres of plants.

Höchst was hot on the trail. Though production did not begin until 1901, its process proved cheaper than BASF's and remained for decades the standard for indigo synthesis. Meanwhile, the company expanded into heavy chemicals and pharmaceuticals, producing offspring like benzene, naphthalene, and phenol—in addition to anesthetics such as chloroform and ether. It also proved expedient to manufacture antiseptics, since compounds from the phenol family to which these invariably belonged were common dye-making ingredients. Then, in the 1880s, came the joint antipyretics and analgesics—which fight fever and pain. By the nineteenth century's last decade, the old dye house had taken a leadership position in sedatives as well. It put out the popular anesthetic Novocain and emerged as the only big dyemaker to pursue the immunological discoveries of Louis Pasteur and Robert Koch—launching large-scale production of vaccines and serums to control cholera, typhus, diphtheria, tetanus, and other diseases. Among these trailblazing offerings: Salvarsan, the first effective syphilis treatment, invented by future Nobel laureate Paul Ehrlich.

The last of the big three German dyemakers was Bayer. The firm initially took a less progressive approach to research—and did not even rank among the first to assemble a scientific staff. Nevertheless, under Carl Duisberg, who rose from research chemist to managing director and guided Bayer well into the twentieth century, the company established itself as a research pioneer—overtaking its rivals to become Germany's largest chemical concern by unveiling a host of innovative chemical and drug products. These ran the gamut from synthetic dyes to artificial rubber to medications for tropical diseases and the pain-reliever Aspirin. While building his powerful research arm, Duisberg would formalize many of the industry's previously haphazard scientific practices and set the tone in motivational techniques, compensation of staff, laboratory design, and organizational structure that distinguished him as one of the driving forces in corporate research.

Bayer was initially a small concern deep in the shadow of BASF and Höchst, though still the biggest name on the lower Rhine. It started out in Barment in 1861, but within a few years moved to Elberfeld, the city later called Wuppertal about 30 kilometers east of Düsseldorf. The founding fathers were Friedrich Bayer, affluent owner of an imported dyewoods business, and chemist Johann Westkott, who cooked up the enterprise's first fuchsin dyes in his and Bayer's kitchens, often depriving their families of hot food. In 1863, the business incorporated as Friedrich Bayer & Co. Early the next decade it followed BASF into alizarin production; these dyes soon became mainstays.

As with BASF and Höchst farther upriver, Bayer started as a manufacturing concern and undertook virtually no research. Three years after its founding, the company hired a university-trained chemist to improve efficiency in synthesizing fuchsin—and for the "invention of new colors." But the man was dismissed after six months, and no replacement hired.

For the next decade, Bayer relied almost exclusively on four *Meisteren* to run its plants and tackle what research was conducted. Two of these foremen proved so indispensable they became partners; and under their tenure Bayer expanded into several new colors. But while production innovations helped create these successes, if Bayer's chemists tried to concoct their own colors, they failed: not one new dye arose from within the company.

The embrace of industrial research came haltingly. In 1873, within a year of starting alizarin production, Bayer began hiring additional chemists. These scientists worked not in a laboratory, but on the factory floor, concentrating on problems of quality, yield, and cost. Still, the company thrived. By 1883, Friedrich Bayer three years dead, the original partnership dissolved, and the business emerged as a public entity, Bayer had closed the gap on BASF and Höchst, ranking just behind the latter in a German dye industry that had already captured nearly two-thirds of the world's dyestuffs trade. That gave the firm the financial capital to invest more heavily in the future.

Somewhat feebly at first, Bayer revisited its old "invention of colors" idea. As early as 1881 it hired a young chemist who devoted the bulk of his time to concocting dyes. Two years of research produced no significant findings. However, on the heels of this disappointment, director Carl Rumpff engaged three additional freshly minted Ph.D. chemists. Instead of putting them on Bayer's Elberfeld staff, he arranged for each to work at a different university on tasks expected to hold promise for the company.

The reasoning was inspired. Not only did these "fellowships" keep the men from being sucked into production tasks, they gave Bayer access to cutting-edge organic chemistry research and a cheap "in" to leading professors. However, largely due to the overly optimistic assignments doled out by Bayer's inexperienced directors, the plan failed miserably. After less than a year, the trio was reassigned to Elberfeld—again leaving the company largely bereft of a research program. But all that was about to change. Fortunately for Bayer, one of its three green chemists was Carl Duisberg.

A native of Barmen, an industrial textile center next door to Bayer's Elberfeld headquarters, Duisberg had been determined as a boy to pursue chemistry rather than go into his family's ribbon-making business. He had taken his doctorate at the University of Jena, then spent a year in military service before joining Bayer in 1884 and being assigned to the University of Strasbourg.

Duisberg was a gifted, though not brilliant researcher. However, he had ambition and confidence—and seemed blessed with a keen organizational sense and the ability to motivate and lead. Upon his arrival at Elberfeld, the twenty-four-year-old scientist was appointed assistant chemist to the head of a small Bayer manufacturing line. He settled into a recently vacated whitewashed laboratory and within two short years, through good luck and sharp observation, discovered three important colors in a relatively new and important class of artificial organic dyes—the azos. The first to have commercial impact was a superior rival to AGFA's Congo red, which had actually been created by Bayer chemist Henry Böttinger, who had skipped the company with his secret.

That single discovery beautifully illustrated the power of research. Not only did the new color generate new earnings, until then Bayer had been entrenched in litigation over its claim to Congo red. Duisberg's creation seems to have stimulated both parties to enter into a cross-licensing agreement. From that point, Bayer carved out a specialty in azo dyes with little competition besides AGFA. Duisberg was rewarded with a raise—from DM 2,700 to DM 6,000, or about $1,400—and 2.5 percent of the resulting net profits. First year royalties alone totaled DM 9,000 marks, which he invested in Bayer stock.

With little planning, a team of assistants gathered under Duisberg's wing. By 1886, the young chemist had earned the title *Prokurist*, enabling him to conduct business in Bayer's name. His nucleus of assistants grew from four that year to a dozen two years down the road, breaking off as a separate research and patent division. While operating units continued to improve production processes, the division under Duisberg concentrated solely on discovery and quickly broadened into pharmaceuticals. A fever-fighting drug called Phenacetin hit the market in 1888. Bayer next followed Höchst into the sedatives, producing Sulfonal and Trional. Eventually, the company created a special lab at Elberfeld exclusively for pharmaceutical studies.

Almost from the beginning, Duisberg labored to assemble the ingredients of a modern industrial research organization. As early as 1886, he instituted university-style seminars designed to encourage interaction and keep staff members up to speed on scientific developments. Around the same time, Bayer recognized research as a lifelong, specialized occupation within the firm—and Duisberg oversaw the creation of a uniform training program that helped identify those best suited to the field. Recently hired chemists spent a year in the various plant divisions and laboratories, learning to apply dyes and test for fastness, tackling synthesis, and studying competitors' colors. Depending on aptitudes and bent, the men then drew assignments in either production or research.

A modern home for research came next. Bayer's first labs were makeshift affairs, arising wherever room could be found: in hallways, old bathrooms, an abandoned carpentry shop. Tables covered with lead sheets served as workbenches, and chemists had to go outdoors instead of "under the hood" to conduct experiments involving noxious fumes. Few of the overcrowded rooms even had a sink and drain.

Largely at Duisberg's urging, directors agreed in 1889 to put up 1.5 million marks to build a central laboratory at Elberfeld. The new home included storerooms, a rare chemicals room, machine shop, and central analytical lab—a break from the past practice of letting each manufacturing division do its own testing that increased efficiency and prevented plant managers from falsifying poor results. Supplementing these operations came an army of technicians, lab boys, dyers, and glassblowers, all existing to free researchers from drudge work. The center also boasted a technical library, to which Duisberg added significantly over the years by purchasing the libraries of recently deceased chemists.

The heart of the action—the research labs—fulfilled Duisberg's vision of an open, stimulating environment that still afforded the solitude scientists typically desired. Ultimately copied the world over, it had been inspired by the spacious labs found in many German universities, where students labored side by side at long tables. But the Bayer research chief added his own stamp—a precursor to modern cubicles. Each of the dozen chemists assigned to a given room occupied a U-shaped niche, stocked with its own equipment and supplies. The sides of the niches consisted of long work benches crowned with shoulder-high reagent shelves. The low partitions separated a scientist from his neighbors while encouraging communication, cross-fertilization of ideas, and a healthy competition. The constant spotlight also made it harder for researchers to keep a discovery secret in hopes of selling it elsewhere for personal gain—as Böttinger had done a decade earlier by hawking Congo red to AGFA.

Bayer's Central Scientific Facility opened in August 1891 and marked a milestone in the industrialization of invention. Plenty of obstacles remained to be overcome before Duisberg fulfilled his vision for research. Despite their relative freedom, his chemists continued to engage in routine analyses and patent work, even some product development. In those early days, as well, the division often served as a training ground to qualify and socialize chemists before the company farmed them out to production labs and other venues—and the high turnover hindered ongoing research. Still, for the first time the research division had broken free from technical control work and assumed a place in the corporate hierarchy alongside sales and production. As the German historian Georg Meyer-Thurow notes, "The locus of invention had been transferred from external, mostly academic institutions to industry."

Throughout his career, Carl Duisberg worked tirelessly to bolster the status and efficiency of research. In 1900, Duisberg joined Bayer's board of directors. A dozen years later, he took over as managing director. During this period, leading almost to the outbreak of World War I, Bayer grew to 10,600 workers and swept past old rivals BASF and Höchst. The scientific method permeated the corporation. Chemists replaced foremen on production lines; the scientific control and tending of manufacturing processes became institutionalized. The number of Ph.D. chemists swelled from a handful when Duisberg joined the firm to 262. Of these, nearly a quarter—57 in all—held the title research chemist. Research spread beyond the Central Scientific Facility to five other labs—covering pharmaceuticals, inorganics, alizarin, artificial rubber, and photographic and technical products—all areas into which the company diversified.

Chief among his gifts was Duisberg's demonstrated genius for industrial organization. He strove to integrate research into the scheme of Bayer's business—not simply by building a lab, hiring good people, and placing them in a stimulating environment, but by physically organizing operations to facilitate communication between scientists and business groups and enhance the firm's overall efficiency. The realization of this goal took shape only a few years after the Central Scientific Facility opened, as Bayer contemplated plans for remaking an alizarin works it had acquired on the east bank of the Rhine, eight miles north of Cologne.

The factory had been owned by one Carl Leverkus, so the site was called Leverkusen. In 1894, urgently needing to expand operations but wanting to hold down costs, Bayer devised a plan for the extensive tract that combined new construction with the retention of some existing facilities. Duisberg considered this false economy and set out a bolder vision in a January 1895 memo entitled: *Denkschrift über den Aufbau und die Organisation der Farbenfabriken zu Leverkusen,* or Memorandum on the Construction and Organization of the Leverkusen Dye Works.

The memo won adoption and established the thirty-four-year-old Duisberg as an industrial pioneer. It outlined a sprawling complex that aimed at maximizing efficiency by coordinating the flow of goods from their arrival at the factory, through production, all the way to packaging and shipping. Starting at its big wharf, the works progressed back from the Rhine in a series of large rectangular blocks. Each block housed operations relating to one major phase of manufacture—ultimately Leverkusen consisted of seven departments—and was set off from the next group of buildings by a paved street about 120 feet wide; narrower roadways separated structures in a given block, while canals carrying the water required for processing ran along the wide streets, as did gas and electric power lines, and some 40 miles of meter gauge railway tracks.

Everything was designed to facilitate communication between related operations and integrate the fruits of research into development and manufacturing. A new central scientific lab was eventually laid out near the administrative building in the plant's mid-section, with the main physiological lab and dye house close by. Duisberg considered it critical that each production department maintain its own centralized development lab—focusing chemists on the tasks at hand and encouraging mutual stimulation. Next to these labs, each department housed a separate engineering operation, so "that works chemists can at any time get into direct communications with the works engineers." The main work was complete in 1907, with various extensions taking another seven years. By that time, Leverkusen covered 760 acres, housed the corporate headquarters, and was manned by 7,900 workers: it would serve as a basis for many twentieth-century chemical factories.

Even as Leverkusen went up, research swelled steadily in size and stature, mirroring the company's own explosive rise. Duisberg continued pressing to free scientists from distracting routine tasks and patent obligations. In 1897, Bayer's patent bureau was reorganized as its own department, building up staff to relieve chemists of paperwork. In 1904, with Duisberg on the board, the company opened a training lab solely for researchers. By further establishing research as a specialty, the move helped ensure that chemists weren't spirited away to other departments: turnover fell by two-thirds.

The very success of research, however, summoned a fresh set of demons. State-of-the-art facilities helped industry match or surpass the overall conditions of academe, increasing the appeal for recruits. But at the same time, the growing army of scientists heightened bureaucracy and red tape, and narrowed communication between management and researchers. There was even the risk of squandering scientific talent by promoting top chemists into newly created layers of management. It all led to greater efforts to plan and control the invention process. These controls, in turn, coupled with the increased complexity involved in making dyes, drugs, and various chemicals, led to more team projects and raised issues of motivation, recognition, internal rivalry, and reward. On all these fronts, Duisberg and Bayer blazed trails still well-trodden nearly a century later.

The regular conferences introduced in the mid-1880s to keep chemists abreast of scientific trends and facilitate esprit de corps represented just a beginning. Duisberg continued to beef up Bayer's library. By 1906, with close to three hundred periodicals on the shelves, a new literary department was circulating abstracts of relevant articles. On the opposite end of the spectrum, scientists began submitting regular progress reports, so that the growing number of managers increasingly distanced from day-to-day work could better follow lab activity. Starting out daily in 1887, these were shifted to weekly accounts about the time the central lab opened four years later, and a supplemental end-of-year summary was instituted in 1906.

To motivate chemists and keep them focused on market needs, Bayer began offering prizes and bonuses for solving specific problems. In accordance with German law, the company also paid researchers a fixed percentage of the profits from any patents. Initially the payout ran at 5 percent, though the figure was cut almost in half by the mid-1880s. Still, the rewards proved substantial. In 1906, of seven Bayer research chemists studied by historian Meyer-Thurow, four at least doubled their base pay. The top chemist's annual salary was a very respectable DM 9,500, or about $2,300, but he took home a staggering DM 28,685 in royalties.

Competition and the need for constant innovation predisposed Bayer against iffy, far-out research projects. Additionally, the company's system rewarded managers and scientists for showing steady progress at regular intervals. So in lieu of a glamorous hunt for breakthroughs, most of a chemist's time was spent conducting the exacting experiments necessary to discover dyes and rarely involved anything more than two years down the road. That didn't mean Duisberg forsook "real" science. The company continued to survey the latest literature, woo academic consultants, and dispatch researchers to university labs. Bayer and other dye houses also played a leading role in creating the chemistry institute of the Kaiser Wilhelm Gesellschaft, the great scientific research establishment founded in 1911 to advance pure and applied studies in a variety of fields. However, science *per se* had little place in daily operations. As the Bayer board explained in 1910 to a chemist leaving for a career in academe: "We esteem highly every branch of pure science, and we appreciate every activity on any field of theoretical chemistry or teaching, but from the chemists working for our company we have to demand industrially useful results of their work."

Those results came chiefly from the meticulous accumulation of thousands of facts. In Bayer's case, one outcome was a monumental accumulation of dyestuffs patents—from 36 in the decade 1877 to 1886 to 512 in a like period leading up to the century's end. By 1914, at the onset of World War I, the corporation held more than 8,000 German and foreign patents. Similarly, in 1900 Bayer sold some 1,000 dyes and 40 pharmaceuticals. Fourteen years later, the dyes had doubled in number, while the company offered more than 150 pharmaceutical and photographic products.

Out of Bayer's various research shops and laboratories came a stream of important, even life-changing innovations. Beyond its dyes, Bayer's chemists developed Antinonnin, a synthetic chemical for crop protection; the anti-diarrhea medicine Tannigen; the leprosy drug Antileprol; and methyl rubber, the first synthetic rubber produced on an industrial scale. Then there was Aspirin. Pioneered at Elberfeld in 1897, the miracle drug debuted about a year later with an unprecedented publicity campaign: thirty thousand German doctors were invited to petition for samples. Sales soon skyrocketed; and Aspirin became the world's most famous pain reliever.

By early in the twentieth century, as other companies launched similar massed assaults on knowledge, the industrialization of invention permeated the German technological scene. A new word—*Etablissementerfindung*, or collective invention—appeared in the public lexicon. To the corporations practicing industrial research, such collectives advanced existing product lines and aided expansion into other areas—as from dyes to pharmaceuticals and photographic products, or plastics, varnishes, insecticides, resins, perfumes, cellulose fibers, fertilizers, and munitions. This expanding wall of technology rendered it extremely difficult for competitors to break into the field: in fact, no new German dyestuffs firm successfully emerged from the mid-1880s all the way up to the Great War.

The Americans, in particular, took note of the industrial-scientific juggernaut created by Bayer and other German dye houses. Duisberg even became something of a celebrity in the United States. In 1902, the Bayer executive was invited to tour the recently opened research lab of pharmaceutical maker Parke, Davis in Detroit; he lavished praise on its personnel and facilities. In New York a decade later, Duisberg made headlines by demonstrating a pair of Bayer's synthetic tires. "These tires are made of artificial rubber which I make in my plant," he boasted. "I have had them on my automobile and tested them over four thousand miles before coming to America. The only other synthetic tires in existence are on the Kaiser's automobile."

Down the road, as defeat in two world wars dampened German economic power, U.S. firms would flex their muscles and establish their worldwide dominance—in business and industrial research. But well before Duisberg boasted to the New York scene about the tires produced by his chemists, American firms had been fanning the flames of corporate science.

In one critical way, the United States in the decades surrounding the opening of the twentieth century mirrored Japan in the 1970s. A trail of technology—and not a strong scientific base as in Germany—led to the country's rise as an economic power.

In place of science came a fervor for technology and invention. The country's great natural resources, vast spaces, and chronic manpower shortages all put premiums on machinery and labor-saving devices. The last two decades of the nineteenth century witnessed the advent of typewriters, agricultural equipment, sewing machines, barbed wire, and automatic canning factories. Assembly lines spread from the Cincinnati slaughterhouses to every corner of manufacturing. America became a world leader in the production of lathes, milling machines, gear cutters—and the utilization of small, interchangeable parts for mass production. These advances allowed for a sharp reduction of costs that spurred the formation of global enterprises. The Aluminum Can Company of America, Dow Chemical, Eastman Kodak, Singer Sewing Machine, Westinghouse Electric, and General Electric represent just a few of the dominant firms created in the late 1800s on the basis of mechanical skill and ingenuity.

The wave of innovation that swept the United States proved so transforming that the last three decades of the nineteenth century are often called the age of "heroic" invention. It was creative practical-minded inventors like Thomas Edison, Eli Whitney, Samuel Morse, and Alexander Graham Bell who emerged as national heroes; few even possessed college degrees. The country had given rise to well-respected physicists such as Albert A. Michelson. But none carried the weight of scientific counterparts in Europe. And there was no one like William Perkin from whose science whole industries took root.

That did not mean business completely eschewed science. From 1834 until his death thirty-four years later Samuel Luther Dana served as resident and consulting chemist for the Merrimack Manufacturing Company, a leading textile firm. He made several important technical and scientific contributions to dyeing and bleaching processes. In 1870, steel tycoon Andrew Carnegie hired a German chemist, one Dr. Fricke, to replace the "quack" who ran his company's blast furnace. Writes Carnegie: ". . . and great secrets did the doctor open up to us. Ironstone from mines that had a high reputation was now found to contain ten, fifteen, and even twenty per cent less iron than it had been credited with. Mines that hitherto had a poor reputation we found to be now yielding superior ore. The good was bad and the bad was good, and everything was topsy-turvy. Nine-tenths of all the uncertainties of pig-iron making were dispelled under the burning sun of chemical knowledge." Because rival companies had not figured out the game, Carnegie snapped up their "low-grade" ores and turned around and sold his own "high-grade" product at tremendous profits. "What fools we had been!" he laments. "But then there was this consolation: we were not as great fools as our competitors."

But these were the exceptions. For the most part, it remained Yankee ingenuity—not Yankee education—that spawned wealth and prosperity. True, German industrial might had also emerged on the backs of independent inventors and entrepreneurs like Friedrich Bayer or Werner Siemens, founder of the electrical giant that bears his last name. But the gulf between science and technology spread far wider in the United States. Despite the early scientific impetus from Benjamin Franklin and Thomas Jefferson, America had spawned nothing like the famous *technische Hochschulen* or the outstanding university labs that attracted the best and brightest into science. Business executives, scrambling to acquire control of resources, move materials, and find reliable labor, typically did not even consider science of pressing economic relevance.

The various barriers that served to keep science out of industry started to erode as the nineteenth century wound down, however. While Americans bent on a scientific career had traditionally been forced to take advanced degrees in Europe, the flourishing of U.S. universities reduced the necessity for such an expensive and often difficult step—

making science accessible to a growing number of people. Beyond their efforts to promote "pure" science, Purdue, Wisconsin, and the Massachusetts Institute of Technology started mechanical engineering departments in the 1880s, with electrical engineering not far behind.

The convergence of education, science, and industry was fueled by a national merger mania. As firms scrambled to achieve market dominance, some 340 companies were gobbled up in 1899—against just three dozen the previous year. In 1900, the number of mergers surpassed 1,200; and all told between 1897 and 1904, some 4,227 firms congealed into just 257 corporations. Only the 1904 *Northern Securities* decision, which challenged market control via horizontal mergers, brought an end to the fury. In contrast to the German dye industry's widespread cooperation—increasingly not an option as the U.S. enforced antitrust laws—this raging marketplace battle spurred more vigorous competition than in Europe. Competition, in turn, required better manufacturing controls and efficiency in production, distribution, and everything else. With the proliferation of the electric motor and related technologies, such burdens accentuated the need for mechanical and electrical engineers to operate factories, fostering demand for executives schooled in science and technology.

Here the parallels with Germany heightened. As best shown by economic historian Alfred D. Chandler, Jr., when scientists and engineers at last came to corporations in appreciable numbers, they generally appeared in the same science-linked fields as across the Atlantic: chemicals and electric power. What's more, they tended to arrive in the three waves seen in the Reich—first as control and testing technicians, then as academic consultants hired to improve processes and products, and, finally, full-fledged industrial researchers.

A last step in this evolution was the establishment of formal, centralized research laboratories. Again as in Germany, this move occurred only after companies achieved economies of scale. It came after the creation of large production and marketing arms and the infusion of technology and applied science into the factory floor and control and development labs—once firms established a measure of dominance and leaders determined that future growth not only depended on finding additional markets for existing products but also on a steady stream of innovation. "A man might believe that new scientific discoveries were of no value to him," explained C. E. K. Mees, longtime research director at Eastman Kodak, "but he could not entirely forget that his active competitor might take advantage of these discoveries—might, indeed, even be secretly making discoveries and might come out some day with a new line of products that would take his business away from him."

The inklings of large-scale industrial research appeared in America before the nineteenth century ended, usually among consumer chemicals and food processing concerns. The Standard Oil Company of Indiana took up research in 1890, when William Burton was hired to look into desulfurization. That same year, M.I.T. alumnus David Wesson—for whom Wesson Oil was named—took over as chief chemist and manager of American Cotton Oil's plant in Guttenberg, New Jersey, where he organized the company's central library. In 1896, long before Mees arrived, George Eastman established an Experimental Department at Eastman Kodak. "If we can get out our improved goods every year," he noted, "then nobody will be able to follow us and compete with us. The only way to compete with us will be to get out original goods the same as we do."

By 1900 at least thirty-nine corporate research facilities existed in the United States. As far as can be determined, all focused on piecemeal improvements to existing products, with little or none of the more far-reaching studies that would eventually establish U.S. labs as paragons of industrial research. In the 1880s, the Bell Telephone System had briefly attempted to push the bounds of science, delving into fundamental aspects of telephone transmission under mechanical department director Hammond V. Hayes. But Hayes, Bell's first Ph.D. and holder of just the second physics doctorate awarded at Harvard, soon called off the endeavor, asserting: "I have determined for the future to abandon theoretical work for this department, devoting all of our attention to practical development of instruments and apparatus. . . . I think the theoretical work can be accomplished quite as well and more economically by collaboration with the students of the Massachusetts Institute of Technology and . . . Harvard College."

Roughly a quarter-century later the company would rethink that position. For the moment, however, Ma Bell protected its future chiefly through Alexander Graham Bell's strong patent portfolio—bringing more than six hundred successful infringement suits against various competitors until the inventor's basic patents expired in 1893 and 1894. Instead, the distinction of creating the first large-scale corporate research facility in the United States fell to the General Electric Company, a rival the Bell System would butt heads with many times over the coming decades.

It was all taking shape by early 1901. On a typical Friday evening a well-dressed man in his early thirties rode west in a Pullman car of the Boston and Albany Railroad. Though outfitted like any young executive, he was a scientist by training, an assistant professor of chemistry at the Massachusetts Institute of Technology. He had been a regular on the Schenectady-bound train since the previous December, when he began spending weekends conducting research for GE. The man was Willis Whitney—America's Carl Duisberg. The still-struggling enterprise he labored to put together was destined to become the world's most famous and important center of industrial research.

Schenectady boomed around the turn of the century. Perched on the transportation nexus formed by the Erie Canal and the New York Central Railroad, it served as a mecca for manufacturers of locomotives, rail cars, farm equipment, and sundry other machinery. The biggest of those calling the place home was GE. In 1886, Thomas Edison had transplanted his Manhattan machine works into a deserted locomotive plant about a mile outside downtown, on the Mohawk River flood plain. A few years later, after a spate of consolidation that culminated with the Edison General Electric Co. merging with Thomson-Houston to form General Electric, the site had emerged as the manufacturing and administrative hub of the expanding concern.

Already by the time Willis Whitney began his commute, GE was devouring Schenectady. It had taken control of the city's street rail system, directing its course to help open neighborhoods for a swelling work force. A mile west of downtown the company fashioned a shady preserve for executives. Farther out, contractors developed housing for engineers, accountants, and middle managers. To the south, among hills lining the river bank, came row upon row of workers' quarters, two- and three-family units known as Schenectady Flats. Finally, down on Liberty Street, along the Erie Canal that bisected the business district, stood the boarding houses for bachelor engineers.

Whitney would make his way toward one of these, a place nicknamed Liberty Hall. There, in a "cluttered, overdecorated Victorian nightmare of an apartment," lived the man who had gotten him going in industrial research—the mathematician Charles Proteus Steinmetz. And behind Steinmetz's rented digs, past where his old housemates had kept a backyard menagerie of raccoons, cranes, a monkey, an alligator, and crows named John and Mary, sat the carriage barn that harbored GE's research. It was the same curious structure where Steinmetz convened his Saturday night poker club, The Society for the Adjustment of Differences in Salaries.

Whitney and Steinmetz had first encountered each other that fall when the M.I.T. scientist journeyed to Schenectady to consider the research post. It had been a meeting of contrasts. Experimentalist Whitney, handsome and erect, was a nonsmoking Republican from a pious American household who had an undergraduate degree in chemistry from M.I.T. and a doctorate from the University of Leipzig. Then there was the theoretician and grad-school dropout Steinmetz, a hunchbacked German immigrant with a scraggly Russian-like beard—an abrasive practical joker and avid socialist who chain-smoked cigars. Eventually, these differences contributed to a rift between the men. But in those initial days the two shared the vision—or at least the hope—that science could be put to work for industry. At age thirty-five, just three years Whitney's senior, Steinmetz already stood as a giant on the American technological scene. Using his clout as GE's chief consulting engineer, he had pushed repeatedly for a research laboratory, finally convincing the company in 1900 that in order to maintain leadership in electric lighting and broaden its reach beyond light bulbs and electric power it was time to act. Evidently, he had persuaded Whitney as well.

Whitney saw the arrangement as an experiment. He started work in December 1900, but had bargained to keep his academic post and conduct research only on weekends. It was still an experiment several months later. However, by the spring of 1901, Whitney had upped the ante to three days a week. In the beginning, the staff consisted of just himself and John T. H. "Tom" Dempster, an able lab assistant inherited from Steinmetz. The men possessed little in the way of amenities; the tiny barn on the banks of the grimy Erie Canal, hardly inspiring creativity and innovation, marked a sharp contrast to the already massive dye industry research complexes along the mighty Rhine. Still, glibly ignoring the German dye houses that had beaten it to the punch by some two decades, General Electric liked to call its old carriage house lab "the barn where industrial research was born."

As Whitney and Dempster settled down to invent the future, General Electric already enjoyed a rich history of discovery that had begun with Thomas Alva Edison. In 1876, a year before his thirtieth birthday, the man whose name would become synonymous with practical invention opened a private laboratory in the hamlet of Menlo Park, New Jersey. It was this lab that gave birth to many of Edison's record 1,093 American patents, including the phonograph and the incandescent light bulb—the lifeblood of GE.

Reveling in his successes, Edison boasted he could turn out "a minor invention every ten days and a big thing every six months or so." Yet he cultivated his image as an old-fashioned Yankee inventor. "Oh these mathematicians make me tired," he gibed at scientists. "When you ask them to work out a sum they take a piece of paper, cover it with rows of A's, B's and X's, Y's . . . scatter a mess of flyspecks over them, and then give you an answer that's all wrong." Still, though not part of any corporation, his was probably America's first large-scale applied research laboratory. The Wizard of Menlo Park possessed a variety of scientific instruments, maintained a chemistry lab, and kept a library teeming with the latest technical books and scientific periodicals. He even hired scientists, including the mathematical physicist Francis R. Upton, and employed teams of inventors that in many ways set the tone for industrial research. As historian Howard Bartlett has noted: "Before many industries had even given thought to research, Edison was keeping seventy-five men busy conducting experiments, designing and building new electrical apparatus for them, and devising methods of measurements so that he could make the use of electricity practical."

Edison initiated his incandescent lamp experiments in 1877, about the same time he started developing a lighting system for cities. Two years later, he invented the carbon filament incandescent lamp—the first commercially feasible light bulb. That New Year's Eve, he illuminated Menlo Park to demonstrate the metropolitan lighting scheme; his Pearl Street station in Manhattan went on line in 1882, drawing headlines as the world's first central power facility.

Throughout the decade, Edison's inventions spawned a host of businesses that bore their creator's name and churned out everything from electric tubes to lamps. Other concerns, such as the Sprague Electric Railway and Motor Company, would be acquired or absorbed by the growing enterprise—consolidated in 1889 as the Edison General Electric Company. Three years later, the firm merged with the Thomson-Houston Company of Lynn, Massachusetts, a profitable arc lighting company and alternating current pioneer. The combination created an overnight giant with a work force of ten thousand and annual sales of around $20 million. Thomson-Houston supplied the first president, Charles A. Coffin. But for a name, the upstart corporation looked to its other half, dropping the reference to Edison and becoming the General Electric Company.

The new General Electric possessed a strong heritage of invention. Not only was there Edison, from the other side of the merger came Elihu Thomson, a British-born engineer who had pioneered arc lighting and alternating current power systems. But the glory days of these inventors were already in the past. Edison had quit the electric industry in the late 1880s, among other things focusing on the problem of moving pictures. Thomson, though retained as a consultant, had turned his main attention to teaching.

With Edison's absence and the diminished presence of Thomson—and with both its power and electric light operations under attack from rival technologies—General Electric faced an innovation problem from its first days. Yet the company made no attempt to reconstitute an Edison-like invention factory. Instead, it attempted to secure the future by purchasing patents from independent inventors and engaging part-time consultants.

That strategy served it well in the 1890s by fending off a challenge from arch-rival Westinghouse Electric & Manufacturing. From his base in Pittsburgh's Garrison Alley, George Westinghouse had snapped up valuable alternating current technology—including Nikola Tesla's induction motor and the transformer patents of British native William Gibbs and Frenchman Lucien Gaulard—and begun supplying electric power at lower costs than Edison's direct current system. But by 1897, after purchasing some technology of its own and bringing in hired guns like inventor Charles Bradley and physicist Louis Bell, GE had improved DC systems enough to redress the technological imbalance. That set up a far-ranging patent swap—GE's inventions in traction, lighting, and an AC-DC converter for Westinghouse's transformer and induction motor—that benefited both sides.

Even as this AC-DC fight played out, however, other holes began appearing in GE's dike. Challenges to Edison's key 1880 patent for the carbon filament incandescent lamp had created a state of limbo that enabled a host of independent lamp makers to spring up. The invention was finally affirmed in 1892, the very year General Electric formed. The company briefly wielded the ruling like a scythe to mow down rivals, but a technicality caused the patent to expire in November 1894, nearly three years ahead of schedule. Suddenly, competition once again became a fact of life.

General Electric counterattacked forcefully, slashing manufacturing costs and blocking foes by gobbling up patents: chief lamp engineer John W. Howell abandoned his bride on their honeymoon to sail to Italy and buy up Arturo Maligniani's patents on improving the vacuum in an incandescent lamp. Company president Charles Coffin then used his formidable manufacturing and patent positions to convince two key rivals to form a federation of "independent" lamp manufacturers that was controlled by GE. By 1900 this National Electric Lamp combine had cornered more than 90 percent of the U.S. lamp market.

But purchasing innovations, leasing scientific talent, and forming questionable trade combines only went so far. With annual research and development expenditures for lamps running at roughly $25,000, less than half what Edison had spent twenty years earlier, GE remained vulnerable to outside innovations. Critically, the company's formidable defenses all centered around the carbon-filament, high-vacuum incandescent lamp. Throughout the 1880s and 1890s, a host of alternative technologies appeared—the mercury vapor lamp, the osmium lamp, and, particularly, a ceramic filament lamp from German chemist Walther Nernst that was longer-lived and operated at upwards of 50 percent greater efficiency than the best carbon-filament product. In 1894, even while fighting GE over power distribution, Westinghouse bought the U.S. rights and prepared to bring the Nernst lamp to American consumers.

By the turn of the century, the various attacks on the incandescent lamp—all drawn largely from scientific investigations—had primed the pump for General Electric to rethink its research strategy. Charles Steinmetz had been hounding the company to create a research laboratory since 1897, arguing that the lab should help GE get into the fast-rising electrochemical field, which utilized electrical energy for such profitable ventures as the reduction of aluminum from its ore. Undeterred by management's refusal to bite on his two initial proposals, he had launched a third assault in September 1900. This time, in a letter to vice president Edwin W. Rice, in charge of manufacturing and engineering, Steinmetz quit pushing aluminum and instead couched the lab as the way to parry the threat of mercury vapor lamps and other attacks on the lighting business.

Aimed straight at GE's soft spot, the third arrow proved the charm. Steinmetz sent copies of his letter to Elihu Thomson, Rice's former high school teacher and longtime mentor, as well as top company patent attorney Albert G. Davis. Both embraced the idea. As Davis wrote Rice: "It seems to me therefore that it would be wise for a considerable sum of money to be spent in the active development of the mercury lamp. If someone gets ahead of us in this development, we will have to spend large sums in buying patent rights, whereas if we do the work ourselves this necessity will be avoided." Thomson painted a bigger picture: "It does seem to me that a Company as large as the General Electric Company should not fail to continue investigating and developing in new fields: there should, in fact, be a research laboratory for commercial applications of new principles, and even for the discovery of those principles."

By that October, president Coffin on board, the ground rules had been sketched out. Fending off attacks on GE's lamp business took top priority. The men also determined that the lab had to be distanced from day-to-day developmental work, and that in order to attract topflight researchers it needed to pursue a broader range of activities than lighting. As Rice later recalled, "We all agreed it was to be a real scientific laboratory."

One of GE's academic consultants, M.I.T. physicist Charles R. Cross, suggested Willis Whitney as the perfect man to launch the endeavor. The chemistry instructor met that October with Thomson, Rice, and Davis. Reluctant to take the job for fear of "burying my possible individuality in a large soulless industry," he promised to seriously consider the position only if it could be part-time, allowing him to stay on at M.I.T.

The General Electric team quickly agreed. For two days' work a week, Whitney was to receive $2,400 a year—more than three times the average American family's annual income. It wasn't all about money. Whitney had earlier turned down a high-paying job from consultant Arthur D. Little. He also had his M.I.T. salary, plus substantial income from his own consulting work. But M.I.T.'s refusal to grant his recent request for a $75 annual raise still rankled. After mulling things over for a few weeks, and visiting Schenectady to meet Steinmetz, Whitney accepted the offer. He started on December 15, 1900, quickly becoming a regular on the westward-bound train out of Boston each Friday, and a familiar sight on the streets of Schenectady, wending his way to the carriage barn behind Liberty Hall.

Willis Rodney Whitney would direct the General Electric Research Laboratory from 1900 until his retirement thirty-two years later following a nervous breakdown. Though lacking Carl Duisberg's organizational genius, Whitney found renown as a corporate and scientific leader admired not only for the lab's accomplishments but for his enlightened style. The sign above his always-open office door—"Come in, rain or shine"—became legend. So did a cheery call as he wandered through his realm: "Are you having fun?"

For months after launching the General Electric lab, Whitney kept up his rigorous schedule: four days at M.I.T., three at GE, two nights on the train. But progress on the Schenectady end came agonizingly slowly. In early 1901, fire damaged the barn, prompting the move that spring to a one-story wooden building at the main works. A few weeks later, the rampaging Mohawk flooded the facility. Within a year, Whitney's operation would uproot again, moving to Building 6, next to the central office building.

To pick up the pace, Whitney courted first-rate scientists. Ezekiel Weintraub, a highly regarded Russian-born physicist, got double the going rate of $1,500 a year, summer vacations in European labs, and an enviable clause in his contract that said: "conditions affecting the position he now holds should not be changed so as to make it exceedingly distasteful or intolerable for him." Weintraub was joined by three fellow M.I.T. graduates.

Even with the infusion of talent, though, the lab produced precious little. In fact, Whitney had done more for industry as an academic than he was accomplishing as a corporate research head. He and M.I.T. mentor Arthur Noyes had created a process that recovered valuable solvents used in making photographic paper. By early 1901, just as he was getting going at GE, Whitney's share of the net profits totaled more than

$20,000; future proceeds would pay for his first house in Schenectady. The failure to replicate anything like that success for GE plagued him. "These men of mine are all at work," Whitney complained to his parents, "but what they do is just fail, fail, fail." To a friend he confided a burning desire to put his flailing operation on the corporate map: "The only thing I want now is to accomplish some great thing for 'General Electric.' They are giving me free hand here to spend and experiment as well as I am able, and I shall die with a ten-ton shadow on my opinion of Whitney [*sic*] if I don't do some good work here."

General Electric also seemed unsure of the lab, burying its birth announcement in the 1902 Annual Report. After summarizing the previous year's manufacturing and engineering highlights, Edwin Rice noted: "Although our engineers have always been liberally supplied with every facility for the development of new and original designs and improvements of existing standards, it has been deemed wise during the past year to establish a laboratory to be devoted exclusively to original research. It is hoped by this means that many profitable fields may be discovered."

Two years later, up to forty people and an annual budget of $60,000, the lab had yet to produce anything of note. By then Whitney was rethinking the role of research. One of Steinmetz's original tenets had been to keep the lab from being sucked into the day-to-day concerns of development and production. Whitney now decided that his lab needed to earn its keep by tackling more pressing problems brought to him by factory engineers. As early as the summer of 1901, he started accepting such assignments; the practice gained steam in those uncertain first years.

Whitney also reconsidered the research environment that had evolved. As an academic, he had modeled the lab after the traditional German university style of individualistic, even secretive investigations. But infighting among jealous researchers grew so bad patent attorney Albert Davis labeled the lab the "bear pit." One staffer took pity on his boss, writing a friend: "I am almost sorry Whitney ever took hold of the German method . . . because he's having an awful time with it."

In a quest for cooperation, Whitney started culling out troublemakers and recruiting more teamwork-minded researchers. To lessen secrecy and isolation—and create a reliable record for patent disputes—he began insisting that his charges maintain detailed notebooks and submit monthly progress reports. He also introduced a Thursday evening colloquium that brought staff together to hear outside speakers or discuss recent scientific discoveries.

Around the time Whitney left M.I.T. and moved to Schenectady in 1904, the lab finally showed signs of turning the corner. That year, Ezekiel Weintraub's article on mercury lamps appeared in England's prestigious *Philosophical Magazine*—the first scientific publication from GE research. Almost simultaneously a spinoff of that work—a mercury rectifier that changed direct current to alternating and vice versa—hit the market. Alternating current had won the battle for transmitting great amounts of electricity, but some cities still ran Edison-produced DC systems, and trolley systems were invariably direct current—so the rectifier, without the moving parts of existing mechanical systems, held great promise as a means of conquering the divide. On its heels came a carbon filament lamp invented by Whitney himself that operated with 50 percent greater efficiency than a conventional incandescent lamp while offering the same eight-hundred-hour life. This General Electric Metallized, or GEM, lamp went on sale in 1905 and proved the most valuable technological contribution of Whitney's career.

Even before the GEM reached store shelves, however, a new threat appeared. Over in Germany in 1904 Siemens & Halske chemist Werner von Bolton had conceived a lamp based on the element tantalum that produced more light per watt than even Whitney's creation. It did not immediately displace the GEM lamp. But as its advantages became clear, Whitney scrambled to launch studies of other elements—especially tungsten—that hovered near tantalum on the periodic chart and might exhibit similar properties. In the meantime, GE was forced to shell out $250,000 for the rights to buy tantalum wire from Siemens & Halske and another small fortune for two other rival European processes.

This was perhaps GE research's darkest hour. Being blindsided by upstart technologies was exactly the situation the lab had been created to avoid. Widening an earlier split with Whitney, Steinmetz declared it "shameful . . . that the successors of Edison should cease to be considered as the leaders in the development of incandescent lighting, and European engineers get the credit which our Company should have retained, if its engineers had not been asleep." When metal filament incandescent lamps hit the market in 1907, the robust 30 percent annual growth in carbon filament models screeched to a halt, then slipped into decline. On a broader front, concerns about a general economic malaise set off a Wall Street panic that left firms scrambling to cut costs. With the GE lab swollen to some 150 workers, Coffin ordered Whitney to draw up contingency plans to slash expenses by as much as 30 percent. No sooner had the besieged research director put the finishing touches on these plans than he collapsed near death with untreated appendicitis.

Whitney spent three months convalescing in Florida, and at the small farm he had purchased in Schenectady. He briefly contemplated a new career as a surgeon, but instead returned with fresh resolve to increase the lab's value to General Electric. The famous sign appeared over his door, welcoming visitors come rain or shine. He changed the policy on colloquia, requiring researchers to discuss their own efforts, noting: "it affects the personnel by effecting closer and more friendly intercourse." Of greater import, Whitney had come to see the dangers of betting everything on a single line of investigation—in this case, lamps. As soon as conditions permitted, he diversified into realms outside lighting, drawing life from a breadth of projects and expertise that would buttress the lab through all its ups, downs, and various iterations over the rest of the century. In the meantime, he cut back even further on long-term projects like the hunt for a better metallic filament lamp—and began taking on more factory assignments. As he wrote his parents in the spring of 1908: "The times are so hard that I don't dare take a long road. That is, I don't dare to try one of those long shots which may turn out good in a few years but is not good now. I see how hard the rest of our organization is pushed to hang on to the value of a dollar."

In redefining his laboratory's scope and mission, Whitney concluded that a lab's spirit was far more important than the individual scientists it contained. In particular, everyone from the director on down had to proceed with chins up through the mist and pitfalls that surround research. As he wrote in 1910, "With active optimism, even in the absence of more than average knowledge, useful discoveries are almost sure to be made." That did not mean a laboratory had no need for stars. Whitney continually angled for top talent, scientists who would not only raise research quality, but act as beacons for legions of other able professionals. And indeed, through his efforts some extraordinarily gifted individuals descended on Schenectady, contributing greatly to the optimism that finally began to permeate the facility.

One of the first of this ilk to arrive was Whitney's former student William D. Coolidge. A distant cousin to future President Calvin Coolidge, the young chemist only agreed to leave M.I.T. for double his academic salary and freedom to devote a third of his time to private experiments. Almost any deal would have been worth it. A brilliant equipment designer and rigorous scientist, Coolidge possessed an extremely modest and quiet demeanor that complemented perfectly Whitney's extroverted and charismatic personality: he quickly rose to assistant director, taking over the lab while the boss recuperated from his 1907 illness. And it was Coolidge who provided the first big breakthrough—a way to produce tungsten filament with much greater tensile strength and endurance than its rivals. His "hot swaging" process, combined with the purchased European patents, enabled the company to reassert its ironclad control over the lamp industry.

Four years after Coolidge joined the lab, Irving Langmuir began his long tenure in Schenectady. A Brooklyn native, Langmuir's dream had been to emulate Lord Kelvin and be "free to do research as I wish." Returning to the United States after completing his doctorate in Germany, he opened his professional career in the fall of 1906 as a chemistry instructor at Stevens Institute of Technology in Hoboken, New Jersey. But he had struggled at Stevens, unhappy with both the students and the tiny paycheck. GE proved his escape route. In 1909, Langmuir took a summer job at the lab. He figured the odds were good that would lead to a full-time position, but planned to accept only while "looking around for a really good position in a university." Instead, the twenty-nine-year-old Langmuir turned out to be the quintessential industrial researcher, a hero for scientists and managers alike who contributed mightily to GE's prestige and pocketbook. For over forty years, he stayed in Schenectady, tackling everything from pure science to home appliances. He totaled sixty-three patents, averaged some five research papers a year, and would win a Nobel Prize for his insights on surface state chemistry.

When Langmuir arrived in July 1909, Whitney told him to look around for a few days and pick a project. The new man found staff members struggling over their tungsten bulb's tendency to blacken after short use, while the filament itself grew brittle and crumbled. Langmuir's doctoral thesis at the University of Göttingen involved the behavior of gases around the heated filament of an incandescent lamp. So he zeroed in on the bedeviled tungsten device. He suspected the problem might result from the absorption of gas by the heated element and discovered that as the lamp warmed its glass walls gave off water vapor which reacted with the tungsten to produce hydrogen. With Whitney's blessing, he put the lamp aside and began basic experiments on the behavior of gases near heated filaments.

Three years later a practical result appeared. It turned out that if water vapor got into the lamp, these molecules would dissociate in the presence of the filament. Freed oxygen atoms would combine with tungsten atoms evaporating off the filament to create tungsten oxide—a compound that fairly flew to the lamp's glass walls and stuck there. Langmuir also showed how the addition of nitrogen could dramatically slow down this process by retarding the evaporation of tungsten. These discoveries—the basis for his Nobel Prize—pointed the way to a much improved lamp filled with nitrogen. Langmuir's patent covering the use of gas in a bulb led to the Mazda incandescent lamp—named for the Persian god of light—and formed a cornerstone of GE's soon-dominant lamp position.

General Electric research was finally ready to enter its golden age. It had taken years of struggle, and stemmed more from the efforts of a few talented individuals than Bayer's collective inventors, but Whitney had finally moved GE into the world's top tier of industrial research. The grateful director spoke in especially reverent terms of Langmuir's work: "He has caused new factories to be built on things too small to be seen by the naked eye. . . ."

CHAPTER THREE

HOUSES OF MAGIC

"Large corporations do not hire technical men because they are large. Rather they are large because they hired technical men when they were small."
—CHARLES KETTERING

"I often think, though probably wrongly, that men in academic positions ought not to be paid as much as men in industrial positions. On the average the men in the latter case work harder, overcome more obstacles, worry more, and compromise with their natural desires much oftener."
—WILLIS R. WHITNEY

THE CHICAGO CROWDS poured past the sparkling fountain standing guard over the Electric Industries pavilion at the Century of Progress Exposition and World's Fair of 1933. Inside the massive structure, they streamed through a wide doorway under the brightly lit sign that had captured their attention—and imagination: "House of Magic." It was showtime for General Electric and its vaunted research arm. Pavilion exhibits revealed off the latest marvels of the electrical age—all-electric kitchens, refrigerators, washers and dryers, and especially air conditioning. But through those doors, inside GE's theater, even those must have seemed mundane. An emcee—a research staff member—was pulling out all the stops. The house went dark—allowing him to demonstrate how black light could illuminate the previously invisible company logo emblazoned on a shawl draped over his shoulders. Next came a speech-controlled train that through a trick of converting syllables to electrical impulses backed up, stopped, and advanced on command.

"So now you may ask me why does our company indulge in this expensive research work?" the host queried his audience. He answered with barely a pause. "Pure science of the past, when applied today, is giving us better quality and new products such as the radio, the refrigerator, and, now, the air conditioner. And we know that just as surely will the pure science of today, when applied in the future, give us more new products and still better quality. And if we are to remain leaders in quality of product in the future, then it's just up to us to be leaders in pure science research today."

A lot had gone down at General Electric research in the past two decades. Willis Whitney had retired in 1932 amidst ill health and a severe bout of depression that led to his breakdown, retreating to his farm outside Schenectady and leaving the lab in the hands of able successor William Coolidge. Long before he departed, however, the once-floundering enterprise had transformed into a wellspring of innovation. In 1913, Coolidge invented a tube that, with key suggestions from Irving Langmuir, revolutionized X-ray treatments. Coolidge's swaging process for metal tungsten paid dividends in making auto ignition contacts that displaced the unreliable old platinum technology. Yale-trained physicist Albert W. Hull pioneered the thyratron, an important element in industrial power control. Whitney's people also contributed greatly to Calrod, the specially sheathed coiled heating wires that made electric ranges practical by eliminating the risk of shock. Meanwhile, Langmuir teamed up with GE engineer Ernst F. W. Alexanderson and others to improve independent inventor Lee De Forest's triode—and their collaboration proved critical in establishing the patent tangle that led GE in 1919 to join Westinghouse and Marconi Wireless Telegraph Co. in forming the Radio Corporation of America.

The end result was that General Electric moved beyond lamps and light bulbs into communication, medical equipment, and appliances—emerging as the world's leading electrical manufacturer. By playing a critical role in this evolution, Whitney's research arm had soared to international renown. In the summer of 1930, popular author and radio host Floyd Gibbons, who had traveled with Pancho Villa during the Mexican Revolution and wore a white eyepatch after being wounded as a World War I correspondent, had toured the lab and come away captivated by the latest X-ray techniques and demonstrations of electric "eyes" that opened doors at the wave of a hand. "This is truly a House of Magic," he had proclaimed during a broadcast that evening.

The name had stuck—turning the lab into a public relations bonanza even before Irving Langmuir won the 1932 Nobel Prize in chemistry and became the first American corporate researcher to capture

science's highest honor. So many visitors had since flocked to Schenectady that the company decided to set up the House of Magic exhibition
that debuted at the Chicago fair. Over a million people would see the
show in the Electric Industries building during its two years in the Windy
City. Within a few more seasons, some 325 campus newspapers would
carry monthly articles showcasing lab activities. GE kicked off a weekly
radio broadcast called *Science Forum* that opened with a short scientific
profile and then provided answers to listeners' mailed-in questions. Millions more people packed theaters and halls around the country, as three
separate House of Magic shows hit the road. These performances eventually included demonstrations of how magnetic wire could record the
human voice, a levitation machine that suspended hats on a bed of air,
and, seemingly always, the voice-controlled electric train.

GE's magic set the standard for industrial research. Its diverse base
highlighted what the endeavor was all about—erecting a strong foundation that both cushioned down times and carved out future avenues of
growth. But long before Whitney's organization had made its mark, a variety of other American corporations were also embracing research.

In 1902, less than two years after Willis Whitney arrived in Schenectady, drugmaker Parke, Davis unveiled its Detroit research laboratory.
That same year, DuPont opened the Eastern Laboratory in Raunoke,
Delaware, with the partial goal of creating safer high explosives. In 1910,
Westinghouse erected a two-story structure at its East Pittsburgh plant to
house physical, chemical, and magnetic studies—an extension of the research department it had operated for eight years. Kodak Research Laboratories opened in 1913 with a staff of twenty.

By the time World War I broke out in 1914, similar efforts were underway at AT&T, Corning Glass Works, General Chemical, B. F.
Goodrich, and E. R. Squibb. All told, though no hard figures exist, America nurtured perhaps one hundred industrial laboratories—staffed by as
few as a thousand research professionals. Overwhelmingly, these labs
continued to appear in the competitive, rapidly changing, science-
dependent chemical and electric industries. A handful of others sprouted
up in related fields such as pharmaceuticals and food processing.

Often those choosing the research route consciously followed the
German and General Electric models. In Kodak's case, Carl Duisberg
had even provided a prime motivation for taking the research plunge.
Kodak relied heavily on German firms for intermediates and photographic chemicals, so on a European business trip in late 1911 or early
1912 chief executive George Eastman had visited his Bayer counterpart.
Duisberg had casually mentioned his labs were staffed by several hun-

dred Ph.D. chemists and wondered how Kodak's research arm stacked up. Eastman went back and started one. To head it, he corralled English chemist C. E. Kenneth Mees. One of Mees's first acts was to seek out Willis Whitney, by then well known, who convinced him to spice up the regimented structure of British labs with a dash of American freedom. Mees then moved quickly to broaden business lines and lessen dependence on Germany, while also delving into such issues as the formation of latent images, optics, and the theory of color reproduction. Research soon proved so critical to Kodak's product portfolio that Mees once boasted: "Almost everything the company makes has been affected by the laboratory investigations."

The initial job of research typically rested in improving existing products or processes—critical factors in the violent market share battles common to technology-rich fields. Often, too, a research arm was created as a defensive action when a firm's bedrock patents expired or rival technologies emerged—just as had happened with General Electric during the constant assaults on its electric lamp business. The government's toughened antitrust stance, going strong into the 1900s, continued to compound these forces, as some companies found innovating their way into new businesses a safer strategy than risking Uncle Sam's wrath by gobbling up competitors or dominating marketing and distribution channels: if you can't buy it, grow your own.

All these factors came into play when American Telephone and Telegraph Company, which had absorbed the Bell System by 1899, started its first real research laboratory. The lab formed in spring 1911, largely at the behest of chief engineer John J. Carty, who urgently wanted a telephone "repeater," or amplifier, to rejuvenate fading long-distance signals and extend the range and quality of phone conversations. Both marked important steps in helping AT&T grow its business and fend off competition from the infant field of wireless radio telephony, which threatened to render the company's massive investment in telephone lines obsolete.

Initially a small branch of AT&T's Western Electric manufacturing arm at 463 West Street in Manhattan, the operation expanded rapidly after 1913. That year it took up work with Lee De Forest's triode, which had become important to Ma Bell after the discovery the tube could amplify as well as detect signals and might therefore serve as the basis of the sought-after repeater. In addition to the benefits in telephony, developing a repeater promised to give AT&T a leg up on building the communications and circuit expertise needed to control radio. Indeed, the triode work eventually brought AT&T up against the Langmuir-Alexanderson collaboration at General Electric and established it as a belated, but critical player, in the RCA story.

If these concerns weren't enough to drive a rapid expansion of research, another incentive came that same year, when the Justice Department and the Interstate Commerce Commission took aim at AT&T's monopolist tactics of buying out competitors. Under the resulting settlement, Ma Bell ceased its acquisitions, shed its Western Union stock, and opened phone lines to competitors. The company then faced the choice of growing its business in the confines of phone service areas it already controlled or expanding into non-telephone fields. Either scenario demanded innovation.

Under the guidance of top Carty assistant Frank B. Jewett and lead researcher Harold D. Arnold—both University of Chicago physics doctorates—the Research Branch of the Western Electric Engineering Department grew rapidly into a patent and research machine. Within two years, AT&T's once-modest research arm, just a handful of workers with a $3,500 budget when triode work began, had swollen to seventy-three staff members with $208,800 in annual funding. As Carty explained to stockholders: "This condition has been brought about by new demands for research in the fundamentals of the science of telephony; together with larger and very important activities in new branches of the telephone, telegraph, and wireless arts."

For industry, at least, the lone inventor was being displaced by teams of scientists and engineers. Arnold's group quickly improved the triode enough to provide the desired amplification. Already by the afternoon of January 25, 1915, when the first transcontinental telephone line opened with a connection between Bell's Manhattan headquarters and the Panama-Pacific Exposition in San Francisco, the cross-country network consisted of seven repeater stations and 130,000 telephone poles splayed across fourteen states. Before patching in President Woodrow Wilson from Washington AT&T hooked up a venerable, white-bearded Alexander Graham Bell in New York with former assistant Thomas A. Watson in California, enabling the legendary inventor to replicate the first words spoken over the telephone back in 1876.

"Mr. Watson, come here, please, I want you!" intoned Bell.

"It would take a week to get there now," came the reply.

AT&T's transcontinental line was inaugurated just as the Great War threatened to engulf the United States. The conflict fanned the flames of industrial research. In January 1915, the very month of Ma Bell's big public relations splash, *Scientific American* launched a six-part series under the main banner: "Doing Without Europe." The opening line of the opening article trumpeted: "If the war will teach us the value of industrial research as an aid in the development of our industries, if it will induce manufacturers to engage technologists to devise processes for manufacturing the goods which hitherto they have purchased abroad, if it will open up foreign markets hitherto closed to us, it may prove a national benefaction."

The statement proved prophetic. Governments who lined up with the Allies—Britain, France, Italy, and the United States—swiftly expropriated German production and marketing interests on their home soil and turned them over to national firms. When the war opened, U.S. corporations doled out an estimated $50 million annually for European technology. The British blockade of Germany severed the flow of dyes, drugs, chemicals, chemical and optical glass, various metals and alloys, and scientific instruments into the country. War-pressed Britain also used its domination of Chilean concerns to soak up the raw nitrates needed in munitions and fertilizers.

Both factors sent American firms scurrying to develop their own goods. Consultant Arthur D. Little, by then president of the American Chemical Society, used his pulpit to lambaste the U.S. chemical sector for not keeping pace with the Germans in coal tar and organic chemistry. "The plain underlying reason is to be found in the failure of our manufacturers and capitalists to realize the creative power and earning capacity of industry research," he declared. Uncle Sam pitched in with programs to promote the ability of chemical and pharmaceutical makers to take up the slack: both industries benefited from the Trading with the Enemy Act, which enabled the appropriation and open licensing—for only a small fee—of some 4,500 German patents.

Industrial research became a watchword for a strong and vital nation. Companies in a host of sectors took up the gauntlet. Kodak and its fledgling lab, along with hard-driving DuPont, which had consolidated research under the chemical department in 1911, served as templates for how to diversify product lines. General Electric, the shining star of industrial science, began developing a portable X-ray unit and delved into making night flares through a novel means of manufacturing magnesium powder, until then a German monopoly. In 1916, Willis Whitney and former M.I.T. classmate George Ellery Hale, the astrophysicist and Caltech founder, took the lead in fashioning the National Research Council, which included an Advisory Committee on Industrial Research. Proclaimed Hale early the next year: "War should mean research. . . ."

Corporate research continued to explode after the November 1918 armistice—and not just in the United States. In Europe, the stiff reparations imposed on Germany under the Treaty of Versailles weighed down the Reich's companies and opened a world of opportunity for competitors outside the Axis. Swiss electrical equipment manufacturer Brown,

Boveri; compatriot chemical makers Geigy, Sandoz, and Ciba (Chemis-
che Industrie zu Basel); Swedish telephone equipment leader L. M.
Ericsson; Dutch electrical concern Philips; and Britain's General Electric Company—no relation to the one back in Schenectady—all expanded dramatically during the war, with research a key to their growth.

Still, American companies led the way. Their rising economic
power, growing links to science, and the squeeze on German rivals all
contrived to give the Yanks a formidable leg up. So did Uncle Sam. The
war had spurred the government to create the National Advisory Committee on Aeronautics, precursor to the National Aeronautics and Space
Administration, which generated an impressive body of aerodynamic
knowledge important to government and industry. Following the conflict, the Naval Research Laboratory pushed electronics and communications, as well as weapons development.

Meanwhile, *Scientific American* and other cheerleaders for industrial research found wide audience. After all, airplanes, trucks, and various chemicals had revolutionized warfare and were rapidly reshaping the
civilian sector. Radio shrunk the wide open spaces, bringing Washington
to the people, along with sporting events and the latest news. Automo-
biles and streetcars eased the burdens of daily commuting. Electrical and
machine technologies transformed the household through washers, vac-
uums, refrigerators, lighting, and even synthetic fibers like rayon. All
told, the United States was spending some $200 million a year on scien-
tific research coming out of the war, with industry contributing twice as
much as Uncle Sam. Scores of articles proclaimed the superiority of corporate studies over the old Edisonian cut-and-try methods. "Science is
not a thing apart," asserted one account in the *Saturday Evening Post*, "it
is the bedrock of business."

It was in this climate that the General Electric Research Laboratory
found fame as the House of Magic—but Whitney's pioneering organization hardly stood alone anymore. The national frenzy for R&D had revamped the American industrial-scientific landscape. By 1921, according
to the National Research Council's inaugural survey of American corporate labs, about three hundred firms practiced research. Located almost
exclusively in the northeastern industrial belt, these operations employed
about 9,300 staff members—including 2,775 research scientists and
engineers—already probably four or five times the prewar figures and
rising.

While the electrical machinery and chemical sectors dominated the
list, research arms were springing up in virtually every major sector of the
American economy—the more so as managers realized that centralized

research held tremendous leveraging potential. Physicists and engineers could tackle everything from vacuum tubes to light bulbs, from telephony to medical equipment. Chemists could churn out paints and varnishes, plastics, and novel fibers. Researchers could work on the far-off future in hopes of launching new businesses, or they could ride in on white horses to resolve a production problem. Even better, they could do both, shifting emphasis as dictated by the times—just as General Electric had demonstrated. In this way, concludes Alfred D. Chandler, Jr., "The research organizations of modern industrial enterprises remained a more powerful force than patent laws in assuring the continued dominance of pioneering mass production firms in concentrated industries."

AT&T was one of the prime movers. Following a wartime hiatus, the company decided research and development warranted its own identity. On January 1, 1925, new president Henry B. Thayer opened the Bell Telephone Laboratories—a separate corporation with Frank Jewett the first president—at the same West Street location that had long housed research. Although commonly associated solely with research, Bell Labs was from the start a research *and* development operation, with the "D" side dominating its activities. Still, the active research arm under Harold Arnold expanded into all areas of electrical communication, from general physics to vacuum tubes, radio, and electro-optical phenomena. By 1935, the department had swollen to eight hundred staff members— about the size of Whitney's operation—of whom fifty-one held doctorates and more than three hundred wielded bachelor's or master's degrees.

Though always in the shadow of Bell Labs and GE, Westinghouse recovered from a 1907 bankruptcy filing and looked to centralized research to gain firmer footing. In 1916, with lamp studies broken off as a separate activity in Bloomfield, New Jersey, the Research Laboratories got new digs on Ardmore Boulevard, about a mile from the company's East Pittsburgh headquarters. Throughout the interwar period Westinghouse continued as a force in electric lighting: it installed the generating equipment at Boulder Dam. The company also broadened into home appliances like irons, percolators, and waffle makers. On November 20, 1920, the first scheduled radio broadcast, of the Harding-Cox election returns, aired over Westinghouse Station KDKA. By the late 1930s, the expansion encompassed television, X-ray equipment, and nuclear physics—moves reflected by research into chemicals, insulation, metallurgy, and the construction of a gigantic, pear-shaped atom smasher on lab grounds. Each year, ten young people won Westinghouse Research Fellowships, which included a $2,400 annual stipend and an invitation to come to the lab and study problems of their own choosing.

Chemical concerns were also on the move. At DuPont in 1914, only

3 percent of sales came from outside its line of explosives. Within seven years, partly to counter monopoly charges, the company had diversified into dyes, films, rayon, pigments, lacquers, and paints—and by 1939 the original mainstay provided less than a tenth of total income. American Cyanamid, an early maker of synthetic fertilizers from calcium cyanide, opened a research lab in 1912 and expanded steadily after World War I into mining chemicals, pharmaceuticals, and coal-tar products. Monsanto Chemical, formed in St. Louis in 1902 to produce saccharine, fanned out into phenol and other fine chemicals—then sidled into intermediates and heavy chemicals. For Dow Chemical, the Great War's disruption of vital European supplies stimulated production of phenol, indigo, and other organic compounds. In 1919, with prices plunging and Dow facing huge building and machinery losses if it could not find ways to exploit its capital investments, a contingent of organic chemists initiated a long period of feverish research that spilled over to inorganic chemistry, biochemistry, physics, and metallurgy. By helping push the company into aspirin, dyes, perfumes, a wide array of chemicals, and magnesium metal alloys for airplane parts, machinery, and portable tools, the effort established Dow as a tower of industrial research.

Good things in research did not come solely through in-house activities. American Cyanamid broadened out from its calcium-based products largely by acquisition. Among its postwar buyouts were the Calco Chemical Company, which had a strong coal-tar chemistry lab in Bound Brook, New Jersey; and Lederle Laboratories in Pearl River, New York, which remained Cyanamid's center for the study of vaccines, serums, and biologicals. These complemented the company's new research facility in Stamford, Connecticut, where scientists investigated mining chemicals and pharmaceuticals. Similarly, Monsanto got much of its research in 1936 by merging with Dayton-based Thomas and Hochwalt Laboratories, to which it had turned increasingly for contract work.

Whatever the route, for the chemical sector diversification held the key to growth—and research and development provided the pathway to that diversification. Those that didn't make the necessary investments faded away. Columbia Carbon, Publiker Commercial Alcohol, and United States Industrial Alcohol, all among the nation's top two hundred firms in the 1920s and 1930s, stuck to narrow product lines with minimal research investment. By 1948, all had fallen off the list.

Rubber and tire manufacturers got R&D religion chiefly in the two decades that followed 1910. B. F. Goodrich had formed the industry's first lab back in 1895, under chemist Charles C. Goodrich, eldest son of the founder. Still, it wasn't until the automobile market picked up that research took off. In 1912, United States Rubber chemist Raymond B.

Price set up a Development Department—including central research facilities—and moved into crude rubber processing and product improvement. From the same intermediates used to stamp out tires, the company pioneered latex thread and sprayed rubber. Goodrich also leveraged research, taking the lead in plastics, polyvinyl chlorine resins, and a variety of rubber-based chemicals. The NRC's initial R&D survey in 1921 put the industry fourth in total technical personnel, after electrical machinery, chemicals, and metals. It trailed only chemicals in research intensity—the ratio of scientific and engineering staffers per thousand workers.

The oil industry, famous for its iconoclastic moguls, showed up as a relative latecomer to research. Scientists and engineers had played a role in the sector since the late 1800s, but almost always in an operating capacity. Even as the Great War ended, virtually all research remained centered on testing and refinery control, though things had been shaking loose in the increasingly competitive climate that followed the 1911 antitrust breakup of Standard Oil Company of New Jersey. Finally, in 1919, president Walter Teagle of post-breakup Jersey Standard wrote board chairman A. Cotton Bedford: "The General Electric and other concerns of a like character lay great stress upon their research department. They consider this department on a parity in importance with the manufacture and sales end of their business. Our research department up to date is a joke, pure and simple; we have no such thing."

Jersey Standard sparked the industry's move into widescale research that same year—bringing together twenty-six analytical and research chemists and three chemical engineers to create its Development Department. Other majors dove in throughout the first interwar decade, spurred on by the auto boom, the discovery of vast reserves, and the rise of international oil companies, all of which heightened competition and put a premium on improving efficiency through innovation. Standard Oil of California launched a research division in 1920; its Indiana counterpart, which had opened a research lab in 1906, dramatically expanded scientific studies in 1922, incorporating the laboratories of three refineries into a research department. Two years later, Atlantic Refining Company started an R&D enterprise, originally called the Process Division. Shell Oil sallied forth into research in 1928, creating the Shell Development Company in Emeryville, on the eastern shores of San Francisco Bay: initial staff consisted of twelve university-trained researchers and forty-five others. By then, Jersey Standard had consolidated research and development into the specially created Standard Oil Development Company. Within two decades, this body would grow to 2,500 workers and become the industry's largest R&D operation.

Early oil industry research focused on refining—separating the stew of hydrocarbons found in crude oil by molecular weight for chemical processing. Lighter fractions go to make gasoline, the intermediates beget kerosene, waxes, and lubricating oils, while heavy crude is cracked—or split—into kerosene.

The first successful gasoline cracking process was invented in 1910 by Standard Oil of Indiana's William Burton, future chief executive of the Standard Oil Corporation. By the late 1930s, thermal cracking techniques like Burton's gave way to catalytic methods that enabled still higher octanes. Simultaneously, various industry research groups helped spawn novel branches of chemistry by devising ways to make synthetic rubber, fertilizers, detergents, alcohols, and insecticides. Notable was the Gulf Research & Development Company, which grew out of a series of Gulf Oil Corporation fellowships set up in 1927 and 1928 to investigate chemical, metallurgical, geophysical, and engineering problems at the newly incorporated Mellon Institute of Industrial Research in Pittsburgh. Three years into the fellowships, the work was transferred intact to a nearby facility, kickstarting the research department of the Gulf Production Company, a Gulf subsidiary. With initial divisions covering chemistry, corrosion, engineering, geology, geophysics, physics, and materials engineering, the operation grew so fast it soon merited special status as a separate Gulf subsidiary. In 1935, a huge research complex went up on a 57-acre tract near Harmarville, Pennsylvania, about 13 miles northeast of Pittsburgh. A prototype for the university-like research haven, the facility encompassed three large lab buildings and did much to establish the oil giants at the forefront of industrial science.

Of all major American industries, R&D probably played its smallest role in steelmaking. The Illinois Steel Company had taken a modest plunge into research with the 1891 hiring of Albert Sauver, a Ph.D. chemist, to undertake a microscopic study of steel: his work lasted five years. Both Republic Steel and U.S. Steel Corporation, formed in 1901 via a mammoth consolidation that made it the largest company the world had ever seen, waited until around 1928 to establish central laboratories. Though at least four U.S. Steel subsidiaries had operated scientific facilities since 1915 or earlier, the company viewed research almost as a showpiece. Directors tried to recruit Caltech president Robert Millikan to head the central lab, but board minutes show some executives worried more about the Nobel laureate's ability to play golf than direct research. When Millikan refused to be considered, the job was offered to Bell's Frank Jewett, who also demurred. Finally, at Jewett's urging, the company selected Yale chemist John Johnston, who arrived to find his division consigned to old quarters in Kearney, New Jersey, with far less funding than promised.

The pharmaceutical and auto sectors supplied most of the remaining players in American corporate research between the two world wars. Parke, Davis ranked as the drug industry's research leader throughout the early 1900s. The scope of investigations increased during World War I, as it moved into vitamins, hormones, and medications for leprosy, epilepsy, syphilis, and various allergies. By the Second World War, under research director Oliver Kamm, its laboratory would expand to sixteen divisions. In 1938, E. R. Squibb & Sons opened the Squibb Institute for Medical Research in New Brunswick, New Jersey, launching studies of organic chemistry, virology, bacteriology, pharmacology, and experimental medicine. Also active in research was Abbott Laboratories of Chicago, which had been formed in 1900 by Wallace C. Abbott, a doctor whose initial "lab" for making alkaloids had been an annex to the family kitchen.

Among early carmakers, General Motors emerged as the only viable research force. Formed in 1908 by William C. Durant, three years after he created the Buick Motor Car Company, GM stood as the only major competitor to assembly line pioneer Henry Ford. Durant's shaky management led him to twice seek financial assistance from the DuPont Company. At the second bailout, in 1920, Pierre S. du Pont was installed at the General Motors helm. Along with protégé Alfred P. Sloan, Jr., who became GM president in 1923, the chemical company executive honed Durant's array of poorly organized operating units into a well-tuned, multidivisional enterprise—in the process defining many standards of modern management and turning GM into the world's biggest automaker. The men revamped production, distribution, purchasing—and research, creating a centralized department aimed at delivering consistent improvements to their many road machines. The job of leading this unprecedented effort went to Charles Franklin Kettering.

One of America's hallmark inventors, "Boss" Kettering had teamed up with businessman Edward A. Deeds in 1909 to organize the Dayton Engineering Laboratories Company (Delco), intended as a research organization to serve the emerging auto industry. His inventions there included the automotive self-starter, the first electrical ignition system, and the earliest practical engine-driven generator—the "Delco-Light"—which provided electricity for hundreds of thousands of farms. But the organization quickly mutated from a research house into a manufacturing concern; not long before the United States entered World War I in 1917, Kettering sold out and established the Dayton Research Laboratories Company, intending to return to automotive research.

Kettering's desires dovetailed perfectly with Sloan's. Since 1911, General Motors had maintained a small research department—mainly for materials analysis and testing—set up in Detroit by Arthur D. Little.

By buying out Delco, and then, in the wake of the Great War, Kettering's new operation, the GM executive raised the stakes. In June of 1920, the General Motors Research Corporation was constituted at Moraine City, Ohio, six miles south of Dayton, with the Boss as president. Kettering set up shop in the old Dayton Wright Airplane factory, carving up its cavernous expanse into forty-four individual labs, each 20 feet by 30 feet and outfitted with vacuum facilities, compressed air, and an array of electrical outlets. The lab was unique in the industry, and Kettering made so many off-the-cuff remarks about breakthroughs—air-cooled engines and 60-mile-a-gallon cars—that senior execs worried about industrial espionage. The Boss countered that openness attracted people and ideas—and made for great publicity. As a compromise, though, he set up a special "tourists'" lab populated by fancy instruments and white-coated technicians engaged in seemingly important but actually trivial work; the good stuff stayed behind closed doors.

The lab quickly proved its importance to General Motors. But the great distance between Dayton and the Detroit headquarters kept Kettering on the road and strained lines of communication. Some five years after it formed, by November 1924, the operation transferred to the Motor City and absorbed GM's original research unit: unhappy with the move, engineers Charles Thomas and Carroll Hochwalt quit and founded their own lab, the very operation that later formed a key part of Monsanto's research. Meanwhile, Kettering's shop flourished. Early in 1929 it moved into the upper floors of an eleven-story brick building on West Milwaukee Avenue in downtown Detroit, where the Boss had a huge office with rubber furniture, fluorescent lights, and his own vise-equipped workbench. The lab achieved full divisional status in 1938, by which time it had nearly four hundred staffers, including one hundred scientific and technical workers—among them eighteen chemists, nine metallurgists, four physicists, and sixty-nine engineers. The place ran so informally a *Fortune* reporter claimed he could hardly distinguish department heads from lab technicians or garage mechanics. But before Kettering retired in 1947, he had built up an impressive technical library that included 23,246 volumes and 1,398 scientific reports. Meanwhile, the operation delved into everything from transmissions to metallurgy and fuels.

Although heavy hitters dominated American industrial research, the little guy still had a few options. In 1886, Arthur D. Little had started a firm to conduct chemical analyses for companies lacking in-house facilities. Within a few decades, this had grown into a strategic consulting house that even took on R&D projects. Two types of alternative organizations had also emerged. One was the trade association lab, pioneered by

such organizations as the Institute of Paper Chemistry, a joint venture between paper concerns and Lawrence College in Appleton, Wisconsin.
Although popular in Britain, however, these found little favor in the
United States. Instead, companies seeking research assistance turned either to consulting labs like Little's or the remaining alternative: technological research institutes that contracted to study specific problems.

The two most significant of these organizations to emerge before
World War II—the Mellon Institute of Industrial Research and the Battelle Memorial Institute—sprang from private fortunes. Mellon traced its
origins to University of Kansas chemistry professor Robert Kennedy Dun-
can, who in 1907 created the country's first industrial fellowship: a project aimed at improving laundering technology for a Boston company.
His advocacy of university programs to study industrial problems impressed Pittsburgh bankers Andrew and Richard Mellon, who invited
him to start an industrial research department at the University of Pittsburgh. The first fellowship began in 1911. Sixteen years later, the organization shed its university ties and incorporated as the Mellon Institute of
Industrial Research.

In the mid-1920s, Mellon's scientists "put cotton shirts on wieners"
by creating the edible cellulose tubing that enhanced frankfurter attrac-
tiveness. The next decade the institute spun off the Gulf Oil research
arm. In 1940, while trying to create a silicone rubber, researchers discovered a moldable material that astounded visitors when it was bounced off
laboratory walls and ceilings: after World War II, the material was marketed as Silly Putty. In 1967, with 1,600-odd patents and some 650
processes or products to its credit, the institute would merge with
Carnegie Institute of Technology to create Carnegie-Mellon University.

Battelle fulfilled the last testament of Gordon Battelle, heir to the
Ohio coal, iron, and steel fortune of his father John G. Battelle. In 1923,
at the age of forty, Gordon died following an appendectomy, leaving
some $3.5 million to endow an institute. The Columbus-based nonprofit
opened its first lab in 1929, and by concentrating on metallurgy grew
into the nation's biggest independent institute. Early projects included
developing corrosion-resistant steel and pioneering a variety of uses for
copper. During World War II it secretly assigned several hundred staffers
to study uranium for the atomic bomb project—and would later create
an important ion exchange process for extracting uranium from its ores.
The landmark payoff, though, came from helping Chester Carlson perfect his 1937 xerography invention; for its work, the institute eventually
received more than $40 million in cash and stock from the resulting
Xerox Corporation.

■

Barely zinged by the Great Depression, the magic flowed out of American corporate research labs throughout the interwar years. In 1937, five years after Irving Langmuir won his Nobel Prize, Clinton Davisson at Bell Labs received the physics Nobel for his co-discovery of the wave nature of the electron. From General Motors, where Kettering reigned, came four-wheel brakes and the lightweight engine that made diesel locomotives feasible. GM researcher Thomas Midgley, Jr., labored for years before hitting on tetraethyl lead to control engine knock, then took three days to come up with Freon.

None of these could match DuPont, however. Through a combination of its own research, acquisitions, and various licensing and patent deals, the chemical titan turned out a flood of innovation that included the commercialization of Midgley's discoveries, viscose rayon, moisture-proof cellophane, Teflon, neoprene—and nylon.

The last two were done entirely in-house. In 1927, partly in response to Herbert Hoover's strident summons for more basic research, but even more because plastics, paints, cellophane, and other company products were polymers or high-molecular-weight compounds, DuPont had established a fundamental program of polymer chemistry under Charles M. A. Stine.

The venture operated out of Purity Hall, a special lab erected at Experimental Station, the company's main research campus of ivy-laced yellow-brick buildings on the site of a small textile mill looking over the Brandywine River in Wilmington, Delaware. Beyond the program's intrinsic value, Stine hoped that a commitment to fundamental research might attract first-rate chemists. He was partly right. Established academics didn't bite. However, a year after the effort opened, a brilliant up-and-coming organic chemist—thirty-one-year-old Wallace Hume Carothers—deserted a Harvard instructorship to head polymerization work. Like Willis Whitney before him, Carothers worried about the price of corporate life on his soul. However, he soon reported, "a week of industrial slavery" had not crushed his "proud spirit." Moreover, funds flowed like wine, and he pleasurably spent his time "thinking, smoking, reading, and talking."

In March 1930, the group Carothers presided over made two major discoveries. The first was a polymer he dubbed chloroprene. It possessed remarkable rubber-like qualities but proved far more resistant to degradation by oil and gasoline: later renamed neoprene, it marked the first all-synthetic rubber. Carothers next created a polyester that could be spun and specially stretched into a superstrong filament. Although the substance lacked stability in water and melted too easily to be marketable, it represented the world's first completely artificial fiber and sparked

Carothers to hunt for more viable alternatives. The search led to his discovery of the polyamids, a special polymer class. In these in 1934 he found Fiber 66: nylon.

DuPont announced nylon hosiery in late October 1938. On the 29th, the same Saturday that Orson Welles sparked a nationwide panic by broadcasting "War of the Worlds," a *New York Times* editorial questioned whether women would be content with stockings that lasted for eternity. Nylon went on to earn DuPont some $25 billion by the 1990s. But Carothers never saw its tremendous payoff. Increasingly, the talented chemist began to feel he had deserted science, growing susceptible to bouts of depression. In April of 1937, just forty-one years old, he poisoned himself with cyanide in a Philadelphia hotel room.

As the fruits of corporate labs deeply impacted everyday life—in radio, telephony, cars, hosiery, and beyond—and brought fame and fortune to inventors and corporations alike, the message came through loud and clear that research paid. Charles Kettering, a gas-pump oracle on technology and business, assured executives that research not only boosted profits, it won markets. Making the promise more tangible, the Revenue Act of 1936 declared corporate research expenditures tax-deductible. The National Research Council pitched in with a good cop–bad cop campaign, hawking industrial research as the "royal road to riches" while predicting doom and gloom for those failing to take the plunge. And it seemed to work. In 1938, the NRC established the Industrial Research Institute, an association for companies and research managers that soon became an independent corporation. Meanwhile, the share of U.S. patents granted to firms rose steadily from 18 percent at the turn of the century to 58 percent in 1936: it would remain around that level well into the 1990s. In certain sectors such as chemicals, corporations garnered upwards of 80 percent of all patents issued. A barn, a tool chest, a Bunsen burner, and a little imagination no longer cut it.

Rising European competition fueled the fire. By the mid-1920s, Germany had turned the corner on recovery. Within a decade its dye houses and electrical concerns—Bayer, Höchst, BASF, Siemens, AEG—had surged back into international markets. Meanwhile, in Britain, two rising stars catapulted onto the scene. Taking the chemical and drug sectors by storm was Imperial Chemical Industries. On the radio, broadcasting, and appliance front came Electrical and Musical Instruments. Both resulted from postwar mergers and could square off with continental and American giants on virtually every front—including research. By the early 1930s, ICI's science machine rivaled DuPont's, turning out scores of pesticides, pharmaceuticals, detergents, synthetic resins and lacquers, dyes, and rubber goods—culminating in its December 1935 invention of polyethylene. In a similar manner, EMI challenged RCA on the global stage.

All these factors—the promise of riches, tax incentives, and rising international competition—spurred American firms to embrace research and helped make "R&D" a familiar term in the public lexicon. In 1936, though industrial lab directors liked to complain that colleges bombarded students with nuclear studies at the expense of more practical knowledge, M.I.T.'s physics department received so many calls for graduates it gave up trying to fill the demand. From 1927 to 1933, the number of corporate research professionals in the United States rose from 6,320 to 10,927. Seven years later, on the eve of World War II, the field encompassed some 2,200 labs employing at least 27,700 professionals out of a total work force some 70,000 strong. Scores of companies began featuring research in annual reports.

The winds of the coming war whipped up the flames. By the fall of 1939, with fighting underway in Europe, the specter of American involvement in another conflict gave fresh impetus to those pushing for greater research commitments from industry and government. Long before the United States formally entered the war in December 1941, often fueled by government contracts, industrial research and development labs were hard at work on weaponry, aircraft, communications systems, electronics, sonar, and radar.

General Electric's research effort continued to flourish for a time, taking on a slew of military projects. Meanwhile, the House of Magic tour hit every Army training camp during the Second World War, putting on 2,308 shows for over one million soldiers, sailors, and marines. The show roared into the peace. Within just a few years of the Japanese surrender, the company claimed that, all told, upwards of 12 million had witnessed its roving spectacle of science and technology. Audiences packed theaters for an updated show that combined old "tricks" such as the voice-activated train with fresh acts like the one based on new phosphorescent materials that allowed their host to shake hands with his shadow.

But things were changing. Amidst several shifts in leadership and more than a few corporate wrong turns, GE's research house would lose much of its luster in the coming years, forcing the enterprise to reinvent itself several times before emerging in robust health towards the century's end. In the meantime, standout innovations from DuPont, General Motors, Gulf Oil, and others brought a host of rival acts into the limelight.

Nowhere was the changing of the guard more evident than on the last day of June in 1948, as scores of New York media descended on the

yellow-brick Bell Labs headquarters in Manhattan to view the latest illus-
tration of industrial light and magic. Press members didn't know it at the
time, but they were witnessing as well the dawn of a new era of industrial
research.

An elbow-to-elbow crowd swarmed into the small Bell Labs auditorium that summer's day. Onstage before the pack of reporters stood bow-
tied research director Ralph Bown. The placard at Bown's feet told the
nutshell story: *"The TRANSISTOR."* But behind him an array of props
and paraphernalia provided the fuller picture. Along with an oversized
schematic diagram and a 100-to-1 scale model of the transistor on
wheels, a pair of towering posters showcased the device's presumed applications in radio, television, and telephony. The research czar would even
speak through a telephone handset into individual receivers to elucidate
the invention's amplifying prowess.

Bown's first words, though, did not relate to the transistor but rather
to industrial research. "Scientific research is coming more and more to
be recognized as a group or teamwork job," he began. "What we have to
show you today represents a fine example of teamwork, of brilliant individual contributions and of the value of basic research in an industrial
framework."

Corporate research was primed to enter its golden age—and those
opening remarks from Bown spoke volumes about the direction it would
take. In the next few decades, the number of corporate labs exploded
from a few thousand to more than ten thousand: these organizations employed nearly half a million workers. The order of the day was teamwork—scores of engineers and scientists from a variety of disciplines
pulling together toward a single end. To an unprecedented degree, science drove the teams.

This was, after all, the age of Big Science. Research had taken hits
in the 1920s, when some charged it was helping protect monopolies, and
again during the Depression, amid accusations that scientific advances
had driven people out of work. Suddenly, in light of the stunning
wartime achievements like the bomb and radar, scientists emerged as sav-
iors who not only knew the future but possessed the wisdom to handle it.
Alumni of Los Alamos and the M.I.T. Radiation Laboratory, a secret facility set up to develop microwave radar during World War II, took center
stage in setting American scientific policy.

Their central tenets were laid out by Vannevar Bush, wartime director of the Office of Scientific Research and Development. A national
hero who had overseen virtually every civilian military-science project
during the war—spanning radar, the bomb, and all medical research—

Bush set the tone in his July 1945 report to President Truman: *Science—
The Endless Frontier.* The landmark document argued that federal support for basic science held the secret to global leadership and ongoing
economic wealth and security. Although much of the role Bush
envisioned for the National Science Foundation was usurped by the
Office of Naval Research, and later the Atomic Energy Commission
and NASA, the report foreshadowed the U.S. government's decision to
invest billions to keep the nation at the forefront of fundamental science.
"After World War II, the rest of the world was more adversely affected
than us," explained Dean Eastman, a former IBM Fellow who became
director of the Argonne National Laboratory near Chicago. "We em-
barked on the policy that science and technology were the keys to the
future."

Apart from a decade-long flat period that began in the mid-1960s,
when a series of inflationary pressures that included the 1973 Arab oil
embargo dampened the U.S. economy, the spigot stayed open through
the Cold War. The first Soviet atomic bomb test in 1949, the Korean War
early the next decade, the 1957 launch of Russia's Sputnik satellite,
heralding the age of intercontinental ballistic missiles and the Space
Race, and Ronald Reagan's Strategic Defense Initiative of the 1980s—
again raising the specter of the Evil Empire—all kept funds flowing.
Most of the money streamed into academic research. But especially
when it came to military systems and defense-related areas such as electronics, computers, communications, and aerospace, Uncle Sam spent
big on industrial labs as well. In 1962, American industry's total research
and development budget reached $11.56 billion. Some 58 percent came
from the Feds, the first time in history that Washington provided the
lion's share of industrial R&D funding.

Labs popped up at a rate surpassing one a week throughout this period. In the physical sciences alone, activities proliferated on a staggering
array of fronts: color television, wireless telephony, semiconductors,
lasers, optical fibers. In the 1950s and 1960s, many U.S. firms also established or acquired labs in Europe. The global push either anticipated or
responded to the 1957 creation of the European Economic Community,
as companies moved to line up foreign talent, stay abreast of university
science in Britain, France, West Germany, Italy, and other nations, get a
leg up on modifying products for EEC markets, and lower research
costs—the same considerations that in the late 1980s and early 1990s
would drive them to open arms in Japan. If that weren't enough, more
than a hundred applied research institutes were formed in the aftermath
of World War II to support the work in corporate labs, as well as pursue
government projects. Among them: the Franklin Institute Laboratories

in Philadelphia and the Stanford Research Institute in Menlo Park, California.

In concert with this great expansion, the mission and style of corporate research also shifted. Early efforts focused chiefly on improving products and processes, with companies sustaining most of their growth by finding additional markets for existing technologies. But in the wake of the Second World War product innovation became as important as product development.

One factor in this metamorphosis was the fortification of antitrust policy. In 1950, mounting criticism over the concentration of economic power in large firms led to tough enhancements to the Clayton Antitrust Act of 1914 that made it extremely difficult for big companies to acquire or merge with firms in related fields. In the 1950s, the government launched a major antitrust suit against General Motors and DuPont, which owned about a quarter of GM's stock. The next decade IBM came under fire; in the 1970s it proved AT&T's turn again.

Competition also spurred this evolution—though in somewhat contradictory ways. On one hand, the dearth of international competitors resulting from the defeat of the Axis forces, especially the great weakening of Germany's electrical machinery, chemical, and pharmaceutical giants, allowed American concerns to break into foreign markets, providing the revenue stream to bankroll more research and development. At the same time, growth in research was also fueled by *rising* American competition. Often thanks to research, the boundaries between industries became blurred as corporations expanded into adjacent areas—as when the oil industry barged into chemicals and synthetic rubber, or when communications company AT&T and electrical giant General Electric converged on the emerging radio market. The barrage of competitors sometimes caused market conditions to fluctuate rapidly—and research became a vital tool for coping with the turmoil. "In the face of constant change," writes historian Howard Bartlett, "industries maintain their stability only by being prepared for the next advance."

In light of such pressures, establishing or beefing up central labs became almost a craze comparable to goldfish swallowing—one demanded by stockholders and Wall Street. Lee Davenport, then at the research helm for General Telephone and Electronics, still shakes his head over the postwar boom. "In the fifties and sixties, R&D was God as far as the world of Wall Street was concerned," he relates. "If you had an R&D program, even if they didn't understand it, you were a progressive company."

It was a time full of magic shows like the one at Bell Labs announcing the transistor. Davenport's operation, housed in Bayside, New York, did vital work. General Telephone owned Sylvania, and during his tenure the lab patented the bright red phosphor later ubiquitous in color television displays. However, the company was not above gee-whiz image building. In the pre-optical fiber days of 1963, purely as a publicity stunt to demonstrate light's capacity to carry signals, Davenport appeared on the popular TV show *I've Got a Secret* with a scheme for transmitting television pictures via lasers. When panelists failed to guess his "secret," he revealed the laser. Smoke blown across the stage allowed viewers to see its red light. Davenport then interrupted the live broadcast's video portion by placing his hand across the beam.

Riding the science wave, companies often elevated research to divisional status for the first time. Labs sprang up in pastoral splendor, far from the madding crowd but outfitted with elaborate facilities rivaling or surpassing those in the best universities. A few years after the war, General Electric research moved to a hilltop outside Schenectady with a stunning view of the Mohawk River: barely was the facility complete than officials deemed it inadequate—and by the mid-1950s the lab had expanded another 50 percent. Labs were constructed with movable walls to ensure future flexibility and equipped with high-intensity fluorescent lighting and key support services such as gas, compressed air, and water. Separate structures were erected to house hazardous work like nuclear studies and X-ray research.

Benevolent, science-friendly management ruled the day. Many firms sponsored colloquia and dispatched researchers to one or more scientific conferences each year. Coffee areas and pleasant lounges punctuated lab hallways; the grounds often included recreational facilities and park-like reserves for informal conclaves or quiet contemplation. It was not unusual for young scientists fresh out of college to be staked to a year or more to decide what avenues to pursue. When physicist Henry Ehrenreich arrived at the General Electric Research Laboratory early in 1956, armed with his doctorate from Cornell University, he was greeted by the head of the semiconductor section. Relates Ehrenreich, "I asked him to suggest some research topics that might be germane to the interests of the section. He said that what I did was entirely up to me. After recovering from my surprise, I asked, 'Well, how are you going to judge my performance at the end of the year?' He replied, 'Oh, I'll just call up the people at Bell and ask them how they think you are doing.' "

Big Science had sired Big Industrial Science. To be sure, the bulk of activities were still heavily oriented toward applied projects. But giants like AT&T, DuPont, General Electric, not to mention newer forces such as Kodak, General Motors, and IBM, put the defining stamp on the times by diving deeper into fundamental research. Papers from industrial labs sometimes dominated scientific journals, most notably the *Physical*

Review Letters. Researchers could advance either through their contributions to the company—or their science. Notes the historian of science David Hounshell, "The shibboleths of this new age were that basic science and well-funded scientists produced dramatic new technologies and that scientists knew better than generals, engineers, or industrialists what science to pursue, which new technologies to develop, and how best to deploy those new technologies."

A myriad of forces beyond steamrolling congressional support for fundamental research created a vortex that pulled companies toward this strategy. Recent advances in atomic energy and solid-state physics—leading via the transistor to the age of semiconductors—encouraged firms to try to master the science underlying these areas. On a more down-to-earth level, as product lines kept expanding in scope and firms stampeded into each other's turf, it became harder and harder for one company to dominate. To gain an edge, many corporations began bankrolling longer-term projects and directing research activities closer to the forefront of science, where fundamental principles applicable to many products might be uncovered. Accentuating this trend, growth in research came as part of a broader expansion of all business operations, including development, which took up much of the shorter-term work previously handled in central laboratories. General Electric, for example, had created a few development labs by the early 1900s. The figure reached about a dozen before World War II, including five in Schenectady alone—for materials testing, general development, engineering consulting, illumination, and insulation. As the postwar era brought the number of development labs to twenty, central research found increasing room for fundamental investigations.

The sentiments of scientists also shaped the course of corporate research. For nearly a decade after World War II, it had not been particularly hard for labs to find researchers. But things began to change as the Cold War heated up. Government research and development outlays soared from $3 billion in 1956 to $15 billion nine years later, of which 14 percent was allocated to basic research. Reveling in the windfall, universities stepped up enrollment. A fresh flock of students entered science, kicking the average growth in the number of physics doctorates awarded upwards of 5 percent a year. Apparently drawn by Uncle Sam's largesse, this latest crop of physicists chose overwhelmingly to stay in academe. Only 180 freshly minted physics Ph.D.s joined corporate labs in 1965; 500 stayed in universities.

Firms seeking to land first-rate recruits had little choice. Even when offered salaries substantially beyond those of academe, many scientists did not exactly warm to the idea of working for an industrial lab. To lure standout graduates, corporations had to offer wide liberties in choosing projects, the freedom to publish in academic journals, and more opportunities to work on the cutting edge of science. Stories are rampant throughout much of the 1950s, 1960s, 1970s, and even the 1980s, of do-what-you-want cultures, projects without hard goals or timelines, and a general research free-for-all. Relates historian Margaret B. W. Graham, "The new breed of researchers would neither be harnessed nor directed, as their predecessors had been. Rather their ideas had to be contained and channeled with the hope that an occasional breakthrough discovery of immense value would compensate for much that was not needed or not useful in the commercial context."

The big three industrial research pioneers—General Electric, DuPont, and AT&T—continued to point the way. Thanks primarily to Irving Langmuir, GE had maintained a small core of fundamental studies leading up to World War II. Amidst the postwar fervor for science and technology, and even before the new lab on the Mohawk was ready, research director C. Guy Suits boosted those efforts significantly. Suits had joined the lab in 1930, then taken leave during the fighting to head Harvard's Applied Physics Laboratory, home to the lion's share of U.S. electronic countermeasures development. Back at General Electric in 1945, he initiated basic meteorological studies, mainly for the military. Langmuir himself studied precipitation and icing conditions. On November 13, 1946, his assistant Vincent Schaefer seeded a supercooled cloud with dry ice and demonstrated the first artificial precipitation. In sync with GE's entry into nuclear power, the lab opened fundamental atomic studies. Basic or quasi-basic semiconductor and materials research also got off the ground. The metallurgical forays led to superior alloys for jet engines and the 1955 creation of the first synthetic diamonds.

DuPont picked up where it had left off with nylon. Way back in 1935, the company had unveiled a new slogan: "Better Things for Better Living Through Chemistry." Barely two months after World War II ended, in October 1945, chemical director Elmer K. Bolton reported to senior management that the way to live up to that commitment was by aggressively expanding fundamental studies. To "retain its leadership," he argued, DuPont needed "to undertake on a much broader scale fundamental research in order to provide more knowledge to serve as a basis for applied research."

The company adopted the idea with more gusto than Bolton intended. DuPont at the time consisted of eleven largely autonomous operating departments. Each ran its own marketing, manufacturing, development, and research arms—in addition to Bolton's central Chemical Department that pursued basic studies. Crawford Greenewalt, who took over as DuPont's president in early 1948, proceeded to establish parallel fundamental programs in all eleven departments, decreeing that each would turn out its own "new nylons." The move came over the vigorous objections of Charles Stine, risen to vice president for research, who felt that since departments were charged with serving the needs of existing businesses, they would not be able to devote sufficient resources to longer-term projects. His advice went unheeded. To avoid duplication of fundamental studies taken up in the various departments, the central research arm pushed into even more basic arenas, including those where the company had little, if any, expertise or interest. Meanwhile, to accommodate their ambitious research campaign, senior officials uprooted the company golf course and authorized a $30-million, nineteen-building expansion of the Experimental Station campus. The addition was dedicated in May 1951.

Over at AT&T, the third member of the basic research triumvirate, the transistor proved only the beginning of a massive expansion into fundamental science. Even before World War II ended, the lab had abandoned New York City's noise and grime for an artfully landscaped plot of land in Murray Hill, New Jersey. It was in that haven that then–research director Mervin Kelly ramped up the small prewar program of solid-state research that led to the transistor. But for young Bell Labs researcher Walter L. Brown the glory days of science really didn't start until Sputnik went up in 1957. When the Harvard-trained physicist arrived at Bell in December 1950, almost three years to the day from the transistor's birth, he found researchers studying semiconductors at all sorts of levels—but primarily with an eye toward commercial applications that would make an impact on society.

Then came the Russian satellite. Suddenly, science, as opposed to technology, took off. The number of disciplines represented at the lab grew tremendously—not just physics, chemistry, and physical chemistry, but biology and astronomy. Bell took on a variety of space studies. Some dealt with the upstart field of communications satellites: AT&T put up its Telstar satellite in 1962. But researchers were also detecting and measuring charged particles in space, seeing possible problems with the particles interacting with semiconductors on Earth. Some probed nuclear physics, still others biological processing—anything broadly related to communications and materials science. "You started seeing people reaching out in all kinds of scientific directions," states Brown. "Sputnik reset the clock."

The bountiful commercial payoffs of fundamental investigations at GE, DuPont, and Bell Labs—revolutionizing consumer items like lights, stockings, and portable radios—helped convince other corporations to follow the big three towards the frontiers of science. All these efforts reflected the effects of World War II, Sputnik, and the Cold War.

Automakers bought into basic research soon after the Second World War. Ernest Breech, former manager of General Motors' Bendix Division, had been hired to help retool the ailing Ford Motor Corporation. In the fall of 1949, he memoed Henry Ford II: "I am convinced that Ford will not have many 'firsts' unless we get a few good thinkers and have a real research department." In mid-1951, Ford opened a "scientific laboratory" at its Dearborn engineering facility that was dedicated to "fundamental research and development in fields broadly related to the basic character of Ford Motor Company—transportation." Departing starkly from Henry Ford's practical bent, the operation soon delved into basic physics and chemistry. Two years later, on May 20, 1953, Dwight D. Eisenhower dedicated the company's Research and Engineering Center via a closed-circuit television broadcast from Washington as part of Ford's fiftieth anniversary extravaganza. The President looked forward to another half-century of achievement, predicting the center would produce "much that brings better living to the United States." Ford followed up with a huge R&D exhibit in its visitor's center. By fall, 600,000 people had toured the display.

Car king General Motors embraced science even more dramatically. A few years after Boss Kettering stepped down in 1947, the auto giant reoriented research along more fundamental lines. Then in 1955, upon the retirement of Kettering's successor, the outstanding engineer Charles L. McCuen, the company tapped well-known nuclear physicist and former Atomic Energy Commission official Lawrence R. Hafstad as its new research head. "The dramatic accomplishments of science in recent years have captured the imagination of everyone, and have caused industry as a whole to move into a 'research era,' " GM chairman Alfred P. Sloan, Jr., explained. Hafstad had no automotive experience. However, Sloan added, "His appointment reflected the fact that the emphasis in the work of the Research Laboratories was moving steadily in the direction of investigation of new, broad research problems."

Since 1944, Sloan and Kettering had planned a sprawling research and technology complex to consolidate aspects of engineering, manufacturing, and styling with the research division, which had spread into pockets of temporary quarters across Detroit. The oft-ballyhooed $125-million General Motors Technical Center, designed by renowned Finnish-American architect Eero Saarinen, opened officially on Wednesday, May 16, 1956, in Warren, Michigan, just 12 miles from the GM building in downtown Detroit. The site spilled over 330 acres of broad lawns and vast woodsy areas. A domed auditorium and two score steel-and-glass buildings, each three stories high with end walls of glazed brick, spanned three sides of a 22-acre, fountain-studded artificial lake.

Some five thousand representatives of science, education, and industry gathered on grandstands by the lake on that uncharacteristically cold and windy gray day. As with Ford's dedication before it, Ike was piped in from Washington. He described the complex as a nerve center for attacking technological frontiers and showcasing the value of democracy. GM president Harlow H. Curtice rode to the speaker's platform in a bubble-topped, shark-finned Firebird II, the company's latest experimental gas turbine car. "We must put more emphasis on basic research, pure research," he declared, later adding: "I hope you will come to regard the General Motors technical center in the same way I do—as one of the nation's great resources—more important even than the natural resources with which we have been endowed." An aging Kettering, no devotee of pure science, joined the parade of speakers and likened the complex to an intellectual golf course: "Here in this institution we have the place where we can make indefinite practice shots and the only time we don't want to fail is the last time we try it."

And this was still just the early phase. All through the 1950s and 1960s, other big corporations began taking chip shots at scientific frontiers. A week before GM opened its center, U.S. Steel launched the largest research complex in its industry. Among its farther-out pursuits during the Cold War: studying moon rocks through fluorescent microscopes. Chemical giant Dow followed in DuPont's footsteps by sponsoring wide-ranging chemical and materials studies. Almost next door, still in Dow's home town of Midland, Michigan, basic research into silicon and polymer chemistry was underway at Dow Corning, a joint corporation set up with Corning Glass Works during World War II to fashion specialized materials to help aircraft operate at high altitudes. Drugmakers mainly avoided wide-scale organized research until after the World War II pharmacological revolution that witnessed the development of antibiotics such as sulfa and penicillin. None dived into basic studies with more gusto than Merck & Company, where Vannevar Bush joined the board in 1949 and began a five-year stint as chairman in 1957. Over in the oil sector, Gulf energetically expanded basic research at its lab complex outside Pittsburgh, while Chevron pursued a similar path out in California. Exxon launched into basic research in 1968, establishing a central laboratory chartered to practice "pioneering research."

Few industries pursued fundamental studies more vigorously than the electronics and electrical sectors, however, where AT&T was showing the way. Pennsylvania behemoth Westinghouse had anticipated the move to basic studies in 1937 by hiring Princeton University physicist Edward U. Condon as associate director of research. Condon left after World War II to take over the National Bureau of Standards; in 1951, he would move again to head research and development at Corning Glass Works. But he had installed a Van de Graaff generator and built one of the nation's leading programs of fundamental research in high-energy physics. In 1955, the company took occupancy of a state-of-the-art research complex—eventually called the Westinghouse Science and Technology Center—in Churchill Borough, 10 miles east of downtown Pittsburgh. A few years later, in early 1960, rival Hughes Electronics created a stunning research facility overlooking the Pacific Ocean in Malibu, California. The lab was charged in part with conducting basic research in solid-state physics. That May, only a month after it opened, physicist T. H. Maiman developed the world's first working laser using a synthetic ruby crystal. Ruby lasers soon formed the cornerstone of a multi-billion-dollar range finder business for Hughes, opening the door to fiber optics and optoelectronics.

Even with accelerators and lasers, it proved hard for these labs to match the stature of RCA, which opened its central laboratory on Princeton's Route 1 corridor in 1941. Up until that time, research had centered on product and business development. It was only on the eve of war, when Uncle Sam began providing major funding for radio-related research, that a central lab was created. During the conflict, RCA focused on military needs such as radar and vacuum tubes. In peacetime, however, the lab expanded into all areas of electronics, hiring a slew of theoretical scientists to provide the foundations for what it later termed "building block research." By 1950, RCA listed 2,723 members of its technical staff—over three times the prewar total. Of these, nearly 500 were based at Princeton. They enjoyed wide latitude in selecting projects to pursue, as well as the opportunity to engage in individual studies: "sustaining research." In 1951, the facility was renamed the David Sarnoff Research Center, after the company president, who erected a suite of rooms for visiting dignitaries. Four years later, theoretical scientists, including physicists, chemists, mathematicians, and metallurgists, comprised almost 50 percent of the staff—as the lab helped pioneer digital computers.

Then there was IBM. Old man Watson—Thomas J. Watson, Sr.—created a Department of Pure Science in late 1944 or early 1945, with the aim of developing the nation's premier organization dedicated to using automatic calculation methods in science. Hired to run the endeavor was noted astronomer Wallace J. Eckert, the company's first Ph.D. scientist, who explained: "By pure science we mean scientific research where the problem is dictated by the interest in the problem and not by external considerations."

The facility's official name was the Watson Scientific Computing Laboratory at Columbia University. Everyone called it the Watson Lab. Eckert and crew found temporary quarters in Columbia's Pupin Hall but moved late in 1945 to a renovated fraternity house at 612 West 116th Street. The five-story townhouse had room for perhaps two dozen professionals, including labs, offices, library, reception lounge, and small machine shop.

The emphasis was on academic freedom. Researchers chose their areas of focus, played bridge and chess at lunch, and dressed casually—though one was admonished for walking around in bare feet. An attempt to get the scientists to punch a time clock failed miserably. The initial focus lay on applied mathematics and mechanical calculation, chiefly with the aim of building a machine to outduel Harvard's Automatic Sequence Controlled Calculator. Within a few years, however, the lab expanded into advanced electronics, digital computing, and solid-state physics, including semiconductors—with smaller efforts in chemistry and metallurgy. Among the early recruits: world-renowned mathematician Llewellyn H. Thomas; John Backus, later famous for pioneering the FORTRAN programming language; quantum mechanics theorist Léon Brillouin; and the brilliant engineer Byron Havens, a veteran of M.I.T.'s Radiation Laboratory. By the early 1950s, the lab had tripled in size to around 120 researchers, expanding into a larger edifice in an old women's residence club at 612 West 115th Street. A fresh wave of recruits included physicist Richard L. Garwin, later a longtime member of the President's Science Advisory Committee, and Erwin L. Hahn, a towering presence in nuclear magnetic resonance.

The lab embodied the tremendous surge in IBM's commitment to research and development. But the company also maintained much larger, applied R&D operations, in Poughkeepsie and Endicott, New York. Between 1950 and 1954, its overall research and development organization grew from about 600 people to just over 3,000—with more than 2,800 in these two facilities. Another 110 were at the Watson Lab, with about 80 more stationed at a small outpost in San Jose, California. In 1955, the company opened an even tinier research arm, in Zurich, Switzerland.

Even as the scope of research grew, outside consultants hired by IBM in 1954 concluded that most endeavors were too applied. Heeding their call for more basic studies, in January 1956 the company broke research off as a separate organization that included the Watson, San Jose, and Zurich labs, as well as the core of Poughkeepsie's physical research department. That fall, it introduced a new research director, prominent physicist Emanuel Ruben Piore, a Russian émigré and former chief scientist for the Office of Naval Research.

Aiming to create the world's most eminent industrial research lab, in the tradition of Bell Labs and GE's House of Magic, Piore wasted little time in reorienting activities toward basic research. Within a few years, his operation got a new home as well. A grand complex went up on a 240-acre site just south of Yorktown Heights, in pastoral Westchester County 25 miles from New York City's northern border. First occupied in the summer of 1960, and dedicated the following spring at a ceremony attended by New York Governor Nelson Rockefeller, IBM's scientific haven was conceived, as GM's before it, by Eero Saarinen. The structure took the shape of a gentle 73-degree arc that ran 146 feet wide. Its shorter rear wall spanned 900 feet in length; the curved glass facade stretched 1,091 feet to overlook rolling wooded hills visited by deer and Canadian geese.

Towards the start of the golden age, in May 1956, the *New York Times Magazine* ran an article entitled, "Key Men of Business—Scientists." "The welcome mat is out for eggheads in industry," the piece began. Although it was pegged to the opening of General Motors' $125 million science and technology center, the article pointed to the bigger, national trend. "Twenty years ago industrial research was a marginal operation in most companies," the magazine intoned. "In only a few was there any systematic assault on the frontiers of knowledge; research departments were tucked away in corners of factories; funds and facilities for long-term studies were rarely available. . . . Where large numbers of engineers were kept for development work, they often sat in regimented ranks in block-long lofts. Today industrial research has ballooned into a $5,000,000,000 a year enterprise conducted in centers that have more kinship to a graduate campus than to an industrial plant. Bright young Ph.D.'s in physics, chemistry and mathematics hold skull-sessions with more seasoned colleagues. Abstruse figures are drawn on blackboards. Ideas fly through the pipe smoke. Informality is the keynote, even to the crew-cut hair and tweedy clothes."

The *Times* took things a step further in its daily pages the next morning, with an editorial about the GM complex headlined, "Molding the future." Asserted the paper: ". . . the technical center is more than a vast institution devoted to the improvement of General Motors' products. It is a national asset. With Soviet Russia proclaiming its intention of leading the world in science and technology, it is to this technical center and others like it that we must turn for aid if we are not to lose the industrial supremacy that we have long enjoyed."

That overriding belief that science—almost by itself—held the magic answer would prevail for decades. "When I became director of IBM research in 1970 we were in an era of science worship," relates Ralph Gomory. The new boss never believed science deserved the idolatry: he knew too much about science. But, generally speaking, most researchers and the academic world believed it—and in those days scientists expected due homage.

CHAPTER FOUR

OUT OF THE PLUSH-LINED RUT

"Scientific leadership is neither a necessary nor sufficient condition for marketplace success."
—JOHN A. ARMSTRONG, FORMER IBM VICE PRESIDENT
FOR SCIENCE AND TECHNOLOGY

"Although it is difficult to persuade research men to admit such a thing, it is also possible to provide too much ease and too great convenience for the researchers. . . . One great research laboratory is said to approach this condition so closely that its inmates commonly speak of it as the 'plush-lined rut.' "

—D. H. KILLEFFER

THE JAPANESE WORKED their visitors to the bone. Led by research director Ralph Gomory, IBM's small technical entourage had arrived in Tokyo in November of 1979 as part of a two-week visit to assess the state of various company cross-licensing agreements with companies in the Asian nation—and their hosts set a grueling pace. Meetings and tours of industrial laboratories lasted all day, with hardly a break. Then came the late-night business dinners.

But it wasn't the hectic schedule that overwhelmed Big Blue's high-powered contingent—it was what they were witnessing. In addition to Jim McGroddy, then the Research Division vice president in charge of semiconductors and integrated circuit technology development, Gomory's group consisted of IBM Fellow Bob Henle, Nobel laureate Leo

Esaki, a native of Japan and a legend in his home country, and about eight representatives from the research, product development, and legal staffs. Their job in a nutshell was to survey the state of Japanese technology and open the process of haggling about who should pay money to whom under the soon-to-expire licensing agreements, as well as lay the groundwork for the next round of deals. But Gomory also had another agenda. Sensing that something remarkable was happening in Japan that portended big changes for the way research and development worked together, he had assembled an all-star cast designed to figure out what needed to be done.

This was Gomory's second such tour; the first had come five years earlier. On that initial trip, he had seen little research going on, with Japan's corporate labs trailing IBM in nearly every regard. But despite his foreboding that things had changed dramatically, the amount of progress in the last half-decade proved staggering. A particular trigger had been a visit to Oki Electric, a second- or even third-tier maker of semiconductor memory chips. IBM at the time ranked as the world's largest manufacturer of semiconductor memories. But the contrast to its research pilot line, where wafers were processed with twenty-year-old equipment, was almost unbelievable. Oki employed well-trained people, had all the latest equipment—and its plant ran twenty-four hours a day.

Even more disconcerting, while Oki marked an especially vivid contrast to the situation back home, it proved much closer to the rule than an exception. "The whole Japanese development and manufacturing cycle was faster than that of the U.S.," Gomory notes. Indeed, Japanese high-tech firms as a whole—in microprocessors, computers, circuitry, and virtually every other aspect of IBM's businesses—had become far more focused on products than Big Blue, raising the bar on quality in such areas as manufacturing and design for manufacturing, just-in-time delivery, and the overall cooperation it achieved between development and production. The result was that while the Japanese still lagged in many areas—especially science—they were implementing far more rapidly. "Believe me," relates Gomory, "we were worried—really, really, really worried. . . . It was clear to all of us that if we did not improve the way we did advanced technology we would be left behind."

Scientific leadership, former IBM research executive John Armstrong likes to say, is neither a necessary nor sufficient condition for marketplace success. If ever that message came through loud and clear, it was on the trip to Japan. That's why, less than two decades after opening its stunning Saarinen-designed research paradise, Big Blue found itself on the front wave of a long hard drive to tame the beast.

Research, as it turned out, had gotten away from just about every-body. Somewhere along the line in the past fifteen years, starting back around the time Arno Penzias says U.S. cars lost their fins and continuing until Ralph Gomory's contingent arrived in Japan in late 1979, the re-search train had become derailed. So while IBM ranked as the world's number one computer company, and in some senses formed the pinnacle of sluggishness and inefficiency, the alarm felt by its research leaders could be taken as a warning shot across the bow of the entire computer-telecommunications-electronics industries. As the 1980s opened, a plethora of big labs found themselves overly fat, awash in multiple proj-ects, lethargic or outright apathetic in transferring technology, and poorly attuned to what customers wanted. In the face of determined and amaz-ingly nimble competitors who attacked the giants at their Achilles' heels, many American companies were primed to get their clocks cleaned.

Awareness of the need to transform research practices dawned on labs gradually throughout the 1980s, as market share was lost not only to the Japanese but to daring start-ups such as Apple, Compaq, 3Com, Cisco Systems, and Microsoft, who became mighty powers in their own right. Typically, research organizations moved first to alter a few practices and institute a handful of initiatives designed to counter the threats. But internal resistance to change ran high, and these initial moves invariably came up short. The crisis picked up steam late in the decade, and then reached a crescendo in the first half of the 1990s in a blaze of budget cuts, layoffs, cutbacks in science, and general turmoil and depression. Some places—like RCA, Westinghouse, U.S. Steel, and Gulf—saw re-search either disappear or become a shadow of its former self. But when the dust cleared late in the decade the best had entered a new era where the rules for conducting research—how projects were managed, the speed at which things moved, the way scientists worked with business units and customers—had all been transformed to produce perhaps the best climate for innovation in fifty years.

This new day had been a long time coming. The era of the corpo-rate Ivory Tower had peaked in the 1950s and 1960s, but kept rolling along for another two decades, spurred on by the Cold War and a flurry of great discoveries. In many senses, it remained a time of wonderfully provocative science and potential breakthroughs—in DNA splicing and genetic engineering, artificial intelligence, optical computing, and exotic semiconductors, to name just a few. In the electrical-communications-computer industries alone between 1956 and 1987, twelve corporate sci-entists won Nobel Prizes—six at Bell Labs, five at IBM, and another at General Electric—for advances in everything from the transistor to chaos theory and high-temperature superconductivity.

Arno Penzias blamed his own lab for keeping the momentum of the academic-style corporate lab going longer than it should have. "The fantastic success of Bell Labs—and it was enormously successful—made a lot of people build similar things," he relates. Even if managers suspected something was amiss with the model—Penzias included—they tended to fight change and stick to the old notion that their job was to protect researchers from the rest of the firm. When the directors did see the light, it was another thing to convince department managers, and then the researchers themselves. Real change often proved ephemeral. As a result, sums up Penzias, "The model itself did a lot of harm, I felt. So many companies had research labs which had very little connection to the business they were in."

But the argument for another approach gained momentum with each passing decade. For one thing, while fundamental research might benefit society enormously, it was proving extremely difficult for companies to reap the economic rewards of in-house basic science, which typically takes years and contributions from many quarters to tame and commercialize. Despite what was sometimes decades of generous support for basic studies, writes the historian of science David Hounshell: "DuPont had no new nylons. Kodak had no radically new system of photography. RCA had lost many opportunities." AT&T certainly made little money off the transistor, which antitrust considerations forced it to license for a modest fee. Meanwhile, already by the 1960s, the invention was being exploited wonderfully by companies like Texas Instruments and Fairchild Semiconductor.

It wasn't just a matter of mimicking Bell Labs that caused problems, however. The very nature of the forces that had driven firms to centralize research—including the need to avoid being sucked into development matters and to work on projects with longer time horizons—also tended to isolate scientists from their company. The retreat of research to academic havens, and the parallel dispersal of development and manufacturing, gave rise to profound geographic and cultural barriers between labs and other operations. Rapid-fire growth that brought in a plethora of young researchers with no commercial experience added to the problem. The stereotypic corporate research center gradually took on the mantle of a country club operation with little accountability—and virtually no understanding of the way markets worked.

Central labs—worse, their companies—could disappear in this haze of disconnection. For industries such as steel, which had never been serious about research, astute observers recognized the danger signs by the early 1960s—well before the lab got into moon rocks. Jerome B. Wiesner, a veteran of World War II radar research later to become

M.I.T.'s president, was serving as science adviser to President John F. Kennedy when he was asked to discuss the state of corporate research at a steelmaker's convention. "I don't know why you asked me to speak about research—you're not doing any," Wiesner bluntly told the group, referring to relevant research. "And if you don't change, you'll be out of business in twenty years."

Outraged steelmakers wired the President to complain. But within Wiesner's promised two decades, American steelmakers, the envy of the world in state-of-the-art equipment and techniques following World War II, had been beaten nearly unconscious. There was, as always, a litany of causes: labor disputes, inflation, policy failures. But ultimately it came down to the management of innovation. "It seemed that the managements at Big Steel had become reluctant to bet money on inventions," writes Richard Preston in *American Steel*. "They stopped taking risks. They couldn't bring themselves to invest in anything they didn't recognize. . . . The chattering grew louder, there was a smell of frying brakes, a wheel hopped the track, and Big Steel toppled over like a freight train and went into a crawling, elaborate wreck that dragged on and on for thirty years, and ended with a chain of dramatic explosions." Indeed, as the 1980s opened and Wiesner's time limit expired, three-quarters of all American steelworkers—320,000 in all—had lost their jobs.

And steel was only the appetizer. In 1965, chemist Charles A. Thomas, who had risen through the ranks of research to become Monsanto's chairman, sounded a more general warning in *Fortune*: ". . . the nation's R&D is now stumbling in a plethora of projects, sinking in a sea of money, and is being built on a quicksand of changing objectives." By early the next decade, the wisdom of his words was becoming apparent. The first big oil crisis underscored the fact that previously laughingstock Japanese automakers like Toyota and Honda had designed high-quality, high-mileage vehicles that were burning up American roads—but not fuel supplies—while Detroit continued to churn out inefficient gas guzzlers prone to recalls and breakdowns. U.S. consumer electronics firms were also under fire from highly efficient Asian concerns that ranked low in science and Ivory Tower research operations, but proved so skilled at development and manufacturing they were killing their moribund rivals in technology and in the marketplace.

Indeed, Japanese companies such as Sony, Panasonic, and Sharp were well on their way to dominating all aspects of appliances and electronics—from the microwave ovens invented at Raytheon soon after World War II to boom boxes and video cassette recorders. Anyone following the tribulations of RCA's David Sarnoff Research Center could see what that portended for go-it-alone American research organizations.

From the mid-1960s through the 1970s, Sarnoff researchers pursued the RCA Videodisc, built on a proprietary technology. But this videoplayer was never accepted by the Consumer Electronics Division, which viewed the lab's work as exotic and pursued its own alternatives. When the research version won out with senior management, the consumer division purchased Japanese technology and introduced VCRs that competed directly with the Videodisc. The Videodisc would finally reach consumers in the early 1980s, only to be withdrawn from the market within three years.

It was in this gloomy climate, with the Japanese coming on strong in semiconductor memories as well, that the contingent from IBM arrived in Tokyo in 1979 to get the shock of its collective professional life. But Big Blue's research division would fare better against this menace than its RCA counterparts—largely because it took fast and decisive action. "We had these Japanese coming up our tailpipe," remembers Jim McGroddy, whose insights proved critical. "I had the view that research had a big role in dealing with that."

The breakthrough came on the long plane ride home, as Gomory and McGroddy dissected things in the spacious first class cabin. It turned out that even before the trip McGroddy had been thinking about ways to improve the R&D process. In particular, he had already helped conceive an innovative means of getting two product laboratories—in Burlington, Vermont, and East Fishkill, New York—to work with Research to accelerate the pace at which memory chip density was improved. Though temporary, McGroddy had seen in the measure the foundations for something more—and he and Gomory had built on the idea to fashion a singular type of program designed to make such cooperation permanent.

The gist from the start was that IBM needed to establish a team of researchers and development engineers to work together on next-generation silicon technology. A big part of the problem, Gomory relates, lay in the nature of IBM's chip-making pipeline. At any given time, manufacturing was busy producing the current generation of chips. Meanwhile, development engineers readied the next version—while research scientists worked on far-out exploratory work several generations down the road. That meant nobody really focused on the generation that would immediately follow what was then under development—a bottleneck that significantly slowed development time.

The new idea hinged on filling this gap with an interdivisional program whose goal was to get Burlington and East Fishkill to work with Research in accelerating the creation of the next generation of various types of logic and memory chips after the generations currently in development. A critical part of the proposal was that it called for a long-term *program*—not a project whose team members went back to their jobs after a specific task had been completed.

Shortly after returning from Japan, Gomory began to flesh out specific goals and structure and determine whether development would go along. The final plan called for making development divisions responsible for half the program funding, giving everyone a financial stake in the outcome and a say in setting the agenda. What's more, outside of a three-year corporate commitment to shield the program from any budget-cutting whims, the participants covered everything from their own existing funds, including the creation of state-of-the-art facilities and the hiring of extra personnel. The payoff: the plan's proponents were convinced they could advance key dates for introducing chip technologies by at least a year, enabling both sides to accomplish more together than they could alone. Indeed, asserts Gomory, "One of the remarkable things about the . . . programs was that they were a mutual benefit."

Toward the end of 1980, chairman Frank Cary approved the plan. Implementation did not go completely smoothly. But the combination of modern facilities, strong status, and shared responsibility helped overcome any difficulties, enabling the group to beat its timetable and speed up chip introductions by more than a year. Gomory then pushed hard to extend the concept, which became known as Joint Programs, to storage technology, packaging technology, software, and other products. One of his central legacies, it would remain an integral part of the company's R&D strategy heading into the twenty-first century.

IBM's research division was not without its problems. A fundamental issue—one that can haunt any research organization—was that despite working well with other parts of the company on existing technologies, it found introducing competing technologies that might upset the business apple cart far more difficult. Even as the Joint Programs got underway, IBM Fellow John Cocke conceived the basis for Reduced Instruction Set Computing, an operating system architecture that greatly boosted computer performance over previous approaches. But research already pursued a dozen exploratory architectures, while product managers were busy focusing on existing computer lines. Despite the fact that the 801, as RISC was called internally, won adoption in the mid-1980s as a product plan in the division that made midrange computers, the change initially proved too large to pull off. In the end, Cocke and Research waited until 1990 before IBM marketed the RISC System/6000 family—and that was through the entirely new business area of advanced workstations. "By then," writes company historian Emerson Pugh, "the power of RISC had been exploited in the marketplace by Sun Microsystems, Hewlett-Packard, and others—forcing IBM to catch up with its own technology."

Still, the creation of the Joint Programs served as a seminal event in the history of IBM research. Critically, for Ralph Gomory, the program's success was not the result of one great idea, but rather a testament to evolution—and all the hard work that had already gone in to bridging the R&D divide. Almost from the moment he had taken over the research arm in 1970, Gomory had worked to overcome this gap. The answer, he had felt certain, lay in somehow getting researchers to seize the initiative for technology transfer. But initially, at least, his options for bringing about that condition were limited, because no base of trust existed between the two groups. "When we started," he recalls, "a lot of my people didn't even know where the other labs were."

Gomory had tried a variety of devices to overcome this formidable roadblock—including sending scientists on "sabbaticals" to development groups and operating divisions. In the end, the Japanese may have provided the catalyst for a major breakthrough, but even their formidable threat would not have been enough without nearly a decade spent persistently building cooperation and trust. As Gomory once noted, "I do not think it would have been possible to launch the first joint program if IBM Research did not already have considerable credibility." Of critical importance, just as the initial proposal was being born, his researchers played a pivotal role in solving a pair of production crises that threatened to bring a major segment of Big Blue's mainframe business to its knees. Those crucial acts enabled the Research Division to vividly demonstrate its worth to development and manufacturing counterparts and pave the way for the triumph of the Joint Programs.

Hammering away continuously at barriers to innovation, and building a ladder of success out of each hard-won step, can be taken as the prevailing theme of corporate research in the 1980s. Indeed, if there's a thread running through the story of top research organizations—especially in the computer-electronics-communications sectors—it's wrapped up in efforts to evolve the same kind of trust with different corporate elements that IBM enjoyed via its first Joint Program.

All through the decade, large corporations experienced similar scares as that facing Big Blue. Often it was the Japanese who exposed the flaws in the R&D process. But attacks could come from anywhere: the Germans, other big American concerns, venture capital start-ups, even from inside one's own company, where new inventions and intrusions onto traditional turf are commonly viewed with suspicion or outright resistance. In any event, research managers at firm after firm tell the same basic tale of crisis and disconnection—followed by long-running attempts to right the ship by bringing down the Ivory Tower. Many times, as would be the case with Bell Labs and Arno Penzias and IBM and Jim McGroddy, the dam only really broke when a steel-nerved manager went against the grain and told scientists what they didn't want to hear: that the problems didn't rest only with "other" sides of the company, but with research as well.

Xerox was reeling in the early 1980s from a monumental research disaster centered around its Palo Alto Research Center, an ivy-laced haven where fundamental studies in computer science dominated the agenda. PARC had been formed more than a decade earlier to invent "the office of the future." Physicist George Pake, a leading force in nuclear magnetic resonance, left Washington University in St. Louis to run the lab, wooing the best and brightest computer scientists with what one recruit, former TV Whiz Kid Alan Kay, called a "wonderful bribe": ten years of blank checks and the freedom to pursue their dreams. And indeed, things had started out great. Within a few years, the digital virtuosos had come up with a host of creations that set personal computing standards for at least the rest of the century, unveiling a system that included the mouse, graphical user interface, laser printer, and the Ethernet local area network.

Then came chaos. Senior management's inability to make sense of computing, Japanese attacks on Xerox's core copier business, and the lab's separation from East Coast headquarters all precluded the firm from successfully marketing PARC's innovations. Researchers left in droves. In 1983, Apple launched its Lisa computer—precursor to the Macintosh introduced the next year—which replicated Xerox's graphical user interface and overlapping windows. With the lab poised to disintegrate, new director William Spencer, a veteran of Bell Labs, delivered the message that research was as much to blame as the rest of the company and began basing performance reviews partly on how well research managers got to know their counterparts throughout Xerox. After all, he noted, these were the "customers" who would ultimately manufacture and sell what research put out.

Somewhat similar maladies dogged General Electric's historic research organization when Walter Robb took the reins in September 1986. Despite Ivar Giaever's 1973 Nobel Prize for sharing in the discovery of superconductive tunneling, the enterprise had never really bought into the postwar science craze. Not only was Giaever's work the result of a largely applied project investigating electronic phenomena that managers felt might be useful in making various devices, under Robb's predecessors Art

Bueche and Roland Schmitt the lab had made some critical contributions to GE's bottom line, most notably in high-performance plastics and medical imaging. Still, as a onetime scientist who had left Schenectady to run the company's plastics operation—and, later, the Medical Systems Group—Robb was the first research boss with any real business experience, and a sense of disconnection nevertheless troubled him.

He started by dramatically cutting back longer-term "knowledge research," which was conducted without specific corporate aims in mind. He also held a two-day meeting with business division leaders to hash out a change in the way the center was funded: from then on, instead of drawing two-thirds of their budgets directly from corporate coffers, research departments had to get the lion's share from contracts with the operating divisions (or strategic customers) for specific jobs that these businesses *wanted* done. Robb next turned to a liaison program instituted by former lab director Guy Suits. Under Suits's concept, three or four scientists had served as the links to GE's dozen or so main business areas. That had since evolved to the point that there was nearly one lab liaison for each of GE's diverse divisions—but Robb took that a step further by increasing the number of lab liaisons and filling the positions with managers instead of researchers.

The same sorts of sagas unfolded at scores of other labs—through Arno Penzias's epiphany at Tanglewood in 1989, and into the 1990s. And they weren't limited to U.S. corporations. Frank P. Carrubba left the helm of Hewlett-Packard Labs in 1991 to take over a vast Philips Electronics central research organization that had always been funded from central monies and deliberately allowed to live life apart from development groups—on the theory that its scientists and engineers should be insulated from shorter-term business needs. The new boss turned everything on its head, paring the corporate contribution down to 30 percent of the research budget and forcing his staff to get the rest through the operating divisions. Carrubba farmed out projects that didn't fit Philips's core competencies to university labs. He also created a special group to evaluate new product ideas, "so lower-level managers don't allow something terrific to die on the vine." Meanwhile, Philips's European competitor Siemens took similar steps, while Japanese giants such as NEC and Hitachi all streamlined their already efficient research organizations.

Positive effects of the various changes were hard to see at first. Not only did resistance remain stiff, such measures needed time to propagate and become fully ingrained in employee attitudes and practices—and even when they proved wildly successful did not ensure overall success given all the other factors that ultimately determine marketplace winners.

The U.S. economic downturn of the late 1980s and early 1990s accentuated the lack of connection between research labs and their business counterparts—and made it harder for labs to recover their equilibrium. During better days, business divisions had had their hands full selling existing technologies in expanding markets and were more content to let research run its course. As the climate darkened, research organizations once ignored as irrelevant were besieged with requests for improvements that could help products stand out from the pack and increase profit margins—and for innovative new sources of revenue as well. But that was easier asked than answered. Researchers and business groups still had little experience working together. At the same time, the financial squeeze was forcing the "research bloodbath" that saw industrial R&D spending in the United States fall in real terms for three straight years—spurring layoffs and straining resources.

By the early 1990s, eliminating redundancy and increasing efficiency had become overriding goals in corporate research organizations, as with just about everywhere else. Gone was the old adage that more R&D is better R&D.

Over at IBM, Ralph Gomory had risen to senior vice president and management board member before leaving the company in 1989 to head the Alfred P. Sloan Foundation in New York. He had been succeeded as research director in 1986 by the physicist John Armstrong, and then, in 1989, by Jim McGroddy, when Armstrong moved to Armonk headquarters as vice president for science and technology. With Gomory's strong support, Armstrong had beefed up the Joint Programs and instituted a variety of other programs to get a bigger bang for research's bucks—steps McGroddy would also continue to pursue under Armstrong. "To use resources more effectively—that's the challenge we have," Armstrong told *BusinessWeek* in 1992. In the same publication a few months later, Bell Labs research boss Arno Penzias spoke for many counterparts by making it clear that the spirit of disconnection would no longer be tolerated—stressing that research strategies, and researchers themselves, must be integrated with overall corporate strategies. "I want to still do fundamental research, but I also want people to understand that they're doing it for a purpose," he asserted. "I said to everyone, 'If you don't like it, please leave.'"

Not every research organization survived the trauma intact. Gulf Oil's impressive center in Harmarville, Pennsylvania, had flourished through the 1970s, but was suddenly disbanded as largely redundant following Gulf's 1984 takeover by Chevron. U.S. Steel—re-christened USX—dramatically trimmed the scope of its research, funneling much of the work to operating units and slashing basic science almost completely: no more moon rocks. RCA's chagrined Sarnoff Center was absorbed into the nonprofit consulting house SRI International, then spun out in 1987 as the for-profit Sarnoff Corporation, concentrating on product improvements in information technology, biomedical electronics, and consumer electronics. Westinghouse's big campus in Pittsburgh, renamed its Science and Technology Center, had employed 2,500 people at its heyday in the 1970s but fell into decline and spiraled down to under 500 workers in the 1990s—as the once-full parking lot transmogrified into a cement field of empty spaces and the business itself was sold to Siemens.

Even for organizations destined to weather the storm, the 1990s began miserably. Morale sank at just about every lab, as top researchers deserted for academe or the rare industrial safe haven such as NEC's impressive new $25 million research institute, which opened to a tidal wave of publicity in 1989 in a wooded park along the same Princeton corridor that housed RCA's Sarnoff Labs. Concentrating on basic research into such areas as quantum chaos and biological computing, the neophyte institute evoked even more resentment toward the retrenching going on in American research houses. One of its star physicists, Richard Linke, a pioneer in storing data on holographic crystals, left Bell Labs amidst the turmoil of 1989 for NEC's more inviting environment. He had been recruited to AT&T by Arno Penzias himself, who had stressed the Bell Labs philosophy that research was about hiring good people and relying on them to tell their bosses what areas warranted investigation. "Well, that was then and he doesn't say that anymore," related Linke, who remained a devotee of the old idea that good ideas spring from the bottom up.

Eventually, even NEC's institute would feel the pinch, though to a lesser degree than most other labs. But in the meantime, at a host of rival organizations, the surviving researchers, developers, and business unit personnel dove into the innovation maelstrom with some combination of fear and determination—and from their growing bond rose a kernel of hope. As IBM's Armstrong once summarized: "People have been so shaken up by the corporate trauma that they realize that they can independently be successful only if they're all successful. Adversity brings people together."

Corporate research evolved in the 1990s from the invention of science fiction to creating scientific and technological fact. That, anyway, is the analogy dreamed up by Wolf-Ekkehard Blanz, a longtime IBM research scientist who joined Siemens's glittering Princeton lab in 1993 and transferred later that decade to its medical division in Germany.

Blanz notes that the wonders science fiction authors predicted for around the year 2000—such as mass transportation to the moon and glass-domed cities on the ocean floor—are within the grasp of modern technology. However, he adds, they are not about to come to pass. The reason: "[writers] forgot the marketing dimension. Nobody is out there that is willing or capable of paying for that."

In the same vein, many industrial research projects in the 1960s, 1970s, and 1980s, while technically feasible, nevertheless lacked grounding in reality. It's okay for university researchers to pursue wild dreams, says Blanz, but not industrial scientists. "I have to educate my researchers to become not only technological researchers and university researchers; they also have to be marketing researchers."

The modern era of corporate research finally found itself in the mid-1990s—and its theme was to make that connection between technology and markets in a far more sophisticated way than in the past. Some business-savvy researchers depict the old days in terms of "technology push," where scientists dreamed up products and almost hurled them at the market. By the early 1990s, "market pull" had become the rage, with customer needs shaping what came out of the labs. Finally, came the age of both push and pull. Paul Horn, the latest in IBM's string of research managers, describes this evolution in more everyday language. The early 1980s saw researchers and developers begin to work together, he says. Relations remained strained, though, and the groups did not really function as an integrated team. "Then by the late eighties and early nineties, we really started to integrate the teams together. And probably by the mid-nineties . . . the strongest of those, you couldn't tell who's who."

To some degree, of course, corporate research organizations have always tried for such synergism. Way back in the late 1800s, German dye companies realized the advantages of fusing research with sales. Chemists were dispatched to branch offices, working with salesmen and sending back a stream of information about what customers needed, or how things might be improved. Clients were invited to spend weeks at the dye houses in order to better understand products. A century later, IBM created its Joint Programs—and there were plenty of other examples in between. But as the latest crisis descended on research in the late twentieth century—and costs rose, budgets fell, and competition increased like never before—the whole idea of connection was revisited in force. Managers moved to unveil a vastly increased arsenal of weapons to help them cope.

For Bell Labs, the early 1990s started off with senior managers pushing concurrent development—the concept of researchers working with developers and customers from the onset of a project in order to make sure the end product had the right powers and features. Similarly, Xerox created codevelopment teams, where researchers, developers, and marketing staff banded together with customers to identify key problems that could be solved with emergent technology. Meanwhile, to beat bureaucracies, a variety of firms began employing small, highly autonomous teams modeled after the legendary Skunkworks unit run by Clarence L. "Kelly" Johnson at Lockheed Aerospace Corporation in the 1940s. Named for the secret moonshine Skonk Works brewery in the Li'l Abner cartoon strip, the initial team took just forty-three days to build the first American fighter to fly at more than 500 miles per hour; speed was even more of a concern fifty years later.

And this was only an inkling of what lay in store for research. The decade's early years saw IBM's Almaden lab institute special innovation grants, where standout proposals for projects outside the core business receive special funding for two years. Siemens kicked off its idea contest. To improve the prospects of finding major new areas of growth, GE launched its Game Changers program, in which researchers propose "home run" projects they believe can result in more than $200 million in annual sales, $20 million in yearly profits—or the equivalent in cost savings. Almost everywhere it became the fashion to spin off technologies that didn't fit well in core business lines and might otherwise have gathered dust on lab shelves—as Big Blue did to much fanfare by forming a start-up company to market its ingenious computerized surgeon, Robodoc. A host of other firms—Xerox, Lucent, Intel, NEC, among them—set up their own venture arms to nurture novel technologies and concepts.

In truth, already by the early 1990s it was hard to find a good lab that didn't employ a variety of strategies to spur innovation along. On the macro scale, unprecedented competition and the skyrocketing price of doing science—industrial research costs had been rising at double-digit average annual rates for much of the past fifty years—virtually dictated that firms find multiple sources of funding for some high-risk, long-term projects. Part and parcel with all the internal programs and incentives emerging to aid the innovation process, a number of intercompany alliances and government programs sprang up to support research—creating a hybrid form of R&D the historian Margaret Graham calls "largely unknown to the previous research generation."

Uncle Sam had gotten the ball rolling back in the 1980s by shifting the focus from basic research to research self-sufficiency and emphasizing applications and shared knowledge. In 1984, Erich Bloch began a six-year stint as National Science Foundation director. A thirty-two-year veteran of IBM, much of it in development, Bloch spurred the creation of a series of research and engineering centers aimed at spreading commercially relevant knowledge to a wide variety of industries and firms. A slew of other programs sought to foster technology transfer inside industries, with antitrust statutes relaxed to encourage such cooperation. The National Cooperative Research Act of 1984 opened the doors for research consortia such as Austin-based Sematech in the semiconductor sector and its crosstown sibling the Microelectronics and Computer Technology Corporation (MCC); such ventures sought to create precommercial technology that could enhance the competitiveness of all the member firms. The Federal Technology Transfer Act of 1986 and the National Competitiveness Technology Transfer Act of 1989 aimed to allow national laboratories and other federally funded R&D centers to spin out previously secret technologies.

The result of such programs was often a profound change in the way large-scale research got done. All told, more than 450 Sematech-like ventures formed in the first decade following the National Cooperative Research Act's passage. Just a partial list of government-backed industrial R&D efforts going strong in 1998 included some three hundred projects under the National Institute of Standards and Technology's Advanced Technology Program; more than sixty regional centers under NIST's Manufacturing Extension Partnership; and at least a thousand ventures under the Cooperative Research and Development Agreements (CRADAs) jointly sponsored by the Department of Energy and the country's national laboratories.

To complement such government-supported initiatives, the industry-university alliance emerged as yet another, extremely popular way to cultivate expertise in early-stage research that might not bear fruit for years. By bankrolling academic projects, firms saved much of the equipment and overhead costs of maintaining their own research group in risky areas. They also got a leg up on identifying and training future hires while cultivating a wider breadth of knowledge about fields outside, but related to, their main businesses. Such alliances were sent into orbit by the rise of biotechnology, as witnessed by Monsanto's unprecedented $23 million research grant to Harvard University in 1974. But the trend expanded throughout the next two decades into just about any field. Hewlett-Packard was just one firm that took up this practice in the early 1990s, initially funding a few studies at Stanford University's Science Center. By 1998 the company supported seventy projects at thirty-six universities in eight countries.

Firms also began positioning their research arms to form the core of global competitive alliances that brought together companies with nodes of complementary technological capabilities. The combination of Microsoft's Windows technology and Intel's semiconductors—an alliance dubbed Wintel—showcased the potential power of such groupings.

The opening-up of their labs carried research houses into a tidal pool of fresh challenges. Alliances proved dicey on a myriad of levels: rights and responsibilities had to be contracted, some partners failed to pull their weight or get what they wanted out of the deal, and even if all that was settled working effectively across organizations proved vexing. Adding the international dimension brought into play an assortment of laws, customs, educational systems, work ethics, and other cultural differences, down to something as basic as what a contract meant. Steve Rosenberg, Hewlett-Packard's manager of external research, struck a deal with an Italian university that involved a graduate student coming to work at HP. When the researcher didn't show, he contacted the professor and was told the student had entered the Army for two years. Rosenberg asked for the money back, but was told, "Well, no. He'll do the work when he returns."

Pitfalls aside, however, even the largest company cannot win on its own anymore. That fact, argues Xerox vice president of research and technology Mark Myers, ensures that various "ensembles of relationships"—which have research as their nuclei but shift as partners' needs and conditions change—will play an increasing role in corporate research. Their very evolution, Myers stresses, vividly illustrates how much the institution of industrial research has changed. Says he: "These are movements of technology-centered competitive posturing that I don't think had any precedent in the days of the formation of IBM's Yorktown laboratory or the formation of the DuPont laboratory."

All through the tumultuous early 1990s, as corporate labs struggled to implement these initiatives and find even more ways of coping with the staggering pressures of layoffs, budget cuts, and dramatic changes in the way things worked, the trauma of getting back on track kept a dark cloud over industrial research.

Initially, labs reacted defensively—with efforts designed to shore up the dam. But as companies pulled through their roughest times, and many of the strategies employed to improve the R&D pipeline and create market-driven products began to pay dividends, a bolder view took hold. The best research-oriented organizations began seeing their size not as an impediment, but as a key advantage. Part of it was the realization that their smorgasbord of potential solutions, many operating on different time horizons, provided cover for the inevitable stumbles—so that even if several projects failed completely, the firm had ample alternatives to draw on. Even more than that, though, the array of talents and technologies inside

their labs became weapons by which to seize the offensive in the perpet-
ual battles for products, customers, and markets—for who better to integrate technologies and offer a full range of solutions than the central lab?

A very perceptible anti-arrogance also began to permeate top research facilities. Adversity, as John Armstrong had noted, *did* bring people together. But more than that, as the Ivory Towers went down some
researchers found that their creative juices were actually stimulated by
collaborations with the outside world—so that interaction became a
choice as much as a necessity. Notes John Seely Brown, director of Xerox
PARC, "When researchers come to realize outside people add that kind
of creativity, it generates that level of respect and willingness to interact.
This changes the sensibilities and moves us into a true co-production
point of view that I think is one of the essences of what really facilitates
technology transfer and moving an invention into being innovated."

It turned out, then, that old dogs were often willing to learn new
tricks. At the same time, another breed of researcher was coming onto
the scene. This was the computer scientist, far different in nature than
the physicist. Members of this group didn't give a hoot about fundamental principles, thrived on seeing their creations in widespread use, and, in
the words of Ralph Gomory, "didn't have the idea they were God's gift to
the universe." What it all boiled down to, the longtime IBM research director astutely asserted in 1997, was evolution: "You can do things with
labs today that you couldn't have done twenty years ago—because everyone's outlook is different."

A bright light shone on the United States as business executives,
academics, and policy leaders—including Vice President Al Gore—converged on the Massachusetts Institute of Technology's glitzy Tang Center
in the spring of 1998 for the National Innovation Summit.

At the time, the country rode gloriously on the shoulders of a boom
period often called the best U.S. economy in history. And while it was
impossible to calculate how much difference the transformation in corporate research had made in spurring this revival, there could be little
doubt the rosier times drew vital nourishment from all the efficiency initiatives put into place at the beginning of the decade—including those
designed to make research more accountable and improve the innovation process. The best electronics, telecommunications, and computer
firms—as well as counterparts in many other industries—all reported a
rising share of their revenues coming from new products (those developed in the last five years), up from about one-third in 1992 to more than
half in 1998. Various studies of R&D effectiveness—the ratio of revenues
to spending—showed skyrocketing increases over this same period.

But still the warning bells clanged. Deep questions lingered in the United States about the long-term effects of the industrial research makeover. By the early 1990s, nearly everyone had seen the need to overhaul the way R&D worked. The severity of the spending cuts and the outright dismantling of many long-range projects, though, had been widely assailed as going too far—shortsighted examples of kowtowing to Wall Street's preoccupation with quarterly returns. As the 1998 innovation summit convened, those same concerns about the future of U.S. competitiveness and industrial research cast their shadow over the proceedings.

Of particular focus was the continued perception that U.S. companies had scaled back research (see box, p. 126)—particularly long-term and more fundamental studies, but all types—while the rest of the world had not. After all, Japan's spending as a percentage of its gross national product had surpassed that of the United States in 1986. If one stripped out defense outlays, things looked even bleaker: Japan and Germany had outspent the U.S. since 1970. And the gap between the United States and other nations had kept widening into the 1990s. Companies in Sweden, France, Canada, Italy, Japan, Germany, and Switzerland were all spending a higher percentage of sales on research and development than U.S. firms. Patent data confirmed the trend. Even though IBM had topped the list of U.S. patent winners for five years running, Japanese corporations had made up the majority of the list for nearly a decade. Japan had even launched a research invasion of the United States. In 1985, its companies operated only about two dozen R&D facilities on American shores: by 1998, despite continued trouble with Asian economies, the number had surpassed one hundred.

To bring such trends into contemporary focus, a featured summit speaker was Harvard Business School professor Michael E. Porter. For the 1990s and beyond, noted Porter, new products, processes, and services would hold the keys to productivity and prosperity. The United States had been doing well, but high wages meant its high profitability had to be continually justified. Partly to evaluate U.S. industry's ongoing prospects, Porter and M.I.T. professor Scott Stern had prepared an elaborate quantitative analysis of innovation by studying a variety of other nations and creating an "innovation index" from such weighted parameters as R&D spending and personnel, share of gross domestic product spent on secondary and tertiary education, and international patents filed. The index, covering the period from 1975 through 1996, provided a window on the future of innovation by allowing its creators to predict patent activity nearly ten years down the road. And, from the U.S. perspective, Porter and Stern didn't like what they saw.

When it came to the current patent scenario, Japan ranked first in

□

Are U.S. Firms Abandoning Basic Research?
Lessons from DuPont and Bell Labs

For more than a decade, policy leaders and academics have lambasted U.S. firms for shortsightedly cutting back on basic research. However, an intriguing body of evidence indicates that shifting emphases between basic and applied research should be viewed as normal—and healthy—oscillations in the research process.

Nowhere is this idea more apparent than in a historical review of research practices conducted at DuPont in 1996. Looking back nearly three-quarters of a century, the study showed that the company had followed a pattern of launching big programs in basic science roughly every seventeen years. Each period of this "discovery research" would then last several years before giving way to a decade or more of applied efforts—at which point the cycle would begin anew.

The first foray into fundamental science started in 1927 and ended seven years later with the creation of nylon, when efforts to commercialize the new material began. The post-World War II era marked the next push, which laid the scientific groundwork for the 1950s commercialization of Dacron, Mylar, and Orlon. The third cycle got underway in 1960, when worries about a withering product pipeline spurred an all-out hunt for new blockbusters: during this period, Stephanie L. Kwolek created the first liquid crystal polymer, which provided the basis for Kevlar fiber. The fourth cycle, beginning around 1980, among other things saw research in the life sciences pave the way for the 1990 formation of the DuPont-Merck Pharmaceutical Company. Finally, after a period of retrenching, DuPont senior vice president for R&D Joseph A. Miller predicted in 1997 that his company was "on the verge of a 'fifth wave' of discovery research." That effort continued into the millennium.

Similar oscillations have taken place in communications, where major physics discoveries have sparked four broad eras of research, according to a study by Bell Labs physicists D. V. Lang and Bill Brinkman. The era of electromagnetism, dating back to 1820, paved the way for telegraphy and eventually the telephone. The discovery of the electron in 1897 enabled researchers to create diodes for detecting wireless telegraph signals, and led to the triode, which could both detect and amplify signals. The age of quantum mechanics and solid-state physics, beginning in the 1920s, opened the door for the transistor in 1947, with much of the rest of the twentieth century devoted to find-

terms of the ratio of patents per million people. The United States ran a close second, followed by Switzerland and Germany—with Taiwan, Korea, and Israel coming on fast. That alone might raise some eyebrows in American circles, though Porter cautioned that patents are merely a first-order guide to evaluating innovation. Far more alarming in his view was that the index indicated the United States would slip to third place

ing further practical applications in this area. The late 1950s saw the birth of quantum optics, foreshadowing such advances as the laser and optical fibers.

Each breakthrough sparked periods of intense fundamental inquiries, as academics and corporate scientists rushed to understand the mechanisms and forces involved. It typically takes the communications industry fifteen to twenty years to find applications for such breakthroughs. But without a long-standing "practice of employing the best physicists to do both basic and applied research," Brinkman and Lang conclude, the lag would have been far longer.

These historic pathways of research provide compelling evidence for concluding that well-targeted basic examinations will always be a part of first-tier industrial labs. Companies also support fundamental studies for reasons other than gaining a competitive edge. One is to create a climate of discovery that attracts top scientists who raise the standards of a research operation and provide a bridge to critical pieces of science in universities. Such work also serves to cover one's bases, since while it is relatively easy to focus R&D on areas likely to affect a company's future interests, it is much harder to be sure nothing has been overlooked. Basic research can therefore keep a hand in the bigger game in case something unexpected turns up—even though the odds remain extremely long that it will pay off directly on the bottom line.

In the late 1990s, at least for the communications-electronics-computer sector, the basic research pendulum was already swinging back—with IBM, Hewlett-Packard, Intel, NEC, Xerox, and Microsoft all cautiously increasing support for far-ranging investigations unlikely to bear fruit for ten or twenty years, or even longer. Much of the work focused on atomic-scale devices, foreshadowing humankind's approach on the limits of conventional semiconductor physics. It's possible that somewhere over this boundary another breakthrough is waiting to kick off a major new thrust in basic studies. Or, one could imagine something farther out, like an era of space flight, sparking a rash of fundamental research in all sorts of novel areas of physics, chemistry, biology, and materials science.

Whether or not that proves the case, natural oscillations occur in all vibrant organisms—and research seems no exception. "There is a time and tide in the affairs of research when the emphasis on fundamental research wanes," notes Robert Spinrad, vice president of technology strategy for Xerox Corp. "Adam Smith would approve—because there is a time to row the boat and a time to bail. But . . . there is a regenerative effect at work, with good research creating new commercial opportunities which, in turn, foster more good research."

□

among western developed nations by 1999—then tumble on down to sixth by 2006, behind Japan, Germany, Denmark, Switzerland, and Finland. "Japan and Germany are not out of the game," Porter proclaimed. "Their fundamental policies suggest otherwise."

But Porter hinted at a monkey wrench in his scenario. It was turning out, he noted, that the actual number of patents granted to U.S. citizens was running ahead of what his innovation index predicted. One explanation, he theorized, was that companies had improved efficiency so much they had moved onto a new curve, despite what might historically be considered an underinvestment in the fundamentals of innovation.

A wealth of evidence—including tours of labs around the world, but also the evidence of the market—indicates that standout corporate research organizations are on that new curve. It's a place where money and growth in funding don't rank as highly as in the past. Science can be vital—but science well-focused on areas likely to benefit the corporation. Arrogance is out. Researchers must know and learn to appreciate their comrades in the far corners of the company.

To get on this curve, a central trick has been to let go of the past. It has not been easy. Big organizations are almost always threatened by change, and corporate labs have seen their senses of order and value come under fire in recent years. But it's a tired canard to think of the great research houses—Bell Labs, IBM's Thomas J. Watson Center, and others—as national assets. They belong to their shareholders. In the words of Kodak's pioneering research director C. E. K. Mees: "Few businesses *can* afford to *support* research. They carry out their research, as they do the rest of their operations, for profit, i.e., to be supported by it." The best are accepting this truth in attitude as well as practice—improving competitiveness and carrying the torch of technological progress around the globe.

Paradoxes abound on this journey to innovation. One, eloquently stated by Xerox PARC director John Seely Brown, is that a corporation has to be fairly large before it can afford a research center. But once it's large enough to create a lab, the company is often so unwieldy it can't really appropriate the value. To overcome this formidable obstacle, argues Brown, research directors have to fashion a malleable environment that combines far-out creativity with the hard-wired facts and demands of the real world, "a milieu that people trust in to really think out of the box" but is "really grounded."

It's a ceaseless battle, Brown relates. "Creating the soft is the hard stuff."

CHAPTER FIVE

IBM: TAKING THE ASYLUM

"At IBM, if you look around, what's happened the last five years is the inmates have taken the asylum—and it's been glorious."
—IBM FELLOW BERNARD MEYERSON

"It is startling. All the evidence—now this isn't easy for a former director of Research to say—is IBM Research is more effective now inside IBM than it's ever been."
—JOHN A. ARMSTRONG, FORMER IBM VICE PRESIDENT
FOR SCIENCE AND TECHNOLOGY

IBM Research Division

Central Research established: 1956 (research department), 1963 (Research Division)
Company sales 1998: $81.7 billion
1998 company R&D expenditures: $5.5 billion
1998 Research Division budget: $550 million (est.)
Funding paradigm: 65% central funds, 25% joint programs with development, 10% outside sources (government, royalties, spinoffs)
Total division staff: 2,950
Research technical and scientific staff: 2,600
Principal research labs: Yorktown Heights, New York; San Jose, California; Austin, Texas; Zurich, Switzerland; Beijing, China; Delhi, India; Yamato, Japan; Haifa, Israel

---

THE GRACEFUL stone-and-glass facade of IBM's flagship Thomas J. Watson Laboratory looks down from its hilltop across wooded parklands teeming with wildlife. The stunning facility opened in Yorktown

Heights, New York, in 1960 as an idyllic scientific sanctuary. Two metal argonauts guard its wide entrance, symbolizing "man's eternal quest for the ideal through research and knowledge."

The inmates, though, have taken the asylum. That, anyway, is how Bernie Meyerson makes the point that it's no longer research as usual inside these confines. You hear this human whirlwind before you see him. Words fire out from behind his office door. Then, suddenly, the man emerges, big bushy mustache and curly hair, sharp eyes, sharp words— moving, always moving.

Meyerson embodies IBM's new, make-it-happen breed of researcher. When other companies and even IBM officially dropped a long-unfruitful line of semiconductor investigations, he conspired with a few key managers and colleagues to go underground, borrowing equipment and calling in chits to keep his project alive. Today, the novel silicon-germanium technology his small team invented is delivering processors for cell phones and communications networks with lower power and double the speed of conventional rivals—leaving competitors eating IBM's dust.

Befitting his new image, Meyerson occasionally sheds his lab vestments and dons a suit—he has two now—to hobnob with the establishment. In addition to his status as an IBM Fellow, the company's highest technical rank, he carries a new title, director of Telecom Technology, and oversees more than one hundred people spread across research, design, development, and production. He knows the semiconductor business, negotiates deals—and he's in heaven. His answering machine holds myriad job offers. But Meyerson hasn't budged. "I could have my own corporate jet and a huge staff and all that other nonsense," he relates. "But it's not what it's about." Instead, he's taking an idea from nothing into a business expected to hit $1 billion in sales in 2001. "Things like that"—Meyerson smiles—"you kind of live for."

His employers wholeheartedly agree. After the mid-1990s upheaval led by Jim McGroddy that saw IBM's prestigious research division barely escape demise—in part by knocking some $120 million off its roughly $550 million annual payroll and scaling back basic science, but mostly by focusing on creating value for IBM—there's been a stunning turnaround. There ain't many like Meyerson. But with new incentives and a fresh outlook that takes researchers into the field to know IBM's customers—and even a cautious uptick in fundamental investigations— the division has rebounded to eclipse its old budget, stepped up recruiting, and opened new labs in Austin, Beijing, and Delhi, making eight in all. Led by Research, the company in 1999 won the most U.S. patents for the seventh year running. Meanwhile, its labs have churned out a smorgasbord of major innovations on which IBM is feasting. These run from Meyerson's chips and other semiconductor advances to the magneto-resistive heads revolutionizing computer storage to "Deep Blue," the SP supercomputer that demoralized world chess champion Garry Kasparov.

Even more, the rise of the Internet and rapid fusion of communications and computers play perfectly with decades of research that, arguably, no company can match. Whether it's in raw computing power, storage, chips, displays, speech recognition, data "mining," or electronic security—IBM's labs are firing on all eight cylinders. Managers admit to some holes in their technological arsenal and struggle to attract and retain personnel, especially in the face of high-flying Internet start-ups. Still, the future seems dazzlingly bright—with chairman Lou Gerstner, Jr., showcasing Research as central to the company's revitalization. "It's the best time to be in IBM Research perhaps in our history," proclaims research director Paul Horn. "We had great thoughts, we had great things. But we never had the IBM Corporation maniacally focused on how they could get our stuff coming out there to the marketplace faster."

It's been a long trip—to heaven and hell and back—to get to this point. The story of IBM Research is one of fantastic heights and abysmal lows—of great technological and scientific triumphs, money-making blockbusters, and lost opportunities. It's a story, too, of a late nineteenth century company initially taking an entirely different approach to research than contemporaries such as General Electric or AT&T. IBM started a central lab decades later than these other giants—and well behind foreign competitors such as Siemens and NEC. But then it went whole hog, committing to scientific and technical excellence. Its litany of landmark achievements—from magnetic storage to the dynamic random access memory to the discovery of high-temperature superconductivity—ranks second only to Bell Labs among industrial research organizations.

IBM's roots trace back to independent inventor Herman Hollerith, a Census Bureau statistician whose punched card technology won the competition to tabulate the U.S. census of 1890. Business waned after the count. But in December 1896, not long after signing a big contract to process freight bills for New York Central railway, Hollerith incorporated as the Tabulating Machine Company.

The firm grew steadily over the next fifteen years. But Hollerith had health problems, and was far more an entrepreneur than a business executive. So in 1911, the fifty-one-year-old inventor agreed to a three-way

---

IBM Research: The Top Twelve Achievements

1956	First magnetic hard disk for data storage
1957	FORTRAN, first general-purpose, high-level computer language
1960s	One-device memory cell: key to dynamic random access memory
1960s	Fractals: revolutionizes computer graphics
1970	Relational data base concept published
1979	Thin-film magnetic recording head introduced in product
1980	Reduced Instruction Set Computing (RISC) prototype completed
1981	Scanning Tunneling Microscope*
1985	Token Ring local area network introduced
1986	High-temperature superconductivity*
1993	SP scalable parallel supercomputers announced
Ongoing	Speech recognition

*won Nobel Prize in physics

---

merger arranged by the financier Charles Ranlett Flint. The deal, which made Hollerith a millionaire, fused the Tabulating Machine Company with time clock maker International Time Recording Company and the Computing Scale Company of America. The new concern was incorporated in June 1911 as the Computing-Tabulating-Recording Company; not until 1924 would it become the International Business Machines Corporation. The thrown-together conglomerate struggled in its early years, however. And in May 1914, the board of directors sought new management to bring the company out from under a mountain of debt. Their choice was the former number two executive at National Cash Register, Thomas J. Watson.

Watson was a firebrand of a manager who would be named CTR's president within a year and guide its transformation into one of the world's most successful corporations. The only son of an immigrant Scots-Irishman, he had grown up near the small upstate New York town of Painted Post. As a boy, he had stood on a muddy roadside and watched Amory Houghton, Jr., founder of Corning Glass Works, ride past in a fancy carriage. That spectacle had fired the youth with ambition, and inspired him to go to Buffalo and seek his own fortune.

The young Watson's first job, at seventeen, was selling sewing machines, pianos, and organs off the back of a wagon for a local hardware store. One day, or so the story goes, he joined some pals in a roadside saloon to celebrate a sale. The youth drank too much—and too long. When the bar closed, he found his horse, buggy, and merchandise stolen. He was fired and dunned for the lost property—and forever after insisted his subordinates forswear alcohol in public settings. After that incident, Watson dabbled in other careers. But at age twenty-one, he landed a job as a salesman for National Cash Register. There, his business gifts blossomed. He was promoted up the ranks, eventually becoming Patterson's right hand until mounting tensions between the two dynamic individuals sparked his dismissal.

From the start at CTR, the most pressing problem Watson faced was product development. In particular, competition was rising from the Powers Accounting Machine Co., founded by Russian immigrant James Powers. Unlike the tabulating machines developed by Hollerith—those sold by Watson's new firm—the Powers machines provided printed output, eliminating the need to manually record the numbers registered in counters or dials. This gave Powers a clear competitive edge Watson had no easy way to counter, especially since Hollerith had farmed out manufacturing to various contractors, leaving CTR with scant in-house technical resources.

It all amounted to a stew of R&D troubles. The only experienced technical person on the Tabulating Machine Company end of the business was Eugene Ford, who worked out of a laboratory in Uxbridge, Massachusetts, near one of CTR's prime contractors. Watson convinced Ford to relocate to New York and set up the company's first full-scale engineering department devoted to developing tabulating machines. The operation took over the top floor of a twelve-story building at Sixth Avenue and Thirty-first Street, near Pennsylvania Station. Largely to fund the lab, Watson asked the Guaranty Trust Company for a $40,000 loan. But three years earlier, Guaranty had put up $4 million to help launch CTR—and it refused to invest another dime until business picked up. To this, Watson replied: "Balance sheets reveal the past; this loan is for the future." He got his money.

That bulldog response—his son Thomas Watson, Jr., later called it one of his father's "greatest sales pitches"—can be considered the beginnings of a modern development organization at IBM. Research, though, was another story, as Watson's outfit took a far different tack than giants like General Electric and AT&T, both contemporaries of CTR's main precursor firm, the Tabulating Machine Company. These other companies formed central research arms in the early 1900s. IBM, by contrast, would wait until 1956 to form its research department. Instead, virtually all its early efforts focused on engineering and development activities that addressed immediate product needs. This reflected the company's relative small size and slower early growth—and especially the fact it did not exist initially as a monopoly.

As it was, Watson's campaign to beef up CTR's technical capabilities was impressive. He did it through whatever means necessary—hiring inventive talent, or buying it. The most decisive factor in warding off Powers lay in Watson's ability to wield CTR's existing tabulator patents to force his rival into paying stiff royalties that ultimately helped push it into temporary receivership during the mini-Depression of 1921–22. But he and Ford also recruited a slew of technical talent, including machine designer Clair Lake and Fred Carroll, an experienced inventor who had worked for Watson at National Cash Register.

In 1917, the battle against Powers still unfolding, Watson closed the Sixth Avenue shop and moved Lake, Carroll, and seven others to Endicott, New York, to work more closely with existing manufacturing operations. He also hired James Bryce, an inventive superstar who within five years would become CTR's—later IBM's—chief engineer. Finally, Watson set about securing rival patents. Probably his most important acquisition came in 1922, when CTR purchased the entire laboratory of independent inventor J. Royden Peirce, who held key patents in automatic time-controlled technology. An intense rivalry developed between Peirce's Manhattan lab and the Endicott group.

Only twenty-six of fifty-nine key patents employed by CTR in Watson's first ten years with the company came from its own inventors. Nevertheless, IBM's growing technical competence primed its pumps for explosive growth. In the 1920s, as its punched card equipment became widely used in scientific and engineering computations, the company moved to find ways to store more information on each card. Here, IBM once again met stiff competition from the revived Powers Accounting Machine Co., which was acquired by Remington Rand in 1927. There ensued a fierce battle, during which the Manhattan and Endicott engineering groups proved critical in maintaining technological parity with Rand. And that, it turned out, meant victory for IBM. That's because the punched cards developed by IBM and Rand were so distinct that those prepared on one system could not be processed by the other. Since IBM had installed more machines, the inability of its competitor to offer clear technological superiority ensured its upper hand in the market.

The Great Depression of the early 1930s—disastrous for so many businesses—proved a boon for IBM. For one thing, the company's practice of leasing equipment rather than selling it kept revenues relatively strong: strapped businesses were far more willing to rent than buy. For another, Watson seized on his good financials to consolidate market position. In 1933, IBM entered the typewriter business—soon a hallmark—with the purchase of Rochester-based Electromatic Typewriter Corporation. The 1935 passage of the Social Security Act brought another windfall: all Social Security checks for retired or unemployed workers were generated on IBM punch cards.

Meanwhile, Watson, the veteran salesman, was turning into a fanatic about engineering and development. In 1933, he shifted his Manhattan engineers to Endicott. With the intent of facilitating technology transfer, a handsome three-story building capped by a clocktower—the North Street Laboratory—was constructed to house the engineering arms of both the manufacturing and product development organizations. The reception room was called the Hall of Products. Around it on the ground floor ran the manufacturing engineering department. The next two levels housed seven of Watson's top-ranking inventors—including Ford, Lake, and Carroll—and their support staffs. The building also contained a technical library, patent office, industrial design shop, machine shop, and electrical shop. "We have realized from experience that the future of our business largely depends on the efforts, brain, and ability of our engineering department . . . ," Watson proclaimed at the facility's inauguration. "That is why today we are breaking ground for this new building, which will be devoted entirely to research and engineering work."

Although Watson spoke of "research," the Endicott organization was aimed chiefly at immediate product needs—and fell far short of the longer-range and more fundamental investigations of a central research arm. Indeed, as the 1940s opened, few signs of a full-fledged research organization could be found floating around IBM. General Electric had its House of Magic. AT&T had Bell Labs—and both had their own Nobel laureate. But heading into World War II, there was nothing to indicate that IBM research would soon be joining their select ranks.

The war changed almost everything for American industrial research and development—and IBM embodied the new era. During the conflict, the company had placed all manufacturing facilities at the government's disposal—turning out fire control instruments, Browning automatic rifles, bombsights, and other ordnance. Far more critical for IBM's future, the success of wartime technologies such as radar underscored the swelling importance of electronics, giving impetus to previously frustrated efforts to make the transition from mechanical to electronic methods of calculation. The late 1940s announcement of the transistor and the explosion of science and technology funding tied to the Cold War fueled the evolution. Within a few years IBM launched massive forays into vacuum tubes, data storage, semiconductors, programming, and more that catapulted it toward dominance of the fledgling mainframe industry.

It was in this climate that research finally blossomed with the late 1944 or early 1945 creation of the Watson lab at Columbia University. The prime motive had been to compete with Harvard's Mark I computer. IBM had helped fund and build it. But management felt the university had failed to properly acknowledge the company's contribution. IBM got its revenge. In January 1948, the Watson lab dedicated the Selective Sequence Electronic Calculator, a one-of-a-kind machine that for a year or so ranked as the world's fastest computer. IBM proudly displayed the symbol of its technological prowess in a specially designed room next to its Madison Avenue headquarters. As one account noted, "The rubber tile floor in geometric pattern, the silver and gold finished walls, recessed fluorescent lighting and black marble columns provided an impressive setting. . . . Behind glass panels set in aluminum frames, the memory units were stored; as they performed, small lights flickered and flashed like fireflies." Not far behind came the Naval Ordnance Research Calculator, an even more powerful machine that remained in service some thirteen years at the Naval Proving Ground in Dahlgren, Virginia.

As Watson researchers pushed the frontiers of computer science, mathematics, and solid-state physics, the lab's high-profile emphasis on academic freedom and excellence turned it into a magnet for top talent. Llewellyan Thomas, John Backus, Richard Garwin, Erwin Hahn, John Lentz—these names would become legends in their fields, if they weren't already. And as one account of their exploits later noted, "it is illuminating to consider the fact—attested to by Lab alumni—that many of the individuals involved would not have joined the company except as member of the Watson Laboratory."

The lab, though, was not the only IBM research effort to spring from World War II's ashes. Within months of the German surrender, executive vice president Charles Kirk called electronics whiz Ralph Palmer to New York for a powwow. Before leaving midway through the war for naval service, Palmer had supervised the company's small electrical laboratory in Endicott—and had pressed for bigger efforts in electronic calculation. His sentiments gelled with those of Thomas Watson, Jr., himself just returned from the war. And Kirk, an experienced executive busy grooming the younger Watson to take over the company, thought such an enterprise might find a home amidst the vast manufacturing and development facilities—first for munitions, and then for electric typewriters—the company had started during the war in an old specialty foods processing plant along the Hudson River in Poughkeepsie.

"I understand you think IBM is falling way behind in electronics," he queried.

"That's right," Palmer responded.

"Why don't you go to Poughkeepsie and do something about it?"

Palmer went. The lab he established in 1946 represented the poor cousin—or alter ego—of the heady Watson facility. It operated largely out of a rundown building—the old pickle works—perched amidst power plants and warehouses near the river, below IBM's main Poughkeepsie factory. The recruits, though often exceedingly talented, typically lacked the academic pedigrees of their more glamorous New York colleagues. The very best—on paper anyway—did not come, noted the late physicist Rolf Landauer, who joined the small semiconductor research effort in the early 1950s. "We were not IBM University," he recalled. "You didn't have to be told. You knew you were there to do something useful."

The lab helped IBM's bottom line almost immediately. Within two years, Palmer's group had produced the 604 Electronic Calculating Punch, which dramatically eased many accounting tasks and proved a hit almost from its fall 1948 release. More to the point, the Poughkeepsie enterprise accelerated the company's investigation into electronic arithmetic elements, storage, ferroelectric materials, diode logic, solid-state devices, semiconductors, and other key areas of computing. It became a focal point of a dramatic expansion of IBM's technical efforts that saw the total R&D work force swell from six hundred to nearly three thousand over the first four years of the fifties. Half the new jobs—1,200 in all—came in Poughkeepsie. In the fall of 1954, a new facility there—named the 701 Building after the company's first electronic stored-program computer product—was dedicated as the IBM Research Laboratory. But Palmer was a results-oriented engineer, not a scientist like those at Watson. To remind everyone that he wanted practical innovations that could be mass produced, and not one-of-a-kind machines, a bust of Thomas Edison was placed in the new lab.

Even before people moved in, however, others inside IBM were pushing for a more detached and independent research organization. A group of consultants hired by IBM agreed, stressing the need for greater efforts in basic science, as well as investigations in fields outside the company's current business lines. These opinions carried the day. In January 1956, Research was broken off as its own organization—reporting independently to corporate management rather than through engineering. In addition to the Watson lab, the new body included the company's roughly one-hundred-person San Jose lab, a facility devoted to storage technology that had opened in 1952, as well as the recently launched Zurich Research Laboratory. Later that year, IBM announced the selection of well-known physicist Emanuel "Manny" Piore, former chief scientist for the Office of Naval Research, as its first director of Research.

Piore aimed to create the world's greatest industrial research organization. Using Bell Labs and General Electric as models, he built the effort around fundamental or quasi-fundamental investigations. The nucleus of his organization came from Palmer's Physical Research department, which harbored top-flight researchers in information theory, magnetics, semiconductors, and other aspects of solid-state physics. The new research director also hired a slew of additional physicists, chemists, and electrical engineers—and launched or dramatically upgraded programs in optics, chemical films, and computer systems design. In 1958, a strong mathematical sciences department was created under Herman Goldstine, a former colleague of the brilliant mathematician and digital computer pioneer John von Neumann. Also under its auspices were IBM's troops of computer scientists, among them John Backus, who the previous year had invented FORTRAN, the first high-level programming language. In 1960, IBM snared another prize by recruiting Leo Esaki. Working at Sony a few years earlier while completing his doctorate from Tokyo University, Esaki had observed the phenomenon of electron tunneling. The discovery would lead to his sharing the 1973 Nobel Prize in physics. His mere presence helped put IBM Research on the scientific map.

Again departing from Palmer's style, the neophyte research organization distanced itself physically from the rest of IBM. Its new home was in Westchester County, some 50 miles south of Poughkeepsie—closer to Yale and Columbia, and only 40 miles from midtown Manhattan. In 1957, the first block of some 125 researchers moved to a grand three-story building on the former Robert S. Lamb estate—future home of the Hudson Institute—outside Ossining. Two other facilities would also be built—one in Ossining itself, the other in nearby Yorktown, New York.

But these three labs were only staging grounds for the real consolidation in Yorktown Heights. There, the stunning new Thomas J. Watson lab designed by Eero Saarinen, with its sweeping glass front and metal argonauts, opened for business in 1960. By that time, the number of Ph.D.s had swollen from 58 when Research became a separate department in 1956 to 243—representing nearly half of all professional staff members. Physicists, mostly, they had witnessed the rise of the transistor and atomic energy. The laser was born that very year—and it was a team from IBM's new facility that built the second and third lasers ever created. The place fairly crackled with science. "The fashion of that time was if you take scientists and give them the money and give them the freedom, marvelous things are going to come out of it," remembered Rolf Landauer. "There was formula in the air."

From these seeds sprang the world's largest corporate research organization in the physical sciences. In the first fifty-odd years after the Watson Lab at Columbia University was founded, a flood of triumphs in storage technology, speech recognition, semiconductors, computer architecture, and basic physics—including back-to-back Nobel Prizes in 1986 and 1987 for the scanning tunneling microscope and high-temperature superconductivity—would place IBM squarely at the forefront of industrial research.

But while some of these advances poured billions into corporate coffers, the division occasionally failed brilliantly to turn its alchemy into gold. In the 1970s and early '80s, IBM sunk well over $100 million into an ill-fated attempt to make Josephson Junction computers: the effort at one point had more than one hundred people on it—a huge number for a single project. Meanwhile, lab innovations in relational data bases and Reduced Instruction Set Computing (RISC) were usurped by competitors such as Oracle, Hewlett-Packard, and Sun. Just as with similar experiences at Bell Labs, Xerox PARC, and elsewhere, such spectacular apparent failures earned Research the reputation of being a "country club" of great thinkers whose members cared little about bringing their creations to market.

The more complicated truth is that for nearly three decades key managers and staff members struggled relentlessly—and often successfully—to make research pay. The innovative Joint Programs launched by Ralph Gomory in 1980 to team researchers with developers and speed technologies to market were critical to building IBM's multibillion-dollar storage, disk drive, and semiconductor businesses. And there were many other valiant efforts—before and after—all of which proved integral to the division's tremendous growth. It's true that as the competitive climate changed drastically, aggressive action was ultimately needed to get the research organization even better attuned to advancing company goals. Nevertheless, not only does IBM Research have to be viewed as a tremendous historical success, these efforts laid the groundwork for the dramatic changes later on—and without them it's doubtful whether the division could have survived when the crisis came.

The first research manager to really home in on the issue of Research's connection to IBM's businesses was probably Gardiner Tucker, an old Watson lab veteran who had moved to Poughkeepsie to head its semiconductor studies. He was named director in 1963, the year Research became a separate IBM division. By then, Piore had moved up the corporate ladder, and his position had been filled briefly—unsuccessfully—by another Watson physicist, Gilbert King.

Tucker found he'd inherited some incredible scientific and technical talent—and a sea of aimless pet projects. "There was a faith that research was a good thing," he recollects. ". . . I think it was true, research *was* a good thing—but not any research." Specifically, he notes that while staff members almost always felt they were working on issues of great potential import to IBM, they had forgotten—or not realized—that while pursuing any breakthroughs they needed to generate more immediate payoffs. For instance, the lab supported a technically beautiful project devoted to automatic Russian language translation. But it was virtually useless to IBM because it wasn't helping develop general technologies and methods of computing. "And that point of view had been lost."

Tucker sought to lay out a well-conceived agenda for advancing corporate aims. He launched a planning effort called Corporate Technology Strategies, which pulled together people from research, development, and IBM business units to ponder the needs of future product generations—and determine what could be handled by existing technologies and what required a radically different approach. The new director cut the big Russian translation effort, but kept an ongoing program in linguistics and speech recognition. He also finally closed down IBM's once-massive cryotron project—an effort to build a superfast, superconducting computer. This long-shot program had gotten way too big, way too fast. In place of such endeavors, Tucker started research into field effect transistors (FETs) and related areas such as processing techniques, chemistry, lithography, and circuit design. The FET work was successfully transferred to development and in the early 1970s formed the basis for IBM's mainframe memories, as well as the logic elements in smaller machines.

After Tucker left IBM in 1967 to become a Pentagon advisor, his generally pragmatic approach was continued by physicist Arthur Anderson. Although the new director of Research would emerge as a major figure inside IBM, eventually becoming the executive responsible for its huge Data Processing Products group, he transferred to the business side after just a few years. Therefore, the real job of trying to channel the Research division toward practical results—while also giving scientists their heads—was left to Anderson's successor, Ralph Gomory. A noted mathematician who proved an eloquent spokesman for his division, Gomory would serve as director of Research from 1970 to 1986—and then oversee its actions as an IBM senior vice president and corporate management board member until his retirement three years after that. More than anyone, Gomory put the defining stamp on IBM Research.

Gomory was born to an upper-middle-class family—his father was a banker—in Brooklyn Heights. After taking his doctorate from Princeton, where he developed a keen interest in mathematics as a tool to understanding the physical world, he spent three years on active Navy duty, working on operations research. He then returned to his alma mater as an assistant professor, finally joining IBM's fledgling mathematical sciences department in 1959. Almost immediately, Gomory distinguished himself by applying his expertise to business problems—saving IBM customers millions of dollars through such innovations as mathematically optimizing the process of cutting large paper rolls and handling the flow of messages in large networks. Only five years later, Gomory was named an IBM Fellow. Four years after that, he became head of the mathematical sciences department—a stepping stone to the director's position.

Gomory took over Research when the Cold War was at its zenith. But while he had to pay some tribute to the altar of science, given the times, Gomory didn't buy into the prevailing view that merely supporting good science provided the key to the future. Indeed, despite building much of the framework that would lead to four of IBM's five Nobel Prizes, the vast majority—more than 90 percent—of his resources were focused on technology. His specifically spelled out strategy had two main components: to work as closely as possible with the business divisions on "in place" technology while at the same time pursuing the most viable alternatives.

In the case of disk drives, this meant advancing magnetic heads while simultaneously investigating optical storage. Similarly, while maintaining the Joint Program in silicon technology, he pursued the Josephson Junction project. Although, like the cryotron effort, this attempt to create a new computing architecture got too big and went on too long, it made sense at the time. "You got to remember that in that epoch . . . fundamental computer technologies had been changing," relates Gomory. "Memory went from being a cathode ray tube to being wound magnetic cores to being bipolars and then to FETs. The notion that the ultimate technology had arrived was not there. We were trying to get to the next technology. Well, it turned out that there never was a next technology."

As part and parcel of these efforts, Gomory beefed up semiconductor research as well as technology transfer initiatives. He worked to overcome the divide between Research and IBM's development groups, which hindered the flow of innovations to the market, undertaking a series of moves that culminated—with Jim McGroddy's critical assistance—in the establishment of the Joint Programs (see Chapter 4). But even this formed only part of a larger policy. Very early in his tenure, he detected a pattern. His researchers would invent something, take it unsuccessfully to development, and come back moaning about engineers who couldn't tell a real improvement when it hit them over the head. Then the invention would be put on a shelf—and the researcher would move on to another project, still grumbling about the development idiots.

"When I was a young director of Research at the beginning, I believed it," Gomory says of his people's disdain for developers. "But I was wrong to believe it." Indeed, as he learned more about development, Gomory realized that many Research inventions were simply not appropriate. Sometimes they didn't take into account manufacturing constraints.

At other times, they arrived at the entirely wrong point in the development process. For instance, it took two years to develop each generation of printers. If Research showed up nine months into the schedule with a better approach, notes Gomory, "it was of no use. There are only certain moments in the development cycle when new ideas can be injected." Not only were the developers not dumb, he adds, "they were right."

In order to halt this practice, Gomory forbade anyone in Research to quit trying to transfer a technology without the approval of the "area managers" directly under him. Then, if the efforts still failed, Gomory promised to argue the case all the way to the management board. This policy, he says, went a long way toward getting researchers to think about development needs from the start. "It smoked out things . . . ," he relates. "It forced the people to be realistic, because they knew we wouldn't give up, and we would be pushing their lousy ideas on the division and they'd be quite embarrassed." Not only that, he adds, it also appeared to make the business divisions a bit more attuned to what was being offered—because they knew arbitrary decisions on their part would not simply be accepted by Research.

Throughout his tenure at IBM, Gomory continued trying to find ways to hone the message that the division had to keep company interests firmly in mind. After a lot of deliberation, he delivered his ideas in a succinct and powerful mission statement. Then, and for years afterwards, every Research manager was given a plastic dodecahedron with the following inscription:

IBM Research Goal:

A Research Division famous for its science and technology and vital to IBM.

— Excel technically
— Know IBM
— Know the technical world
— Provide technical leadership

Ralph Gomory's successor was another of IBM's long line of talented physicists, John Armstrong. Though Gomory moved up the ladder to Armonk as senior vice president and chief scientist and had cultivated the division's strong science efforts over the past sixteen years, in a way Armstrong oversaw the scientific heyday of Research, because it was on his watch that IBM won its successive Nobel Prizes in 1986 and 1987.

Armstrong tells a story about the discovery of high-temperature superconductivity. When he directed the physical sciences department in the late 1970s, he convinced Zurich researcher K. Alex Müller to come to the Yorktown lab for an eighteen-month sabbatical. As Armstrong recalls: "Alex had come into my office at the beginning of the sabbatical and he'd said, 'Well what is it that you want me to do?' And I said, 'Alex, I don't care what you do. As a matter of fact, you can stay home for eighteen months and read books as far as I'm concerned.' And he said, 'Well one of the things I thought I might do was learn about superconductivity.' And I said, 'Alex, that's a dead subject, but you can do anything you want to.' "

Years later, at a university cocktail party, an eminent historian of science approached Armstrong and said: "I want to shake your hand. You are one of my heroes." It turned out he was writing a history of the discovery, which had garnered Müller and IBM colleague Georg Bednorz the Nobel Prize; and Alex Müller had told him the story of how he got into the field. Armstrong winced inside. He felt certain that the lesson the historian wanted to draw was that enlightened industrial research management lets scientists do whatever they want. However, he relates, "[The] freedom to follow your own inclinations in industrial research is a freedom that is earned. It's not somehow a right or the general policy. That was not my policy towards physicists. That was my policy towards Müller."

As IBM's research boss, Armstrong often told people he had two jobs: the first was to create an environment where research was focused on the right areas and in a position where it could thrive. The second: to ensure that even standout developments arising from meeting the first challenge resulted in something that mattered to the company.

Superconductivity and the scanning tunneling microscope were not about to fill that bill—but Armstrong had some ideas about what would. At the top of his list were the Joint Programs. Beginning in 1981, he had spent two and a half years in East Fishkill, New York, where he had overseen some 550 scientists and engineers working on advanced semiconductor logic devices and processes. He was the first director of Research since Ralph Palmer who had actually worked in product development—and had played a hands-on role in cultivating the Joint Programs from the other end. So while giving a few scientists like Bednorz and Müller their heads, he drew on that experience, and on contacts he'd made around IBM, to continue the work of integrating the programs into all aspects of the company's business.

Here he had a lot of help and guidance from Gomory in Armonk, who was also striving to advance the programs and improve IBM's prod-

uct position relative to its competitors. Indeed, by the time Gomory retired in 1989 and Armstrong took his place as vice president for science
and technology, Research participated in nineteen Joint Programs. These
covered virtually every aspect of its activities—from silicon and gallium
arsenide semiconductors to storage to workstations, displays, data bases,
and packaging. Taken as a whole, the Joint Programs still formed only a
small part of Big Blue's vast R&D efforts, but as competition increased,
and the pressures seen in Japan came to the fore, they represented some
of the few bright spots.

In the end, though, even more was needed. In one sense, mused
Rolf Landauer, it might have been a mistake to break Research off as a
separate department back in 1956 and move it away from the Poughkeepsie development groups. After all, that split physically created the divide
between "R" and "D" and set the stage for a host of future problems.
Until then, research projects often went straight to manufacturing—accompanied by their progenitors. Under the new setup, researchers were
supposed to hand inventions off to development, trusting to others—or
not trusting them—to see their brainchildren into production. The Joint
Programs went a long way toward repairing that damage—but would
they ever beat the Palmer days?

The company's dominant market position throughout much of the
period undoubtedly widened any gaps between Research and the rest of
IBM. For one thing, the enormous profits generated by entrenched technologies made it difficult for business groups to accept upstarts. For another, big profits engendered waste throughout the R&D pipeline. By the
late 1980s, IBM's annual research and development budget hit $6.5 bil-
lion. The Research share hovered close to $550 million; that left a lot of
room for waste.

And Research had its flaws as well. Despite all the improvements
made with the Joint Programs and other efforts, it still suffered from arrogance, a distinct underappreciation of business unit needs and schedules,
a tendency to see development as its rival, and an overemphasis on personal credit rather than the common good—things never totally vanquished from any Research organization. Then, too, big budgets made it
easy to fuel scientists' demands for more equipment and pet projects—
acts that further distanced Research from the business divisions. In the
1970s and 1980s, in Alan Fowler's view, so many physicists, chemists, and
materials scientists were hired that it would be hard to justify on the basis
of corporate needs. "Research was fat," he relates, "there's no question
about that." The division launched efforts to detect theoretical particles
such as neutrinos and magnetic monopoles—potential Nobel Prize–win-

ning stuff, but hardly relevant to the information technology business. Fowler himself worked on "localized transport in two-dimensional semi-conductors." It was a fascinating scientific problem, he says, but one with no possible connection to IBM. "And I knew it had no connection with anything IBM was doing."

To some, this marked a failure to uphold the standards set by the Research's own goal. Ralph Gomory's legend declared the division should be "famous for its science and technology and vital to IBM." In reality, relates physicist Wolf-Ekkehard Blanz, who in 1993 left the company's Almaden lab in California (formed seven years earlier out of the old San Jose facility) to join Siemens research, "it was perfectly okay to excel in one of them." Too many researchers ignored the corporate side.

John Armstrong, whose heroes in corporate research were people like Irving Langmuir, who did great science and great things for their companies, sees some merit in Blanz's assessment. "That's one of the things that I fault all of us in management for. We made it easier than we should have to excel in just one. We didn't do enough to encourage the atmosphere that the ideal was to do both." For Ralph Gomory, though, Blanz's comments miss the point. "It *was* all right for an individual to excel in one of them. That was not a misunderstanding. That was the intention. What mattered was that the division as a whole did both, not that each individual did both."

Indeed, Gomory adds, the point of devoting a small fraction of the budget to fundamental studies was to create a climate of excellence and discovery that built prestige for his organization and IBM—and attracted top people who otherwise might not have come. But the vast majority of his efforts focused on technology and business needs, and in this regard Gomory says Research should be considered a huge success. "I got my budget year after year under three CEOs not in some moment of inattention, but in a full-dress hearing in front of the Management committee with all the staffs in attendance and putting in their critiques," he relates. A key element of his annual presentation was a world map showing all the IBM manufacturing and development locations and listing the major contributions Research had made to each. Whether or not he supported some pure science studies was largely irrelevant to this process, Gomory asserts. "The presentations always focused on our direct contributions to the company. They were what got us our budgets."

The truth likely encompasses all these views—and more. The one inescapable conclusion, though, was that as the 1980s wound down, Research as everyone had known it was coming to an end. The crisis would reach a head in the early 1990s, when IBM began hemorrhaging billions annually. However, the writing was on the wall several years earlier. As with many corporations, competition was increasing on all sides of IBM's businesses—from mainframes to PCs to chips. Only for Big Blue, notes Gomory, the situation was even worse—"because IBM was falling apart internally."

These forces dictated different projects and methods of operation. Throughout the 1980s—under Gomory and Armstrong—IBM had still been pretty much in its proprietary age. The biggest issues had been delivery of advanced technology into products and out to the market faster, with almost no contact with the sales side. But in the new day, Research would have to extend the close ties it was forging with development and manufacturing to the rest of IBM—and its customers. All sorts of barriers remained to be overcome. Indeed, recalls Carol Kovac, a future Research vice president, the Watson lab's fieldstone walls made for perfect rock-climbing practice—and researchers had taken to belaying from the roof during lunch and after work. To many inside IBM, she notes, it was known as the place where people climbed the walls—literally. "They thought we were a bunch of wackos."

The job of guiding Research into this new age fell largely to Jim McGroddy. In 1989, Armstrong moved to the Armonk headquarters in Gomory's stead. The new director of Research would report to him—and Armstrong had a major say in choosing his successor from a short list of candidates. He felt strongly it had to be McGroddy, who had the resolve, insight, and experience to make the hard choices ahead. "We were in complete agreement that changes were coming and that Research would have to change," Armstrong remembers. "I never spent a microsecond, nor did he, trying to convince each other that Research somehow should try to be exempt from the changes. If we had taken the view that we were so special and so good that we needed to be spared somehow . . . even though the rest of IBM needed to change—then Research would have been just wiped away."

Armstrong would work closely with McGroddy until his own retirement in 1993. He played a strong supporting role in what went down. But he gives his colleague the credit for picking the right path through the darkness—and pushing for McGroddy's selection is one of the things he's most proud of. "He was clearly the best person for that job. If any of the other people who had been on that short list had become the director of Research when the axe really fell . . ."

McGroddy was a tough, clear-thinking New Yorker who had come to IBM in 1965, on the heels of receiving his physics doctorate from the University of Maryland. For some twelve years, he had worked mainly in the physical sciences department, home to most basic physics. But he had pushed throughout much of his tenure to better channel research toward areas of corporate interest. For instance, in the mid-1970s Mc-Groddy had realized that the Watson lab's world-class semiconductor research often focused on exotic materials, while IBM's products lay chiefly in silicon. So he had started a program called Silicon for Lunch, where a group of about a dozen researchers met every few weeks to hear one of their number deliver a talk on some aspect of the company's silicon efforts. The goal wasn't to get people to quit doing science. However, explains McGroddy, "As they did this first-rate science—do it on materials where the knowledge might have an impact on silicon technology."

Partly because of that effort, McGroddy was tapped to head a new Research department called semiconductor science and technology. In this capacity, he had joined Gomory on the late 1979 technical mission to Japan that spawned the Joint Programs—drafting the essence of the proposal on the plane ride home. His star on the rise, a few years later the young executive had moved to IBM's business side, where he distinguished himself in several manufacturing and development organizations. Among other things, McGroddy had spearheaded the creation of Display Technologies Inc., a Japan-based liquid crystal display maker jointly owned by IBM and Toshiba. DTI's extraordinarily big and bright displays—drawing on Watson lab technology—spurred the success of IBM's ThinkPad line of laptop computers when it was introduced in April 1992. By the late 1990s, DTI ranked as a perennial top-three LCD producer. ThinkPads, meanwhile, accounted for about $5 billion in sales.

These experiences had exposed McGroddy to all aspects of IBM's business and raised his cachet throughout the organization: both proved critical in smoothing the way for the changes he later made in Research. Looking back, he calls the period 1990 to 1992 the first phase of the make-over, before the full trauma of the budget reductions hit. A crystallizing moment came early in the period, when he visited Colin Crook, chief technology officer of Citibank. Crook was a physicist by training, and at the beginning of their meeting, he began talking about the information technology value chain, with atoms and basic mathematics at the bottom, storage, displays, and chips in the middle, and customer solutions at the top. He told McGroddy it was one thing to have great middle- and bottom-level stuff—data bases and mainframes. But what really lit Citibank's fire was up at the top, where it could differentiate itself from Chemical Bank and Chase. Therefore, what Crook looked for when choosing an information technology supplier was not the technology per se, but the applications that came with it.

That experience forced McGroddy to look candidly at his organization. It suddenly became clear that Research activities were heavily skewed toward the middle and bottom layers—and sparse at the top. As he recalls, "Our motto was to be famous for our science and technology, and vital to IBM. Were we really vital to IBM if we were doing almost nothing in the layers of the value chain that mattered most to customers? The answer was obvious."

Over the next several months, McGroddy talked to a lot of IBM customers, learning more about their needs and stressing to his researchers the imperative of providing value at the top of that chain. By spring 1992, he had created a new way of looking at his organization. Rather than the traditional "wheel chart," which essentially listed all major areas of focus, McGroddy reclassified Research activities into strategy areas he called "the Five Boxes." One box, labeled Basic Science, sat in the middle of his chart, where it overlapped and influenced the other four. Another box was Technology, covering semiconductors, displays, and other hardware. Then came Storage—not just devices such as disk drives, but related materials research. A fourth box, Systems and Software, dealt mainly with processors, databases, and operating systems. Finally, corresponding to nothing Research had ever put in any organizational chart, came Services, Applications, and Solutions.

The inclusion of this last box sparked consternation—even outrage. Not only did it mean a lot of people had to change their roles, but "people bitched like crazy and said it wasn't even research," McGroddy recalls. "But it was." What's more, he was committed to driving the percentage of work that could honestly fit in that box from virtually nonexistent to 20 percent by the end of 1995.

As McGroddy refined his plans throughout the summer of 1992, embattled chairman John Akers subjected the division to a stem-to-stern review by ten senior executives from finance, sales, and manufacturing. The five boxes and McGroddy's explanation of the Research value chain enabled the organization to pass with flying colors. But Research was hardly out of the woods. That October, McGroddy called an all-hands meeting—much like Arno Penzias had done at Bell Labs in July of 1990—to formalize his commitment to the changes. He wanted to deliver the news personally to every staff member. So he rented an auditorium at the State University of New York campus in nearby Purchase, where he could reach everyone from Watson in two sessions. From there, he planned to jump on a plane to California to visit the Almaden center. Then it was off to Tokyo and Zurich. The Watson sessions were set for a Monday. The previous Friday, he got word from Armonk that the Research budget would be slashed by 10 percent for 1993. So it was even worse news he delivered.

Nearing the end of 1992, McGroddy became convinced Akers

would soon step down. It was at this time that he began preparing the sixteen-page report detailing Research's contributions to IBM. He was ready, then, on April 1, 1993, when Louis Gerstner, Jr., introduced himself to senior executives and asked for detailed reports on their operations. A few days later, McGroddy had his trimmed-down report on Gerstner's desk—and about a week after that the new chief executive was at the Watson lab.

In preparation for the visit, McGroddy had cleaned up his cluttered office. On his desk was a ThinkPad, with the DTI display. In addition, IBM Fellow Ted Selker had designed the red cursor-control button called the TrackPoint that was built into the keyboard to provide a speedy replacement for the mouse. And there was probably a Research-developed infrared port that enabled printing without cables. "The only thing I had out was a ThinkPad," McGroddy relates. "And I started the conversation by telling him we thought about Research in terms of having an impact, and an awful lot of that impact came together in this product."

A trial by fire still lay ahead for all of IBM. Over the next few years, the company reduced its payroll by more than 100,000 employees, bringing its total number of personnel to under 250,000. McGroddy worked with Gerstner to slash overall R&D funding by a third—from about $5.1 billion a year to under $3.4 billion. His own budget did better, ultimately falling about 22 percent—some $120 million annually—from its 1990 peak. To those in Research, anyway, the fact that the division suffered less than other parts of the company was a tribute to its historic value to IBM.

McGroddy had moved early to trim fat. He saved $3 million annually simply by switching utility suppliers; a series of other belt-tightening actions netted several million more. In 1991, he formed a business development group charged with coming up with an additional $50 million a year by taking on government contracts, and spinning off or licensing technologies that didn't fit into IBM business lines. This led to the commercialization—and eventual sale for $47 million—of the Zurich lab's strong laser group. Another big success was Integrated Surgical Systems, a spinoff company co-founded by Research that produced Robodoc, an automated surgeon. ISS went public in late 1996. McGroddy had kept its existence off the corporate books, and while Gerstner knew there was some connection between it and IBM, he didn't know what. "Lou, three things I need to tell you," McGroddy recalls telling the chairman just before he retired. "You own forty percent of this company, but it's not on the balance sheet. Second, it's going public next month and your finance people don't really know it. Third is I've been chairman the last couple of years, and I never got around to telling you. He took it well."

□

Speech Recognition: The Chinese Connection

In the thick of shaping up a sluggish IBM, chairman Lou Gerstner took time in 1994 for a critical word to his speech recognition experts: China.

A longtime Sinophile, Gerstner knew an untapped computer market when he saw one. Although PCs permeated the world, the complex Chinese language made it extremely difficult for even highly educated workers to go digital. Either keyboards were mind-numbing arrays of hundreds of characters, or users had to clumsily input English letters representing Chinese sounds. Gerstner realized that a good speech-to-text program could bring computing to the masses. Who better to do it than IBM, a speech recognition pioneer? The only catch: with no such project on the books, any upstart effort would likely take years to bear fruit.

Enter C. Julian Chen, a basic physics researcher with no speech recognition experience. Chen's unusual approach to the Chinese problem catapulted IBM past all competitors, leading to ViaVoice Chinese—the first PC application to provide viable recognition of conversational Mandarin.

Chen's story begins at Beijing University, where as a physics major he became obsessed with languages—picking up English, Japanese, German, and French to go with high school Russian. Linguistics went on the back burner as he earned a Ph.D. at Columbia University, then in 1985 joined IBM's Watson lab. In 1994, seeking to help his hard-hit company's bottom line, Chen rekindled his love for linguistics and took up Gerstner's Chinese challenge.

At the time, after two decades of sweat, researchers were finally cracking English speech recognition. Chinese, though, seemed a much tougher nut. The written language is nonalphabetic, meaning words or concepts are repre-

---

Although these maneuvers mitigated the effects of the cutbacks, they could not stem the tide. McGroddy moved forcefully to eliminate unnecessary projects, trying to figure out what to cut based on its potential value to IBM. It was here that basic science really began to feel the heat. The magnetic monopole and neutrino detection efforts got the axe. By 1993, IBM's world-class theoretical physics group had also been reoriented, and began working with leading-edge scientific customers to extend the power of computing in science. But the effects were felt throughout the organization, as McGroddy eliminated some 25 percent of the 3,400-odd Research work force—roughly 900 employees in all. Many were veteran and highly paid researchers, who were pressed to go after university professorships—driven by the choice of changing the nature of their activities to move up the "value chain" or get out. It was a dark time. Morale plummeted—and many top scientists left, including some McGroddy wanted to stay. "A lot of people would have gotten rid of me if they could have," the former Research director sums up.

sented by characters. There'd been some progress with "discrete" recognition, where users speak slowly-with-pauses-between-words. But continuous recognition—converting everyday language to text—still eluded researchers. Chinese has no natural boundaries, or pauses, between words. What's more, the same sound often changes meaning depending on pitch. For instance, various inflections of "maa" mean mother, hemp, horse, or curse. Such differences dictated a completely novel approach from English recognition schemes—until Chen arrived with his unique ability to span linguistic boundaries.

From Chen's practical speaker viewpoint, Chinese and English shared great similarities. Verbs and adjectives in both languages have few special endings. Case agreement—masculine, feminine, neuter—is a nonissue. And nouns are pluralized by adding a single sound. As for pitch, the standard procedure was to treat tones as properties of syllables—creating thousands of possibilities for each phrase. But Chen's breakthrough approach handled tones as English vowels, where any shift also alters meaning, as in "bat," "bit," "bet," "but." Therefore, while others strove to create a distinct Chinese recognition system, he was able to "map the English technology to the Chinese perfectly."

In 1995, Chen hired fifty-four Chinese speakers living in New York, recorded their voices, analyzed inflections, and built an acoustic data base that among other things indicated which words and phrases were likely to go together. He soon created a program that recognized everyday Mandarin with 85 percent accuracy. The work was handed off to the Beijing lab, where scientists fortified the data base—making critical improvements to recognition accuracy—thanks to three hundred more speakers with various regional accents. The trailblazing product was licensed to most Chinese PC makers and preinstalled in millions of computers that began rolling out in late 1998.

□

Not everything was about downsizing, though. Around 1995, Mc-Groddy brought the small research labs in Haifa and Tokyo—both historically run by IBM subsidiaries in their host countries—into the Research fold. He feared their marginalization during the budget crunch, and wanted them part of a more nurturing environment. He also opened two new labs—one in Austin specializing in microprocessor and circuit design, and another in Beijing. He was especially proud of the China lab. The country held a massive potential computer market, and McGroddy wanted to get in early, hire top native talent, and concentrate on tapping it. This proved an extremely expedient move. Putting down roots and signaling that it was there for the long haul tremendously boosted IBM's image as a corporate citizen. The lab also paid dividends on the research front (see box, above).

By 1996, things were really looking up. IBM had successfully introduced its SP scalable parallel supercomputers, based on Research technology. Almaden's magnetoresistive heads were in production, enabling

Big Blue to capture 40 percent of the market for laptop hard drives. And the Services, Applications and Solutions portion of Research had surpassed its 20 percent target—fitting perfectly with Gerstner's push to provide Global Services, which had become the fastest-growing segment of IBM's business. Many researchers even found working with customers and seeing their work impact the real world—not to mention IBM's bottom line—an invigorating experience.

Even before McGroddy's retirement dinner, when Gerstner announced Research had come perilously close to meeting its demise, the outgoing executive was amazed at the number of people who had changed their view. And while he would never find universal praise, his image inside Research continued to rise long after he had left.

"Jim McGroddy was a great Research director," proclaims William R. Pulleyblank, director of the mathematical sciences department as Research headed into the millennium. "I'm amazed at the way he could read the world and see what was happening, and figure out how to position the division to lie right at the center of the new IBM."

Sitting in his expansive office, casually clad in a black mock turtleneck that accentuates the gray in his salt-and-pepper beard, Paul Horn is looking happy to have been saved. And why not? His division is delivering big time on Gerstner's faith—reasserting its standing alongside a similarly rejuvenated Bell Labs as one of the world's two best industrial research organizations in the physical sciences.

Far fewer innovations seem to escape Big Blue nowadays. After twenty-five years' hard labor, the Watson lab's trailblazing speech recognition efforts are taking off commercially. Two home run blows from semiconductor researchers are also paying dividends. One comes from Meyerson. The other, announced in September 1997, involves replacing the aluminum used in microprocessors with copper, a superior conductor that makes for smaller and faster chips. In fall 1998, Almaden lab researchers announced their Microdrive, a 1-inch-square storage device bound for digital cameras and hand-held computers. The following spring, they were helping IBM's Storage Systems Division set a new storage density record of 20 gigabits per square inch using magneto-resistive head technology. Then there's Deep Blue. Even beyond the free publicity associated with demoralizing chess champion Kasparov, the SP line of Deep-Blue-like supercomputers has emerged as a market leader.

These technologies and others are already generating as much as $25 billion annually for IBM—by one Horn estimate, anyway—and poised to do more (see box, p. 153). That's a stunning figure that ac-

IBM Accomplishments—How Good Is Good?

Each year Research scrutinizes its creations and classifies the very best as either Accomplishment, Outstanding, or Extraordinary. These are judged either by their impact on science and technology or on IBM. An innovation can be upgraded over time. Following are some category definitions and results from 1998.

	SCIENCE AND TECHNOLOGY	IBM IMPACT
Accomplishment	significant development in field, some outside recognition	significant product or business impact; customer impact
Outstanding (clearly beyond Accomplishment)	fundamental development, key publication, outside recognition: papers, talks, awards	major product impact; changes technological direction of IBM, key patent, technology
Extraordinary (clearly beyond Outstanding)	worldwide recognition; major awards: Nobel, Turing, Buckley prize	major innovation; sets industry standard; fundamental patent

Extraordinary—Deep Blue elevated from Outstanding. Publicity reputedly generated $3 billion in equivalent advertising. Raised awareness of SP supercomputers.

Outstanding—Copper semiconductors upgraded from Accomplishment as technology begins to impact bottom line. Joins silicon gallium arsenide, which made O levels in 1997. Every chance to make extraordinary.

Accomplishment—
Microdrive: Already heading into digital cameras and other products.

Information Theory: Math computer science algorithm designed by researcher Greg Chaitin had big impact on field.

Silicon on Insulator: Performance-enhancing process for building semiconductors that protects the millions of transistors on a chip from routine electrical contamination.

counts for nearly a third of IBM's 1998 revenues of some $82 billion. John B. Jones, Jr., an analyst with Solomon Smith Barney, says success stems from the quality of IBM's innovations—and its ability to shed once-weighty bureaucracies and turbocharge creations into production. Take semiconductors. "They appear to be outinnovating [competitors]," Jones notes. "And they're also bringing their technology to market faster than the competition—faster than Intel, faster than Motorola, faster than TI [Texas Instruments], faster than Fujitsu." What goes for chips, he adds, also goes for storage and other areas.

A few key factors are driving this spurt. One lies in the improved focus that comes from having survived hard times and being embraced by Gerstner and the business divisions. To give just one example that took place even before the new chairman arrived, Research had made many forays into highly parallel computer architectures in the mid- to late 1980s, with absolutely no impact on IBM's product line. But all that changed with the SP1, which came along in the early 1990s, at a time the company's troubles were so manifest all the parties realized they had to work together.

Pay raises, bonuses, and other incentives—inside Research and company-wide—have been broadened to reflect a team-oriented approach to innovation as well. But researchers and managers insist the big change is cultural, not structural. "What has changed is the ability to do things in a timely manner," says Bernie Meyerson. "We never had the ability to do things in a timely manner around here. Lou's emphasis is, 'Get the job done, I really don't care where you sit.' It used to be everybody would put their hands up and say, 'Stop. That's my job—you can't do it.' " Thomas Theis, director of the Physical Sciences department that spawned IBM's big semiconductor advances, agrees. "We've just taken the chains off people," he says. "It's not that we're more inventive now. This is an environment where people have much more chance to be successful in moving their ideas into the real world."

That world also seems to be playing into IBM's strengths by putting a premium on advances that facilitate the fusion of computers, communications, and commerce—all things Research has been working on for more than a decade. Indeed, with the personal computer turning into a low-profit commodity item, it is from such synergies that IBM expects to fend off more nimble start-ups and niche companies—and drive future growth. In fact, as he prepares to guide Research into the twenty-first century, Horn feels he's taken the Research reins at maybe its best time ever. "I'd like it to be viewed as the heyday," he says.

Horn is yet another of the IBM physicists. He'd been directing the Almaden lab before being tapped as McGroddy's successor. With close to three thousand people in the division, Research is still down some four hundred from its early 1990s apex. But Horn says the current figure includes far fewer support personnel—and that his technical staff is probably higher than ever. Benefiting from IBM's own rebound, the budget, too, is at or near an all-time high of around $550 million. About 65 percent comes from corporate headquarters. Another 25 percent is supplied via the Joint Programs, which now number more than forty. While Research still gets a small amount of government funding, the remaining 10 percent is generated almost exclusively through spinoffs and patent royalties.

Horn characterizes his leadership as evolutionary, rather than revolutionary, as he continues on the course set by McGroddy. But that hasn't kept him from putting his own stamp on things. The most vivid example is his increasingly global view of Research. He opened the Delhi lab in 1997. Two years later, in early 1999, he shifted around the Five Boxes and added the first strategy area outside the United States (see box, p. 156).

As he seeks to weave together this far-flung organization, three broad but intimately related visions guide Horn's efforts. The biggest he calls pervasive computing, the now almost-universal idea that everything and everybody will be electronically networked. The second is deep computing—the ability to extract knowledge from the wealth of data accumulated and stored on-line. Finally comes security, an essential to IBM's concept of e-business.

"We're entering a new era of computation in my mind," explains Horn. "We're really going beyond the sort of PC-centric era into what some people call the dawn of the post-PC era. As communications protocols and data protocols are merged, you start doing computing from all sorts of locations. . . . And that gets into networking, and it gets into electronic commerce, and it gets into knowledge management."

PERVASIVE COMPUTING

MITCH STEIN is a big bearded guy with a big infectious smile. But confronted with the typical home entertainment center, answering machine, or PC, he's not happy. Consider the gizmos and high jinks necessary to navigate these supposedly user-friendly electronics—none of which easily go together. As senior manager of IBM's consumer systems and software group, Stein takes it personally. That's why he's fused many of the stand-alone devices of a modern house to create a face of pervasive computing he calls Life Networking. "I like to tell people that one of my best characteristics is I'm profoundly lazy," Stein explains. "A profoundly lazy person will go to great lengths to solve a problem and never have to worry about it again."

In Stein's Watson lab "family room," phone messages and e-mail, TV viewing, Web-surfing, and other aspects of home control are inte-

An Increasingly Global Research Organization

IBM's eight labs are run by on-site directors. However, Research is also divided into five strategy areas that cover everything from customer solutions to basic science. Roughly equal in size and authority, each area is headed by a Research vice president. Increasingly, researchers form global teams that converge on problems beyond the scope of smaller operations.

Eight Labs

LAB	ESTABLISHED	EMPLOYEES	CORE AREAS
Watson	1961	1,700	semiconductors, math, physical science, computer science
Almaden	1986	480	storage, data bases, and computer science, pervasive computing, physical science
Austin	1995	40	advanced circuit design, microprocessor design and tools
Zurich	1956	180	communications systems, security, physical sciences, e-business, computer science theory
Haifa	1982	280	math, computer science, multimedia, VLSI design verification
Tokyo	1982	170	LCDs, speech recognition, computer science theory, networking, mobile computing
China	1995	40	speech recognition, pervasive computing, e-business
Delhi	1997	30	e-commerce, weather forecasting, deep computing, Internet collaboration

Five Strategy Areas

Services, Applications and Solutions—Works with customers to develop applications and systems aimed at solving specific problems.

Computer Systems and Software—Personal and parallel computer systems, RISC, networking, speech recognition, security, deep computing.

Telecommunications Technology—New in 1999 and based in Zurich. Fiber optics, laser transceivers, asynchronous transfer mode, token ring.

Storage Technology—Almaden-based area covers investigations into traditional magnetic storage, giant magnetoresistive heads, and farther-out areas such as holographic storage.

Systems, Technology, and Science—Heavy emphasis on microelectronics: copper chips and silicon-germanium originated here. Displays, infrared and RF technology, most basic physics.

grated onto a big television screen. Stein feigns a man returning home from work. His presence is detected, lights come on, and his dormant (but never off) system awakes. "Hello Mitch," the computer's tinny voice greets him. "Several messages are waiting for you."

And so the demo begins. Stein preaches that "ease of use" is an overly broad term which has lost much of its meaning. He wants to augment it through such concepts as "ease of learning" and "joy of use." To that end, he can view his e-mail—or listen to it. Similarly, he can dictate electronic messages or ask the system—in normal, everyday language—to dial a friend. Shifting to entertainment mode, Stein asks: "What time is *The X-Files* on?" The response: *"The X-Files* comes on at nine tonight. I will record it as usual."

Virtually every computer lab these days is experimenting with similar systems. But few offer IBM's soup-to-nuts expertise in speech recognition, graphics, storage, processing power, and more. It was the chance to tap that extensive technical base, coupled with incredible marketing clout, that brought Stein, Apple's former director of Human Interface Technologies and a founder of five start-ups, cross country to join Big Blue. At a smaller company, he notes, "I came to the realization that maybe, best case, I'd be able to do one or two point products. But I knew I would never be able to attack the whole platform."

Indeed, Big Blue just might pack more pervasive computing power than any place on the planet. Other Watson researchers show off wearable screen displays implanted in eyeglasses. The pioneering speech group is zeroing in on true natural language interaction with computers. The Almaden lab supports some seventy-five projects aimed at helping people deal with machines intuitively, as they do with other humans (see box, p. 158). Over in Zurich, scientists are developing smart cards that enable owners to securely conduct e-business from anywhere. On the basic science front, Nobel laureate Gerd Binnig of scanning tunneling microscope fame is fashioning nano-electromechanical devices designed to read and write data bits less than 50 nanometers wide—potentially engendering the kind of portable, library-sized storage capacity everyone might need in the pervasive era.

While such ideas may represent more than a little wishful thinking, IBM is starting to probe the market with a few technologies designed to fuel its vision of pervasive computing. Often, these are launched under its novel First of a Kind program, where researchers team with an adventurous customer to tackle specific problems in some unique way. The appeal for customers is to gain an early entry into a new technology. IBM gets to work out kinks in a controlled, real-world test bed. If things don't pan out in one year, the project is terminated. But if all goes well, the idea might be expanded dramatically, opening new areas of growth—be-

---

□

One Face of Pervasive Computing

Omit the steel door and lab coat, and it isn't hard to see Ted Selker as a corporate "Q," 007's pleasantly eccentric gadget man. Until September 1999, when he joined M.I.T.'s famous Media Lab, Selker headed USER—User Systems Ergonomics Research—at IBM's Almaden Research Center. Like Q's, his job was to dream up ingenious devices—only instead of instruments of destruction, Selker and roughly twenty-five colleagues devised ways to allow people to interact with machines intuitively, more like they do with other humans. "The culture here is prototype now, ask questions later," said Selker, an IBM Fellow and co-inventor of the TrackPoint—the red cursor control button now in about four million laptops—who's been gadgeteering since boyhood.

Much like in Q's lab, contraptions abound in USER. Computers turn on as people sit down. A weight machine knows who slides onto its benches, conceives a personalized workout, and bellows, "You Can Do It!" Witness laptop lids that transform into slide projectors. Machines that recognize facial expressions. Easier scrolling. Easier searching. Right on down to Room with a View, a virtual room projected on a wall that allows people armed with laptop keypads to select tomes from the bookshelf, surf the Internet by selecting the globe hovering outside the window, or investigate other activities. The keypads even control chair temperatures and order back rubs.

A whirlwind tour turns up several focal areas. Some highlights:

Physical. The basic idea behind the personal area network is that anyone toting a special transmitter can emit a digital aura that conveys driver's license, credit card info, and access codes without the need to swipe cards through

---

coming what Mitch Stein calls "the Trojan horse that will start the avalanche of capabilities."

The fall 1996 introduction of continuous speech recognition software provides a prime example. The company worked with New York's Memorial Sloan-Kettering Cancer Center and Massachusetts General Hospital to produce MedSpeak, a specialized application for radiologists, whose distinct technical vocabulary made recognition easier. As the technology improved, IBM expanded into legal dictation and then more general products—establishing its ViaVoice line as a leading force in speech recognition. In a similar vein, early in 1999 Stein was working on a home information center application that might lay the groundwork for fulfilling his Life Networking concept. Also under way at Watson was a disaster management effort called TeamBuilder, developed with the city of Orlando to automatically identify and alert the personnel best suited to handle different emergencies.

"readers." It's such technology that automatically logs people onto the workout machine. But a more likely twist involves travel. Researcher Thomas Zimmerman envisions people entering a hotel lobby, walking up to a kiosk, and learning their room number from the screen: doorknobs will be preprogrammed to grant guests access during their stay.

Graphical. USER's "ecological" alternatives to the Xerox PARC-Mac interface more realistically simulate actual desktops—down to desk, bookshelves, and wall hangings. Psychological studies indicate such arrangements—especially in 3-D rather than standard 2-D—dramatically improve the ability to find computer files.

Cognitive. Humans share six "universal" facial expressions: anger, disgust, sadness, fear, surprise, and joy. Project Blue Eyes, an image recognition experiment conceived jointly by USER members and Almaden's Visual Media Management arm, aims for computers to recognize such emotions and adapt behaviors to human moods. For example, the computer might detect pique when a phone call interrupts a person's work and decide to route calls directly to voice mail.

Don't look for context-sensitive sensors involving facial expressions to blossom anytime soon. More immediately, though, personal area network–like creations could handle mundane tasks such as hotel registration—allowing humanoids more quality interactions. "I see a future where we are going to spend more of our time interacting with the personality of the person we are doing business with than taking Visa cards and things like that," predicted Selker. "We're going to have less lines, less forms, less missing each other because we are at the right corner at the wrong time."

□

IBM's first First of a Kind project in pervasive computing applications kicked off publicly in February 1999, when select customers of British supermarket chain Safeway Stores PLC began buying groceries electronically. The concept was hardly original. But IBM's idea was to employ the popular PalmPilot hand-held computer as the cornerstone of an unusually flexible system. Its seven full-time team members spent a year working with Safeway customers to design an interface that allowed items to be readily accessed by category. To this they added new protocols for optimizing data transmission, an IBM Web server, and a DB2 Universal Database management system—largely with this product, IBM had come roaring back from its earlier slip in relational data bases to become a market leader—that automatically determined usage patterns and devised impulse-buy suggestions tailored to individual preferences.

The result was Easi-Order, a system IBM claimed offered unparalleled ease of use from refrigerator to storeroom. Rather than being chained to a desktop, people could compile lists in front of their refrigerator, in their favorite chair, on the bus, wherever. When ready, users simply plugged the Palms into a special modem—and the order was downloaded in under two minutes. During that time, the store uploaded fresh price lists and special offers.

For Safeway, the idea was to lower overhead and build customer loyalty by offering a wider range of services. Even before the initial test—with two hundred select customers—had concluded, it was seeking to bring the technology to other hand-held computers, mobile phones, and TV screens. While IBM doesn't make any of these devices, expanding the market for pervasive computing opens opportunities for its backbone products and services. By the spring of 1999, it had already shown the core technology to Wal-Mart and other retailers—and was predicting future applications in travel and myriad financial services.

Whether it's through these systems or something else entirely, computer technology seems destined to become more pervasive. But that begs another question: with the ability to create, access, and transmit digital information increasing so rapidly, how can people possibly make sense of it all?

DEEP COMPUTING

MACHINE OUTTHINKS MAN. May 11, 1997—Mother's Day. In a stunning turnaround from their Kasparov-dominated initial match fourteen months earlier, Deep Blue finished off the grandmaster in barely an hour. For many, IBM's victory was a testament to pure computing power. After all, the machine's processors—considering 200 million moves a second—whirred away twice as fast as before. But there was far more to it than that. "How do you beat the world champion?" queries William Pulleyblank, director of the Mathematical Sciences Department at Research. Speed is vital, he explains. But it's often not worth much without the right smarts. So IBM had to combine its crunching prowess with improved algorithms that better factored in the ways grandmasters view chess.

Such is the essence of deep computing—applying vast amounts of computing power, plus the right approach to problems, to vast amounts of data to drive decisions. In fact, Pulleyblank says, "Deep computing really came from Deep Blue." And he should know. In May 1999, his duties were extended to running IBM's newly created Deep Computing Institute, an umbrella organization that seeks to pull together deep computing research at four IBM labs: Watson, Haifa, Tokyo, and Almaden.

By fusing these distributed talents in computing theory, statistics, computational biology, financial mathematics, and data mining—expertise previously turned largely on scientific challenges—IBM wants to solve complex business problems.

The Safeway project illustrates one aspect of deep computing—mining data about retail sales and customers. But IBM's initiatives, encompassing roughly one hundred researchers, probe far broader horizons.

One ambitious project underway in 1999 aimed to combine Big Blue's supercomputer weather-prediction capabilities with energy and financial modeling to help utilities meet power demands more efficiently—to the point of buying and selling excess capacity on spot markets.

It's a deep computing problem if ever there was one. IBM isn't known for weather forecasting. But Deep Thunder—an SP computer outfitted with 3-D graphics and powerful modeling algorithms—can often predict weather in localized areas with greater precision than anything available through government or commercial channels. The system debuted at the 1996 Atlanta Olympics, where it aided the scheduling of sailing and other weather-dependent events—and helped save the closing ceremonies by predicting a powerful storm that had officials ready to postpone the extravaganza would stay 10 miles from Olympic stadium.

Because weather is central to determining energy demand, even conventional forecasts can help utilities use their generators more efficiently. In a program with a midwestern utility that ended in 1998, IBM researcher Samer Takriti helped develop proprietary algorithms he says can trim 3 to 5 percent off generating costs—enough to save a typical utility with some one hundred generators upwards of $40 million annually. Add in Deep Thunder's capabilities, Takriti reasons, and the model could get even better. So he started working with Thunder researchers to do just that. At the same time, recent industry deregulation is spawning volatile energy markets, complete with futures, options, and hedging strategies. Takriti also hopes to capitalize on IBM's years of modeling supply-and-demand dynamics and currency trading to create a system that accurately forecasts energy prices.

Developing such comprehensive models will likely take years. But what if a utility could be nearly certain it would escape a cold front while its neighbor would not? It might fire up extra generators anyway—knowing the overcapacity could be sold at a good price. And that's just for starters. Pulleyblank says the fusion of weather prediction and financial modeling could be applied to a variety of agricultural industries, not to mention property insurance and commodity futures markets. Asserts he, "We have trouble even anticipating all the places it's important."

IBM has a poor overall track record in this kind of specialized application programming. If past trends hold true, a wealth of potential rivals serving each particular industry will soon be able to supply the same thing more rapidly, or tailor it better to individual company needs. Still, the company remains optimistic about its future ability to compete on this level.

What's more, even as IBM redirects deep-computing resources from science to business, another vein of research focuses on mining the business of science. An especially hot area is computational biology, which involves analyzing huge data bases of biological information to unearth patterns that might point the way toward creating safer and more effective crops and drugs. IBM dove into this blossoming field in the early 1990s, but began its real push around 1997. It now has some two dozen researchers assigned to it, with scores more in related areas. One key hire was Barry Robson, who in the late 1980s designed a computerized system instrumental in developing a test for mad cow disease. Since then, he has helped launch several computational biology ventures. Now strategic adviser to IBM's Computational Biology Center, Robson says that through clinical drug trials and decades of genetics research, scientists have amassed stockpiles of biological data that today's powerful computers and sophisticated algorithms can finally begin to decipher. "My passion is to really see it become an everyday applied discipline."

IBM's biggest publicly announced effort—an agricultural genetics program launched with Monsanto in January 1998—revolves around Teiresias, a pattern-discovery algorithm that can scour vast protein or gene sequences to find repeated patterns that might code for similar functions in different molecules. Monsanto hopes Teiresias, named after a blind seer in Greek mythology, will hunt through its proprietary data bases and speed the identity of genes responsible for improved yields, higher nutritional content, or pest resistance.

The initial deal concluded in the spring of 1999, when IBM transferred to Monsanto an improved version of Teiresias adapted especially for drug and agriculture applications. Under a new agreement that ran through year's end, IBM was also using the algorithm to look for patterns in the far larger public data bases maintained by the National Institutes of Health and other government organizations—and then construct a novel "biological dictionary" from the data. Whenever academic or government researchers decode protein genetic structures, they post the sequences in these data bases—along with notations of any known functions associated with the genes. These compilations are so vast—20 million seqlets, or mini-sequences from one data base alone—that previous attempts to collate or compare structures looked only at the most promising areas. But using Teiresias and some powerful computers, IBM will put the entire meta-data base into play, hoping to pick up patterns and highlight avenues of investigation that otherwise would have been missed.

That might lead directly to the discovery of new genes. But that's not the only potential payoff. For instance, notes Sharon Nunes, who heads the Watson Lab's Computational Biology Center, Teiresias might turn up unsuspected commonalties across protein families that will enable scientists to design drugs that attack a wider variety of ailments.

Teiresias and other computational biology efforts are still in their infancy—and face the same questions about execution and implementation that confront the electric power work and other elements of IBM's deep computing initiative. But Paul Horn, for one, is particularly enthusiastic about the field, which he sees as a potentially world-changing marriage of IBM's previously esoteric efforts in computer science, chemistry, biology, mathematics, and genetics. "These guys have been doing these very narrow problems—the traveling salesperson problem—when they could be used to cure cancer," he marvels. "When I think of where the next Nobel Prize will come from in our laboratories, those are the most fertile areas."

SECURITY: THE VIRUS WARS

As IBM STRIVES to shape a world of digital culture and commerce, computer and network security become increasingly important—underpinning almost every aspect of its business. The Research division's efforts in areas such as cryptography go back decades, and some security research can be found in virtually every lab. But in the United States, anyway, most of the efforts come together at Watson. One big thread converges around Charles and Elaine Palmer (see box, p. 164). The other centers on the fight against computer viruses.

IBM got into the Virus Wars in December 1987, when an electronic message named CHRISTMA EXEC arrived at Watson over the then-tiny Internet. Theoretical physicist Steve R. White was working on an unrelated computer security problem when the communiqué—once opened—first unfolded on a colleague's screen, slowly tapped out keyboard characters in the shape of a pine tree, then signed off with the salutation "Merry Christmas."

□

The Palmers—A Secure Couple

In 1975, Charles and Elaine Palmer were students at Louisiana State University. Charles handled student punch cards in the computer room. Elaine kept getting her programs wrong. After Charles repeatedly handed her cards back, she broke down and cried. Then, to Elaine's astonishment, "He actually leaped over the counter. He said, 'Don't worry, I'll help you.' So it's been love ever since."

Since that auspicious vault, the computer science couple—married twenty-two years in 1999—have been almost inseparable. After college both took jobs with Standard Oil of Indiana. In 1984, they joined IBM. Six years later, Elaine began dabbling in security issues—and a few years after that Charles founded the Watson lab's antihacker program. These days, the two die-hard Louisianans—their license plates read "EATCAJUN" and SUTHNUH"—have offices across the hall from each other. Charles manages network security and cryptography. Elaine heads secure systems and smart cards. From electronic vandalism to industrial espionage, the buck usually stops with the Palmers.

In soft southern drawls, the friendly couple explain that security is both a hardware and software problem. Charles oversees eighteen researchers tackling the software side. One group—the "crypto ninjas"—works on things like data protection algorithms and encryption keys. Another includes IBM's legendary ethical hackers, who test security by breaking into computers and networks. Perhaps the biggest effort, though, comes in creating evaluation tools and defenses like the Network Security Auditor, which scans for bugs and configuration flaws exploited by hackers and points out ways to fix them. Such fixes prevent the vast majority of attacks. However, warns Charles, "There is no totally secure system, unless it's turned off and filled with concrete." As a

---

As White watched, the visitor accessed his colleague's electronic address book and sent a copy of itself—ostensibly from the colleague!—to the 1,500 or so entries in the data base. A magician doing a disappearing act, CHRISTMA EXEC then erased itself. People were stunned. Loudspeakers blared a warning not to run the rogue program. But it was too late. As it turned out, the seemingly amusing holiday message was a Scrooge-like virus that replicated itself hundreds of thousands of times, clogging up the company's internal e-mail system for nearly a day. Long before all the damage reports were in, however, White had dropped what he was doing to concentrate on the invader. "You realize as soon as this happens that it's something bad," he recalls. "I said, 'That's it, I'm not working today. I'm going to watch this happen because this is a seminal event in history.' "

complementary strategy, the group therefore fashions intrusion detection software like their recently released Haxor technology, which monitors networks for signs of known attacks and other suspicious activity.

No matter how good the electronic defenses, though, computer systems are also physically vulnerable. That's where Elaine comes in. "The key," she explains, "is the key." That is, an encryption system's password or numerical key, when stored on a server or some other hardware, turns that piece of equipment into a target. Attackers will drill through casings and insert hair-sized probes to download stored information. They'll even drop hardware into liquid nitrogen or zap it with X-rays—freezing any destruction mechanisms so they can get at the information inside.

To thwart such nefarious acts, Elaine's seven-person team worked five years to perfect the 4758 cryptographic coprocessor, a small metal box equipped with its own memory and processor that plugs into a server to perform cryptographic operations and store such sensitive data as encryption keys. Under its metal exterior, the 4758 is wrapped in electroconductive mesh. If a probe strikes the mesh, it destroys all essential information. Temperature sensors, voltage meters, and X-ray detectors do the same for other forms of attack. Explains Elaine, "It doesn't self-destruct in a Mission Impossible way, but it does destroy all the sensitive data inside." Just before Thanksgiving 1998, the 4758 became the first product to pass Level 4 (the highest level) of the Federal Information Processing Standard for cryptographic devices. IBM envisions the $2,000 device as an e-business essential—protecting everything from digital postage meters to electronic check writing systems.

So do the Palmers discuss security issues in their spare time? You bet. Charles says if he doesn't have a lead on a specific problem, Elaine will—and vice versa. "It's a networking thing," he says. "Some people have to go to conferences. . . . I just go home."

□

Little did White know it at the time, but that intriguing message would consume his professional life for more than a decade. It brought him—and IBM—to the front lines of a never-ending struggle that affects a million computer users every year and threatens to intensify in the age of Internet communications and commerce, when viruses can be passed rapidly around the globe. With huge bets placed on the future of e-business, and virtually every virus created aimed at its computers and compatibles, no company takes the threat more seriously than IBM. Back in 1987, maybe three digital viruses existed. Today, every day, somewhere between six to ten PC viruses stream into the Anti-Virus Center at IBM's Hawthorne Laboratory, an ultramodern extension of the Watson lab located a few miles down the road from its parent. So far, the IBM group has battled about twenty thousand separate invaders.

And that's not even the half of it. Until recently the enemy at least seemed contained: once IBM's investigators or their counterparts in a few other organizations turn their attention to the virus, it typically takes less than twenty-four hours to decipher the code and divine a cure. But in today's Internet society, with millions of people swapping files and conducting business round the clock, once-sluggish mutant codes can go global in well under a day.

White leads the way to a two-room suite a couple of floors above the Anti-Virus Center. Here, sealed off from the outside world by computer firewalls and other defenses against hackers, resides a prototype of what IBM thinks might be the savior of the Net. It's called the Digital Immune System. The idea is to create digital white blood cells—much as humans develop antibodies to biological agents—that will be permanently available on line. In theory, automatic virus-scouting programs will transmit suspect codes directly to the immune center, where they will be analyzed and debugged and the cure beamed back before mere mortals even know there's a problem. "If the Internet is going to survive," declares White, "we're going to need an automated response on this rapid time scale."

A computer virus is a bit of software code that wends its way into a machine—typically through a disk or electronic message—and, once opened or "executed," co-opts its host's resources to make copies of itself and order up aberrant behavior ranging from posting an innocuous message to wiping out hard drives. Although the theory behind these beasts harkens back to at least the 1970s, viruses did not emerge in the "wild" until the late 1980s. Almost always aimed at IBM and compatible computers running Microsoft's DOS and Windows operating systems, which now serve some 90 percent of all PCs, they speedily became an everyday menace. Annual sales of antivirus software are expected to surpass $1 billion in the year 2000.

Nearly as soon as viruses came into existence, the myths appeared about virus writers. As cast in TV shows and movies, they are brilliant iconoclasts who run circles around hapless corporations. No such romanticism affects IBM's twenty-five-person antivirus team, however. Viral codes are rife with bugs—and none of the writers would land a job with Big Blue. "The writers, they might think that they're showing off their programming prowess," asserts Jeffrey O. Kephart, an instrumental figure in IBM's antivirus fight. "But in most cases they're displaying their ineptitude to the world." Though few good studies of virus writers exist, the available evidence indicates they're overwhelmingly male, usually in their teens or early twenties—and have an attitude. A couple of two-drawer file cabinets inside the antivirus center hold boxes of diskettes bearing copies of every virus the lab has tackled. A few samples from this "morgue" illustrate the point.

One selection is Wazzu, which in its 1997 heyday infected Microsoft Word files by randomly shuffling words and inserting its name into text, as in 'Now is wazzu the country to come to the aid of your time.' Form, another demon, caused keyboards to make an annoying clicking sound on the 24th of each month. Even more telling, in a notes area inside his errant code, where only debuggers would see it, Form's creator made no bones about his motivation: a girl. "Virus sends greetings to everyone who's reading this text. Form doesn't destroy data! don't panic. F— —ings go to corinne."

While such creations may seem like harmless pranks, lab members warn there's no such thing as a benign virus. Take Wazzu. "Well, you might think it's a funny virus unless you're writing the Israeli-Palestinian peace treaty," White explains. Even keyboard-clicking Form is far from harmless, since a virus's very existence means a computer's works have been jiggled, increasing the risk of crashes and corrupted files. And without naming names, White tells horror stories of corporate mergers in which one company infected another because viruses got into spread sheets they exchanged.

Which explains why Big Blue opened the Anti-Virus Center way back in early 1989, barely a year after CHRISTMA EXEC hit. Working initially with IBM corporate customers and later expanding into the consumer sector with its IBM AntiVirus software package, the team has built up an impressive data base of everything it has encountered, classifying viruses according to what they attack—files, operating system, and so forth—and tracking their incidence rate. Setting the stage for the coming showdown in cyberspace, White now identifies five historic epochs to the long-running wizard war.

Cambrian Explosion (1982–1988): The real Cambrian Period saw an explosion of multicellular invertebrate life. Not quite the same with viruses, but close enough. By the mid-1980s, the personal computer had evolved from oddity to productivity tool—and long-theorized viruses became a reality. Towards the end of the period, CHRISTMA EXEC arrived at Big Blue.

Age of Dinosaurs (1988–1992): Invertebrates were joined by great beasts. Some were file infectors, so-named because they infected individual files, or applications. An early arrival was the Jerusalem virus, which on every Friday the 13th put black rectangles on screens and erased any files executed that day. A programming miscue allowed the virus to invade files multiple times, adding 1813 bytes of data with every reinfection; programs infected multiple times no longer fit in memory.

Around the same time boot infectors arose. They were activated when users started up, or "booted," their computers with a floppy diskette. Between 1990 and 1992, the Stoned virus was the world's most prevalent. The only real consequence of the virus was that one out of every eight times users booted up from a floppy, it flashed the message: "Your PC is now Stoned!"

Asteroids Hit the Earth (1992–1995): File infectors were the first casualty. Their decline coincided with Microsoft's segue out of DOS and the growing popularity of its Windows 3.1 operating system, which wouldn't work with infected files and thereby changed the environment for these beasts the way asteroids or comets striking Earth may have wiped out the dinosaurs.

Boot infectors, though, coexisted perfectly with Windows. Early in 1992 viruses were elevated to national prominence by the discovery of Michelangelo, set to be activated on March 6—the artist's birthday. The widely publicized virus—even *Nightline* did a segment—raised the specter of computer Armageddon, since it promised to erase hard disks on a specific day. Lines ran around the block at some stores selling antivirus software.

Michelangelo proved largely a dud. The fevered scanning of disks for signs of the virus it inspired turned up scores of other boot infectors—but these viruses mounted a comeback that lasted nearly three years. Then Windows 95 debuted. Thanks to a design quirk, it refused to spread boot viruses—making the environment uninhabitable for these agents much as Windows 3.1 did for the file infectors. Boot viruses, too, went the way of the dinosaurs.

Mammals on the Rise (1995–1999): What Bill Gates giveth, he taketh away. A new predator evolved: the macro virus. Far more nimble than the file-and-boot-osaurs, it hid inside "macros," the little programs inside Microsoft Word documents and Excel spread sheets that busy themselves with formatting and other subtasks. These viruses, including Wazzu and contemporaries dubbed Npad and Paix, throve with the widespread sharing of files via e-mail—where such exchanges take place far more readily than through disks. For the first time, the tools of executives were affected. A CEO might dash off an electronic memo to all employees and contaminate the entire company.

The first macro virus, Concept, arrived just as boot infectors took their mortal blow late in 1995. Its one overt act was to put up a message box bearing the number "1." A message in the macro code, at the spot that could have included instructions for more damaging action, read: "This should be enough to show the concept."

The period from 1996 through 1998 marked a brief, golden time for antivirus forces. Even the stealthiest new beasts rarely lasted longer than twelve months. The telling factor seemed to be antivirus software, as IBM and others grew increasingly adept at putting out timely updates—and people seemed more willing to use them.

Monster from the Cyber Lagoon (1999–?): Ah, but then the Internet blossomed. As long as viruses replicated chiefly through disk exchanges, it might take a virile creation a year to spread around the world. That left plenty of opportunity to install antivirus updates. In the age of e-mail and macro viruses, infections could become global in a month—still a reasonable time to fortify defenses. But now viruses can travel worldwide in twenty-four hours. "The compression of the time scales from a year to a month to a day changes everything," asserts White. "Basically, the way the antivirus industry is trying to solve the problem right now just breaks down."

In 1990, when the antivirus lab was only a year old, White discussed the forthcoming need to automate virus hunting with Kephart, a newly arrived expert in nonlinear dynamics. This concern ultimately persuaded the young researcher to join the virus wars.

"Patterns and connections have always thrilled me," Kephart relates. "For the last decade or so, I've had lots of fun exploring analogies between large, decentralized computer systems and things like ecosystems, biological systems, and economies."

Almost as soon as White started talking, the analogy bells started ringing. Kephart began musing about how humans ward off biological bugs by building up antibodies, and wondering whether a similar system could be contrived to fight digital viruses—especially ones yet to be seen.

The first step was to understand viruses better. The lab began compiling its elaborate data base of attacks on IBM's corporate customers. Armed with the location and date of each "incident," the number of infected PCs, and the type of virus involved, researchers employed techniques from mathematical epidemiology to figure out how viruses replicated and spread. In addition to watching for alterations in key parts of memory and other hallmarks of virus activity, most antivirus programs today roam hard drives or disks looking for specific pieces of known viral code. These codes typically run a few dozen bytes in length. But drawing on pattern-matching techniques from computational biology, the IBM group was eventually able to spot known viruses from snippets as small as 3 to 5 bytes—speeding up detection. Identifying unknown agents was tougher. But the fact that a single signature often characterizes whole virus "families" proved crucial: since these set code patterns are directly linked to functionality, a wide variety of viruses could be recognized—and cured—even though they had never been seen before.

In the end, it took nine patents and seven years before what came to be called the Digital Immune System finally materialized in the two-room suite at Hawthorne. The inner sanctum holds a series of computers and software components stacked floor to ceiling. In the outer room, equipped with a large sofa, a kickboxing bag, and more computers, a small research group monitors the system and labors to perfect it.

The basic concept is that antivirus clients will be networked directly to the central immune system computer. Monitoring programs on each PC will beam a copy of any suspicious program to the system's analysis engine, which makes a quick guess as to what type of virus might be arriving. From there, the sample is passed to what the team describes as a series of digital petri dishes. This is a separate collection of computers fitted with decoy or "goat" programs that simulate the kinds of environments viruses like to invade. Goats can be run in different languages. And, since certain dates—Friday the 13th, for example—might trigger viruses, it's even possible to simulate rapidly all the days of the year, as well as different hours of each day. Because the starting condition of each goat is known, the system can track the exact path of any infection, figure out the invader's traits, and develop rules for removing the virus—hopefully undoing any damage it has done. The immune system then copies the virus and tests those assumptions to make sure its cure works. The resulting "antibodies" will be transmitted back to the infected client and any other machines on the network and will become a permanent part of their memories. Eventually, every customer will receive a copy through regular updates.

IBM publicly demonstrated the immune system in October 1997 at the International Virus Bulletin Conference, a gathering of the world's top virus fighters and their customers that was held that year in San Francisco. Jeff Kephart and a few colleagues brought a virus-plagued PC onstage, and the delighted audience watched as errant code was transmitted to the Hawthorne center, a cure derived, and the solution beamed back in a little over three minutes. Since that time, field trials have been scheduled, with full commercial rollout planned for early 2000. The previous year, the company announced a deal that gives Symantec Corporation, the California-based maker of sales leader Norton AntiVirus, rights to its antivirus work—thereby fusing Norton's strong market position with IBM's technology.

Without an immune system-like defense, White says, viruses threaten to "stop the forward progress of computing." But that's not to say he feels the days of the computer virus are numbered. Even as the Digital Immune System was being readied for field trials, a renegade program for automating macro virus production was circulated in the virus-writing community. It also appears that today's rapid-fire file exchanges, often involving passing data between various applications, can cause viruses to mutate and take on properties more damaging than even their creators intended.

In short, just as there is no end to the human battle against biological bugs, the campaign against their digital counterparts endures. Sighs White, "This is going to be a problem that stays with humanity as long as we use computers."

■

Shortly after taking over as Research director in 1996, Paul Horn changed the division's motto. Out went the phrase, "famous for its science and technology and vital to IBM." The new legend simply declared: "vital to IBM's future success." Talk about shock. "He took science and technology out of the wording," recalls Randall Isaac, Research vice president for Systems, Technology, and Science. "People were wondering, 'What does it mean, what does it mean?' "

While the motto change itself hardly raised eyebrows outside IBM, what it spoke to has dogged the company since the McGroddy days: namely, science doesn't hold the place it once did. Cherry Murray, Lucent Technologies–Bell Labs director of physical sciences, pulled no punches during a December 1998 talk in Washington, D.C. "IBM lost fifty percent of its physics researchers. When you do that, what you had is lost."

Horn's people counter that axed basic science projects involved long shots like neutrino detection, which even if successful were unlikely to impact the company commercially. Meanwhile, Research continues to support fundamental studies in key areas of physics and materials science, as witnessed by its storm of semiconductors and storage advances and its explorations into quantum computing and other scientifically risky areas where gambling makes more sense. What's different from the past—the point of the new motto—is that science is no longer considered an end unto itself, says Isaac. "We still want to be famous in science and technology—but its goal is to be vital to IBM's success."

In this view, IBM has an ally in John Armstrong. "The real source of the unhappiness," the former Research director relates, "is people's deep anxiety about ever raising the question, 'How much is enough?' " That's what IBM has done, he says. "They've said, 'Look, we have found that less will do.' And that is deeply disturbing to the national scientific community." Meanwhile, he scoffs, people target IBM for scaling back science. "But they didn't scorn Intel for never having had serious research until recently. Same is true for Microsoft, same is true for Motorola."

Horn has worked hard to counter the idea that less science means no science. In November 1998, he invited scores of press to the Watson lab for Science Day. On hand were some of Research's biggest names—from veterans like Nobel laureate Binnig, who described his nanotechnology investigations, to rising star Almaden newcomer Isaac Chuang, an expert in quantum computers. "We've been trying to regenerate the excitement for the long-term exploratory programs, both internally and externally," Horn explains, "because we still continue to believe that they're very important for us."

In the end, though, far more serious than how much science to support are the straightforward challenges of staying nimble and creative inside a worldwide organization. Horn admits Research isn't as competitive as he'd like in networking and certain Internet technologies; though he won't specify which ones, one prominent failing is Internet telephony, an area researchers seem to have missed almost completely. But in attempting to beef up these fields IBM must battle the perception that it is stodgy, and, well, uncool.

Since taking over, Horn has moved aggressively to address these concerns as well. To better compete with start-ups, an unprecedented number of researchers now receive stock options. A Watson out-building was turned into a gym. The Hawthorne lab got a new entertainment room—the Hawlodeck—rigged for video games, go, and chess. In what would have been a sacrilegious act at previously teetotaling IBM, Research now hosts "Summer Fun Days," with live music, beer, and wine. Horn's even hired an activities director to ensure summer interns have fun.

As the twenty-first century nears, Horn says that these efforts are beginning to pay off in improved recruitment. But at least some potential hires have a bigger question on their minds: Can radical ideas thrive in the new environment? Former IBM Fellow Jerry Woodall, now an electrical engineering professor at Yale University, thinks this is Research's Achilles' heel. In their zeal to bolster the bottom line, he says, managers have virtually eliminated spontaneous, curiosity-driven investigations in areas that don't relate directly to current business needs. Such projects, Woodall believes, could be vital in the future, "the high-tech corporation's equivalent to 'seed corn.'"

Lou Gerstner himself had the chance to confront this issue in July 1998, when Research treated seven hundred summer students to a day of music and brainstorming sessions held simultaneously at its labs around the world. At the Watson festivities, a young would-be recruit who listened to the chairman's address said he'd never come to IBM for fear that if he did do something truly different and important it would never see the light of day.

Gerstner didn't miss a beat. Pointing out a curly-headed, mustachioed figure in the crowd, he replied: "Ask Bernie Meyerson about that."

CHAPTER SIX

HOUSE OF SIEMENS

*"I do not like this differentiation between basic, applied, and
fundamental research, et cetera. I'd like only to differentiate
between good research and bad research."*
— CLAUS WEYRICH, SENIOR VICE PRESIDENT, CORPORATE
TECHNOLOGY

*"The future is made by companies like Siemens—it doesn't
just happen."*
— SIEMENS CHIEF KNOWLEDGE OFFICER GERHARD ZORN

Zentralabteilung Technik—Corporate Technology Division

Central research established: circa 1914
Company sales 1998: $66.1 billion
1998 company R&D expenditures: $5.1 billion
1998 Corporate Technology budget: $306 million
Funding paradigm: 50% contracts with business units, 35% central funds,
   15% government and outside contracts
Total staff: 2,100
Research technical and scientific staff: 1,600
Principal research labs: Munich, Erlangen, Berlin, Germany; Princeton, New
Jersey

AROUND 7:30 on the morning of July 10, 1986, Siemens director of
research and development Karl Heinz Beckurts was being chauf-
feured through the Bavarian city of Strasslach in a dark blue BMW
sedan, on his way to work in Munich, about 15 miles to the north. Suddenly, an explosion shook the area. The two security men following just
behind in a second vehicle watched helplessly as a 90-foot flame burst
from under the executive's car, tearing its hood, trunk, and doors open,

and sending it twisted and shredded into a ditch across the street. Beckurts and driver Eckhard Groppler died instantly.

In a seven-page letter found near the scene, the Red Army Faction
claimed credit for planting the remote-control bomb that claimed Beckurts's life. The terrorist group said it had targeted the research official in
part because he supported nuclear energy—but if the faction hoped the
attack would disrupt his agenda, the plan failed miserably. Beckurts was
the first of a new breed of Siemens R&D heads bent on banishing the old
leisurely pace of research and moving lab developments quickly and
profitably to market. Although the company was rocked by the tragedy, if
anything the bombing proved a catalyst to the quest for research efficiency. Beckurts's efforts to make Siemens a more viable worldwide competitor picked up steam after his death—and have kept on gathering
momentum into the millennium.

Although it's hard to imagine any major corporation cowed by an
act of terrorism, Siemens's reaction to adversity is especially telling—
rooted deeply in its more than 150-year history. Besides the normal va-
garies of business, which make it nigh impossible for any firm to innovate
for a century and a half, the company has had to bounce back from heavy
losses in two world wars that saw its plants ruined and severe trade restrictions imposed. In World War II, especially, between massive Allied bombardments and Russian troops ransacking its main plant in Berlin,
three-quarters of Siemens's manufacturing capability was destroyed. The
company lost 25,000 patents to foreign powers, and was banned under
surrender terms from research into nuclear energy, transistors, and other
key areas of potential growth.

Nevertheless, against these long odds, the company soon emerged
as a powerful global competitor. Research, in one form or another, has
played a critical role in driving its success and longevity—and Siemens
for years has ranked among the world's top R&D spenders. Its overall philosophy is palpably German: careful and generally far more conservative
than that of longtime competitors such as AT&T and IBM. Its brief foray
into basic science—despite producing some Nobel Prize–caliber work—
ended long ago. Still, for all its huge size and cautious bent, Siemens
maintains a nimble and resilient central research organization—part of a
remarkably resilient company.

Siemens AG bears the name of a gifted practical scientist and entrepreneur whose inventive talent occasionally rivaled that of Thomas Edison—and who eclipsed his American contemporary in business acumen.
Werner Siemens launched the firm in 1847, at the age of thirty. He pioneered in telegraphic equipment—and later, after co-discovering the

electro-dynamo principle that made large-scale electric power generation and distribution economically feasible, established his company as one of the world's four great electrical concerns, alongside German arch-rival Allgemeine Elektricitäts Gesellschaft (AEG), Westinghouse, and General Electric.

Siemens was born in December 1816, the eldest son of a land leaseholder who sired fourteen children. From youth, he exhibited a gift for the physical sciences—and at eighteen he enrolled as an officer cadet at the Prussian Army's Artillery and Engineering College in Berlin, where he received formal training in mathematics, chemistry, physics, and ballistics.

Upon graduation, Siemens entered the Army as a lieutenant, carrying out a variety of experiments during his garrison postings to provincial towns—and while incarcerated for serving as a second in a duel. In 1846, stationed once again in Berlin, he developed a telegraph of greater speed and reliability than the conventional Wheatstone apparatus—in part because it could be operated by tapping keys, as opposed to turning a crank, which produced uneven current impulses. He then convinced a talented precision engineer named Johann Georg Halske to become his partner in starting a company to exploit the invention. On October 1, 1847, Telegraphen-Bauanstalt von Siemens & Halske—the Siemens & Halske Telegraph Construction Company—began operations at Schöneberger Strasse 19 in Berlin.

With Siemens working out of a ground floor lab, Halske laboring two floors above him, and a workshop in between, the firm quickly established a reputation for reliability and craftsmanship. The telegraph business grew rapidly, as the firm laid cable around Europe, Russia, and the Middle East. It took about a dozen years to exploit Siemens's 1866 discovery of the electro-dynamic principle. But then the company moved aggressively into electric power generation and distribution, as well as electric lamps. In 1879, for the big Berlin trade fair the following year, Siemens built the world's first electric railway. Siemens lighting systems illuminated Berlin train stations, the Royal Palaces, and a growing number of offices and factories. Around the time its founder retired in 1889 at age seventy-four—he would live just three more years—Siemens AG totaled 4,500 employees, including 2,000 outside Germany, and was the nation's leading electrical manufacturing concern.

Siemens dreamed of creating a family-run industrial enterprise that paralleled the Rothschilds' financial empire—and he got his wish. Upon his retirement, younger brother Carl took over as chairman, flanked by Werner's sons Arnold and Wilhelm. Three years before his death in 1906, Carl merged the electrical machinery works with Nuremberg-based power engineering firm Schuckert & Company to form a multinational enterprise called Siemens-Schuckertwerke: controlled by Siemens & Halske, the new concern built power stations, turbines, electric trams and railways, and a variety of other equipment. Meanwhile, Siemens & Halske took up the pioneering X-ray tube discoveries of Nobel Prize–winning physicist Wilhelm Röntgen—who took out no patents—laying the foundation for a major medical equipment line. A revolutionary tantalum-filament lamp invented by company chemist Werner Bolton established Siemens as a major force in the incandescent lamp business. The firm also moved into shipboard communications, radio, telephone systems, electric locomotive design, and railway and subway construction.

In the early 1900s, overrunning its Berlin sites, Siemens began shifting operations to a tract of land along the Spree River in northwest Berlin. By 1914, on the eve of World War I, the vast industrial complex accommodated some thirty thousand workers, nearly a quarter of whom lived on site. With its own school, food and merchant shops, post office, and police station, this city-factory became officially known as Siemensstadt: Siemens Town. Writes business historian Alfred D. Chandler, Jr.: "A similar complex would have appeared in the United States only if the GE plants at Schenectady, New York; Lynn and Pittsfield, Massachusetts; Harrison, New Jersey; and Erie, Pennsylvania, had been placed along with Western Electric's large Chicago plant, which produced nearly all the nation's telephone equipment, at one site in the neighborhood of 125th Street in New York City, or at one near Rock Creek Park in Washington, D.C."

Germany's defeat in World War I saw the expropriation of all Siemens possessions in France, Italy, Belgium, Russia, and Britain—allowing European competitors such as Philips from The Netherlands and Sweden's Ericsson to make inroads in the company's traditional strongholds. To make matters worse, Arnold Siemens died in 1918, near the end of the war—and brother Wilhelm fell into a deep depression and passed away the following year.

At this grim stage, it was left for Werner's last surviving son, Carl Friedrich, to save the day. He proved an organizational virtuoso, overseeing a dramatic expansion of assembly lines and a program to standardize products that lowered costs and significantly boosted profit margins. In 1919, the year he took over, Siemens, AEG, and Auer combined their lamp factories to form the Osram Company, later taken over completely by Siemens. In 1924, Siemens & Halske gained a majority stake in electro-medical manufacturer Reiniger, Gebbert & Schall of Erlangen. Within three years, this led to the formation of Siemens-Reiniger-Werke

AG, one of the world's largest medical equipment makers: along with communications giant S&H and electric equipment manufacturer Siemens-Schuckertwerke, it formed the third great pillar of the House of Siemens. Riding—and simultaneously driving—the nation's economic recovery, these businesses helped Siemens extend domestic operations, then expand again beyond German borders. By the outbreak of World War II in 1939, the company was stronger than ever, with 183,000 workers worldwide.

The Second World War saw the entire arduous cycle of destruction and revival begin anew. Carl Siemens, now ennobled as Carl Friedrich von Siemens, fell ill and died early in the war. The company's factories, many turning out military equipment, were heavily bombarded. In 1945, Siemensstadt was ransacked by Russian forces. Between bombardments, lost equipment, and the seizure of its foreign property, Siemens was reduced to a quarter of its prewar value. This did not include the loss of 25,000 foreign patents and the obligation to publish details on all important new fields of research, development, design, and production. Again, though, the German economy recovered rapidly—and the restoration and expansion of the nation's telephone network, railroads, and electric power grid provided the backbone for rejuvenating the company.

By the 1950s, its headquarters and key production facilities moved to West Germany, Siemens was again reaching into foreign markets. A decade later, the oft-battered concern had almost completely regained its global standing and expanded into all major fields of electrical engineering and electronics, including early digital computers and integrated circuits. Growth continued into the 1990s—so that on the eve of the twenty-first century, Siemens did virtually everything except make batteries and computer displays.

Not everything had gone as planned. The initial foray into computers probably came too early, before markets had sorted themselves out. Ultimately, Siemens dropped its money-losing internal operation and bought computer maker Nixdorf AG, which became the foundation of the successful Siemens Nixdorf line. The semiconductor business also stumbled badly throughout the 1980s. A dramatic catch-up effort spearheaded by Karl Beckurts eventually turned things around enough to show a profit. But the trouble of maintaining it all proved too much— and in a massive fall 1998 reorganization Siemens announced plans to shed its semiconductor and components business for good.

Still, more than 150 years since its founding, Siemens found itself in an enviable position. With some $66 billion in sales in 1998, the company ranked as one of the planet's largest corporations. It had approximately 416,000 employees working in some two hundred countries.

Counting foreign subsidiaries like lighting products manufacturer Osram Sylvania and the telecommunications company Siemens Information and Communication Networks (the successor to what were the Rolm and Stromberg-Carlson corporations) in the United States, 50 percent of its work force and 70 percent of total sales came from outside German borders. As a company saying goes: "Everywhere you find Coca-Cola, you find Siemens."

THE PATH OF RESEARCH

BEGINNING WITH Werner Siemens's private experiments in the Artillery Corps, and reflecting its founder's ideal that the best firms must innovate from within, Siemens has supported some form of research since even before its formal founding in 1847. Indeed, the corporate annals are replete with research triumphs, beginning with the electro-dynamo principle and including the electron microscope and gallium arsenide semiconductors. And without a commitment to preserve this legacy, the company could never have overcome the tremendous obstacles history placed in its path.

For nearly two decades after starting their Berlin workshop-factory, Werner Siemens and Johann Georg Halske effectively *were* the company's R&D operation. Siemens turned out a series of profitable inventions from improved relays to an important method for measuring electrical units of resistance. His meticulous partner, meanwhile, oversaw production and quality control, refusing to let anything ship without his personal approval.

Taken as a whole, the company's creations considerably advanced the developing fields of telegraphy and electrical engineering. Still, as the partnership grew increasingly successful, it became clear that a modern enterprise had to draw on a much wider variety of talents and insights than just its two founders. In 1869, the firm scored a coup by hiring Carl Frischen, an inventive engineer who had discovered—independently of Werner Siemens—a way to send two telegrams simultaneously over the same wire in opposite directions. Four years later, Siemens hired its first physicist, Swiss national Oscar Frölich.

These men and a handful of others formed the nucleus of a blossoming research and development enterprise. Initially, Siemens employed scientists and engineers almost exclusively in measurement, testing, and product development—supplementing these efforts by hiring consultants, contracting with private labs for specific jobs, and acquiring outside inventions, as it did with Röntgen's X-ray advances.

Siemens Science and Technology Firsts

1847	Werner Siemens invents improved pointer telegraph; company founded
1866	Electric dynamo principle discovered; first dynamo built
1879	Electric train demonstrated at Berlin trade fair
1896	X-ray tube patented
1930s	Schottky elucidates theory of semiconductor rectification and structure
1939	Müller invents electron microscope
1952	Welker announces discovery of III-V compounds and predicts their semiconducting properties
1953	Semiconductor-quality silicon produced
1965	Real-time ultrasound diagnostic unit unveiled
1991	Joins IBM in producing prototype 64-megabit DRAM chip
1995	Develops 256-megabit DRAM with IBM and Toshiba

During Werner Siemens's lifetime, beyond the founder's personal lab, the company apparently maintained just two in-house R&D arms. One was a general laboratory on Markgrafenstrasse in Berlin under Frölich, where investigators studied everything from ozone to the torsion dynamometer. The other was a product testing facility known as the Bohneshof experimental station just outside the city in Moabit, near Charlottenbürg.

By early in the twentieth century, as Siemens expanded, these were joined by a host of other facilities—mainly development labs—covering all aspects of the company's business. Lighting investigations took place in a dedicated incandescent lamp lab, as well as separate electrochemical laboratories in Berlin and Vienna. As the Siemensstadt works went up, each unit operated at least one lab for daily materials testing. A number of development facilities were also spawned: specializing in such areas as telecommunications and measuring instruments, these labs eventually occupied an entire wing of the works.

The 1903 merger that created Siemens-Schuckertwerke, and the 1927 formation of Siemens-Reiniger-Werke, brought into the fold a host of additional laboratories and projects. When combined with other R&D activity, the last merger created what one company historian called an "almost unsurveyable host of specialized laboratories" that covered the closely related fields of radio waves, conductors, telegraphy, telephony, television, magnetism, and materials research—and cumulatively employed one in every six Siemens engineers. Some labs did important research: indeed, between 1930 and 1945, investigators from a variety of Siemens facilities published fifty scientific papers on magnetic fields alone. Still, almost without exception these were highly applied enterprises created to serve a specific purpose: amplifier valves, alternating current instruments, pyrometers.

The first lab to take on aspects of a central research organization and concentrate on problems beyond the scope of any existing group's interests was the *Physikalisch-Chemisches Laboratorium*, or Physical Chemistry Laboratory. Established at the rapidly growing Siemensstadt in 1905—in temporary quarters nicknamed the Doctor's Dungeon—the lab's driving force was chemist Werner Bolton. At Wilhelm Siemens's urging, Bolton undertook the arduous series of experiments that led to the tantalum filament lamp—one of the most important inventions ever to emerge from within the company's research organization. The lamp created a worldwide sensation because it worked with all customary voltages and current supplies, was shockproof, and proved capable of burning in any position—even inside tramcars.

Upon Bolton's untimely death in 1912, Wilhelm Siemens appointed his colleague, Hans Gerdien, to be chief of research. It was Gerdien, once a lecturer at Göttingen University, who formally proposed combining several previously independent labs into a central research organization focused on problems significantly removed—the word "frei," or "free" was used—from immediate business group needs. It was a tough sell. As one official account puts it: "amongst the Works Managers, the majority of whom were members of the Managerial Board, there were few who were prepared to admit the necessity for such laboratories." Wilhelm Siemens, though, loved the idea. With his weight behind the proposal, an extensive building plan was conceived. Work on the prestigious new facility began in the summer of 1914, but was not complete until five years later—after the Great War.

Though Wilhelm never lived to see the lab's dedication, the creation of the Research Laboratory of the Siemens Works was one of his last organizational feats. Renamed the Research Laboratory of Siemens & Halske and Siemens & Schuckertwerke in 1924, the lab pursued several lines of investigation throughout the interwar years. One area of focus— acoustics—was spearheaded by research assistants Hans Riegger and Ferdinand Trendelenburg. The work led to a Riegger-developed condenser microphone that helped enable speech and music recordings of unsurpassed fidelity. The lab also studied the physics of electrical discharges and the transmission of electricity. It launched important physical-chemical research into raw materials, leading to highly purified iron and nickel that formed the basis for machine tool steel, as well as sintered corundum for rectifiers and spark plug insulators. In 1935, partly to protect him from rising anti-Semite fervor, the company recruited the Jewish Nobel laureate Gustav Hertz, a nephew of Heinrich Hertz, to head Siemens Research Labs II, an adjunct to the existing lab set up to do physics research. Four years later, under Hertz's auspices, researcher Erwin Müller invented the field electron microscope; far down the road, at Penn State University in 1955, Müller would employ his new field ion microscope to become the first person to "see" an atom.

A last area of scientific investigation—probably the most fundamental the company ever pursued—was ongoing throughout much of this period in the lab of Walter Schottky. The great physicist joined the central research arm in 1927, and throughout the next fifteen years stunned the scientific world with a series of fundamental insights into semiconductor rectification and structure.

World War II interrupted nearly all of Siemens's research. After the war, when it became apparent that the German Democratic Republic would surround Berlin, the management board shifted operations to West Germany. Siemens & Halske moved far to the south, to Munich, with both Siemens-Schuckertwerke and Siemens-Reiniger-Werke, the medical arm, based in Erlangen, a university town just north of Nuremberg. Within about a decade the shared research structure between Siemens & Halske and Siemens-Schuckertwerke was abandoned, as all three businesses built their own central laboratories.

The most fundamental work took place at Siemens-Schuckertwerke. The lab was established in downtown Erlangen, in a reddish-pink building known as the *Himbeer* (Raspberry) Palace. The first director was Ferdinand Trendelenburg, who had earlier studied acoustics in Siemensstadt. Prime areas of focus included solid-state physics (especially semiconductors), plasma physics, physical and chemical analysis, and polymer chemistry, essential to power cable insulation. After 1955, when restrictions on nuclear research were lifted, the company launched an impressive atomic energy program that ran the gamut from fission to fusion.

In tackling these core areas, Trendelenburg's primary interest lay in bringing the lab quickly up to world-class scientific standards—not generating products. "The way to start was to get the most experienced people in the field, well-known scientists, experts—to bring them together," recalls Heinrich Schindler, a physicist who joined the lab in the 1950s and later headed research for the vast Energy Division. Siemens-Schuckertwerke also inherited the services of Schottky, who in 1944 was given a lab in an old castle in Pretzfeld, a village about 30 kilometers from Erlangen: this was a precursor to a formal Pretzfeld lab started under Eberhard Spenke two years later. Probably the best new recruits in that early 1950s drive were solid-state physicist Heinrich J. Welker and nuclear physicist Wolfgang Finkelnburg, who had been a research fellow at the California Institute of Technology under Nobel Prize winner Robert Millikan in the 1930s, then returned to America after the war to work in the U.S. Army's Energy Research and Development Laboratories. In stark contrast to the way Siemens currently operates research, these and other scientists—the lab had maybe fifty Ph.D.s by decade's end—were given academic-like freedom to define their own topics of study.

The biggest payoff came in semiconductors. In 1953, independently of parallel work at Bell Labs and the U.S. Army Signal Corps Laboratories, Siemens researchers discovered a way to produce the extremely high-purity silicon necessary to manufacture semiconductor chips: this "floating zone" refining method for manufacturing silicon crystals was widely used nearly a half-century later. But all that paled in significance to Welker's pioneering work in gallium arsenide and other III-V semiconductors, so-named because they blend elements from the periodic table's third and fifth columns. Welker had been hired by Siemens-Schuckertwerke in 1951 as head of solid-state research, shortly after identifying these compounds as a potential new class of semiconducting materials. He spent most of that year equipping the still-empty Raspberry Palace labs and recruiting personnel—but by fall of 1952 had confirmed his theory. At the time, nobody foresaw the compounds' future applications in optoelectronic devices such as light emitting diodes and lasers. Instead, these semiconductors exhibited unique sensitivity to magnetic fields that Siemens was able to harness for measuring magnetic phenomena in a wide range of electric power applications. Meanwhile, Welker's fundamental work put him alongside Schottky on the short list of potential Nobelists.

In the early 1960s, still out to expand its R&D horizons, Siemens-Schuckertwerke erected a sprawling research campus in parklands on Erlangen's southern outskirts. There, it brought together three main efforts: the central lab, directed by Welker, who had taken over upon Trendelenburg's 1962 retirement; a nuclear reactor development program under Finkelnburg; and an automation facility. For the central research staff, this was almost heaven on earth, with state-of-the-art buildings set amidst fish-stocked concrete ponds and long tree-lined streets bearing the names of famous scientists: Einsteinweg, Wattstrasse, Nernststrasse. The center was inaugurated on May 26, 1965.

Change, though, hung in the air. By 1966, Siemens had largely regained its prewar standing. However, faced with increasing global competition, the first postwar recession, and the rising costs of maintaining separately managed companies, the overseeing board decided to unify its three main concerns as Siemens AG. The various business units continued to maintain autonomous research and development labs. But in 1969, the two main central research enterprises—the former Siemens-Schuckertwerke facility in Erlangen and a series of Munich labs established by Siemens & Halske—were reunited to form the Research and Development Department, with gallium arsenide pioneer Welker as its first director. This decision marked the birth of a modern central research organization that encompassed almost all aspects of electrical engineering—from semiconductors to computers and software, chemistry, materials science, and giant electric power systems. It also brought an end to the relatively free hand given to research since World War II. With business booming in the postwar years, operating group managers had fretted little over the percentage of sales they paid to support research. But in leaner times, managers began to inquire more insistently about what they got for their money.

The emphasis on applied research continued throughout the 1970s and into the 1980s. Still, when Karl Beckurts took over in late 1981, he found a somewhat moribund group that among other things was having a hard time making the transition from its mechanical engineering past to microelectronics. Nowhere was this more true than in the once-pioneering semiconductor work. By 1984, Siemens was the only European producer of dynamic random access memories—but found itself two or three years behind Japanese and American competitors, struggling to recoup massive investments in infrastructure and equipment. Beckurts guided research into microelectronics and information technology, and pushed the bid to play catch-up in semiconductor production by joining with Philips in the MEGA project, a largely successful effort to produce 1-megabit memory chips.

Beckurts's murder by the Red Army Faction in July 1986 threw the R&D Department into disarray, forcing Siemens chief executive Karlheinz Kaske to step in as interim leader. It wasn't until the fall of 1987 that a successor was chosen. He was Hans Günter Danielmeyer—the first Siemens R&D head to possess degrees in science and business administration—who focused research on core technologies and accelerated his predecessor's efforts to make operations more efficient. On his heels came another physicist, Claus Weyrich, whose highly market-driven approach to R&D was destined to lead Siemens research into the twenty-first century.

■

The new research boss was born in Czechoslovakia and raised chiefly in Istanbul, where his German father taught mathematics in a technical university. After spending the first few years of high school in Turkey, Weyrich switched to an Austrian boarding school. He joined Siemens in 1969 after earning his doctorate in solid-state physics from the University of Innsbruck. Between travels and schooling, he brought with him a command of German, Turkish, French, and English.

At Siemens, Weyrich specialized in lasers and light emitting diodes, advancing steadily up the ladder—including a one-year stint in a semi-conductor business group—to senior director for applied materials research. In fall 1994, when Danielmeyer relinquished some responsibil-ities to focus on international research and development, Weyrich took over central R&D activities in Germany. In April 1996, he was appointed head of the renamed Corporate Technology Division—*Zentralabteilung Technik,* or ZT—which differed from the old R&D group in that it in-cluded all central research and development worldwide, as well as re-sponsibility for patenting, intellectual property, and production methods such as robotics and automation. That fall, upon Danielmeyer's retire-ment, he was named senior vice president and a member of the manag-ing board.

Heading into the year 2000, Weyrich oversees some 2,100 employ-ees, of whom about three-quarters work strictly in research. Of these, roughly 950 are based in Munich-Perlach, 365 in Erlangen, and 140 in the Princeton lab. A 50-person Berlin facility specializing in microelec-tronics rounds out the picture. The fiscal year 1998–99 budget for this operation totaled nearly DM 550 million, or roughly $306 million.

This figure represents only 6 percent of the $5.1 billion Siemens spent on research and development during the same period. In addition to extensive development laboratories, the sixteen business groups—counting the Osram lighting company and Siemens Nixdorf computer operation—all maintain research efforts of some kind. In the United States alone, these include an important facility near Chicago for devel-oping nuclear medicine equipment, a radiation therapy systems center in Concord, California, and the research groups of Siemens-owned Osram Sylvania and Information and Communications Networks. But these op-erations concentrate on a narrow range of activities specific to their busi-nesses. By contrast, ZT focuses on core technologies that affect many products and business lines. That doesn't mean business groups eschew work on core technologies, notes Ernst Schmitter, a strategic planner on Weyrich's staff. "But we are from the general part, and they are more product-oriented. We have the responsibility for the future of Siemens."

As part of Siemens's efforts to reduce redundancy and leverage

core strengths company-wide, the labs serve as global centers of excellence. Munich is in charge of information and communications
research, including a wide range of software studies and semiconductor and materials work. Erlangen concentrates mainly on energy,
Berlin on microengineering. The Princeton lab specializes in computer-
related areas such as multimedia technology and imaging and visualization.

To this framework, Weyrich brought an almost single-minded purpose. As he explains, "Our aim was strictly to focus corporate R&D on innovation: innovation means to bring it to the market." In particular, he
felt that only extremely nimble organizations could succeed in the fast-
changing technological, structural, and regulatory environment that
characterized Siemens businesses. Past successes aside, he saw far too
many barriers to cooperation and rapid action between his three main
labs, between research and the various business units—and between Corporate Technology and Siemens customers.

Under Weyrich's concept of Fourth Generation R&D (see box,
p. 187), those barriers could only be brought down by casting aside many
traditional views of research—and challenging the status quo. This philosophy became starkly apparent in the wholesale changes he made in
the way ZT was funded. When he took over in 1994, about 70 percent of
the budget came from central funds, 15 percent from contracts with business units to do specific jobs, and the rest from government funding. As
far as the centrally funded portion was concerned, although business
groups had some say in what was done with the money, they had no
choice but to pay it. Forced into this shotgun marriage, many ignored the
research agenda—sometimes resulting in the perpetuation of fruitless
projects. As one longtime research group member puts it: "Some of the
research topics lived forever."

With business unit complaints about the system rising, the Siemens
board asked for a new plan. Weyrich's answer was to flop the funding paradigm almost on its head—tripling the percentage of ZT's budget that
came from business unit contracts by bringing it to half the total. To off-
set that increase, he slashed the centrally funded portion to 35 percent,
while leaving the government-funded percentage intact.

Talk about culture shock. Faced with having to get half their funding by tackling specific jobs the business units wanted done, research
managers suddenly found that they had to actively hawk their ideas and
expertise around the company. That meant thinking far more intently
about what the business groups wanted and needed. In Weyrich's words,
the plan forced managers to become "technopreneurs"—technology-
minded entrepreneurs—and jolted them out of their complacency.

□

Claus Weyrich: Fourth Generation R&D

"When we started to reengineer corporate research and development," recalls Claus Weyrich, "it was not the aim that we wanted to bring up some substantial changes—because Siemens corporate research and development has always been a very successful part of the company. But you know, you have to change from time to time. Otherwise, when you always run with the same structure, that will result in the reluctance to do something other than the way we did it before."

Maintaining the status quo is not for Weyrich. While every lab director works hard at orienting research to business needs, few have undertaken a more market-centric approach than the head of Siemens Corporate Technology. Driving his changes is the evolution to what Weyrich calls Fourth Generation R&D, a novel extension of the Third Generation concept introduced in the early 1990s by the Arthur D. Little consulting group. As Weyrich explains it, here's how corporate research has advanced since World War II:

First Generation: Spanning the 1950s and 1960s, this was the era of "technology push." Labs were largely centrally funded with a wide array of projects; with companies profitable and growing, research felt little pressure to deliver innovations. Labs threw inventions over the wall to development and hoped the strength of their creations would push things into the market. The result: a "small fraction of ideas and results [were] carried through to the market."

Second Generation: Soaring competition and rising financial pressures dominated the 1970s. Labs responded by reducing the spectrum of research, with a growing number of short-term projects dictated by business needs: "market pull." The innovation rate increased, often thanks to an untapped core of research from the previous era.

Third Generation: During the 1980s and early 1990s, notes Weyrich, "People found out the most important thing is to have feedback—feedback between the technology push and the market pull." With budget woes worsening, the emphasis was on lowering internal barriers to cooperation and forging a system of continuous feedback between research, development, and the market: seamless innovation. Speed was of the essence as the scope of research narrowed further and labs concentrated on developing core competencies that gave them an edge with customers.

Fourth Generation: The rise of "customer-driven innovation." Organizations look beyond where technology is heading to future customer needs. Labs must work closely with customers, strategic partners, and universities—not just to deliver products, but to generate new businesses. Projects are framed through sophisticated planning that considers social and regulatory influences as well as technological trends—with prospective economic success determining budgets. Knowledge management becomes decisive, so R&D is carried out by highly flexible, interdisciplinary task forces supported by a vast information and communication infrastructure. The goal: "We want to make available our knowledge at any time to any place and anybody in the company."

□

"That was a tremendous shift toward becoming market-oriented," he notes.

But changes in how labs got their money—and even in researcher thinking—marked only a first step. To sow knowledge and bolster cooperation between ZT and the business groups, Weyrich wanted the individual core technologies inside his organization run like small enterprises—able to solve problems and launch new initiatives with a minimum of red tape. "You can do this only by giving responsibilities to them," he asserts. Yet when he took over, department heads had to approve nearly every step of a project: how to staff it, what equipment was needed, whether outside experts should be consulted. "Some things even had to be countersigned by myself," he says in amazement.

One of Weyrich's first acts, therefore, was to hand individual project leaders the authority to make most of these decisions. "I said every level wants to have a certain responsibility—and I do not want to touch this responsibility," he relates. "I want results, and to be on time and budget. The way they do this is completely up to them. So I am not asking them whether they are going to conferences, whether they hire students or are working on a Ph.D. thesis."

Weyrich also moved decisively to streamline his operational structure. He reorganized ZT into eight Technology Departments—down from a dozen—each of which nurtured up to seven related core technologies. Then, he cut from sixteen to eight the number of managers assigned to coordinate efforts with the business units—a move that made it much easier to stay abreast of developments in a single sitting.

The last of his major changes zeroed in on planning. In the early 1990s, Siemens business groups began utilizing a technique called Multi-Generation Product Planning to create road maps of specific products needed over the next several years. For instance, talks with automakers might reveal the need to develop a better control device for low-emission motors by the year 2000. Planners then create a "technology tree" that breaks out the core technologies involved—piezo ceramics, hydraulics, and thin-film technology—as well as key components and systems, such as a high-pressure injection valve and mixture control system.

While these plans were always discussed with representatives of Corporate Technology, Weyrich asked his people to bring to the table their own scenarios of where research was heading in related areas of core competency. Termed "technology road maps," these generally look out much further than the MGPPs—up to twenty years down the road—and so are not always in sync with more immediate business group needs.

The two sides often disagree passionately about how to address the two visions. But Siemens research managers praise that discussion for forcing both parties to view every project as part of a continuum of efforts. Individual projects may be short term, argues ZT head of marketing and cooperations Dietmar Theis. "But they are part of a puzzle which goes into the future."

Joint planning also acts to shift ZT's efforts toward more concrete goals. "We don't offer technology any longer, we offer solutions to problems—and we offer them to the people who are interested in the solutions," relates Gerhard Zorn, another top Weyrich aide. Even more important, he adds, "People that discuss develop a common vocabulary—and this common vocabulary is about 80 percent of what you want to get from the whole process, because the vocabulary is usually what limits the transfer of R&D to the business units."

So is it working? Weyrich looks to customer satisfaction—how the business groups think ZT is doing—as the most important measure of the changes he has implemented. "There's a proverb," he notes. " 'At the end of the day, money talks.' We fulfill our budget, and apparently we are offering what the business groups need—otherwise they wouldn't pay for it or want to ask for it. And that for me is the best single figure of merit that we are apparently successful."

With that process in place, a look inside ZT's three main labs shows some of the hottest technologies being pursued.

MUNICH-PERLACH

THE FIRST FEW powder-coated aluminum modules of the new Siemens research and development complex stood stark and lonely in 1978, on a treeless plain along Munich's remote southeastern outskirts, in the city's Perlach district. The initial occupants were mainly researchers and the company's original computer group. So when the winter proved cold and cruel—worsened by construction-induced heating problems and the fact no public transportation to the site was available—people took to calling the place *Data Sibirsk*: Data Siberia.

Twenty-odd years later, apartments, shops, roads, subways, and city congestion lead straight to the once-barren site. More than a dozen additional modules have gone up, as the facility made room for some nine thousand employees working chiefly in Siemens Business Services, computers, chip-making, and central research. The rooftops and trimming, ablaze with primary colors, fairly dazzle the eyes: yellow denotes stairwells and walkways, red shows off gas pipes, blue is for ventilation ducts.

These hues, coupled with the overall modularity, have given rise to another epithet: Legoland.

Buildings in Legoland are named for trees—elm, oak, willow—with their living counterparts planted before each structure. But these relatively peaceful surroundings belie the beehive of activity inside one of the world's largest R&D complexes. Vast underground tunnels, crisscrossed by pedestrians, workmen in electric carts, joggers, and bicyclists link every building, as do the yellow-painted walkways extending out from the fifth floor of each edifice like space station spokes. Above or below, it's possible to traverse the entire campus without emerging into the weather. Siberia has been vanquished—data lives.

Corporate Technology chiefly inhabits the complex's western side, the 950 Perlach-based researchers representing 60 percent of its total. Here, the old Siemens & Halske focus on communications still dominates the agenda. But the telephone has long since been invaded by computers, so hardware and electronics investigations are joined by big efforts in artificial intelligence, speech recognition, virtual reality, user interfaces, and a host of cryptographic and security studies. Inside this vast umbrella, projects range from seemingly mundane improvements in circuit switches or electronic signal filters to imaginative bootleg efforts like SIVIT—the Siemens Virtual Touchscreen.

Much of this activity takes place inside Information and Communications, one of ZT's eight technology departments. It, in turn, is divided into seven core areas—one of which is Bernhard Kämmerer's Man-Machine Cooperation Lab. Consisting of roughly forty-five staff members and another fifteen students and postdoctoral researchers spread over the second floor of Building 10—Maple—the group aims to help people get along better with computers and other machines through such tricks as virtual interfaces, gesture recognition, and speech recognition.

A tour of the lab often starts with SIVIT (see box, p. 191). But a larger—and potentially far more lucrative—endeavor centers on biometrics, where researchers seek to identify authorized users of everything from credit cards to guns through hard-to-alter "natural" traits. These include not only voice, but facial features, iris and retinal patterns, handwriting, and fingerprints. While active on virtually all these fronts, Siemens has emerged as a world leader in handwriting and fingerprint recognition—hoping to capitalize on the need for greater electronic security while also freeing people from the pains of memorizing passwords and PIN numbers.

The epicenter of investigations into biometric methods for analyzing handwriting is Brigitte Wirtz. Surrounded by computers and other biometric paraphernalia, she explains that most handwriting verification

Bathtub Inspiration

Billed as the first modern computer that doesn't utilize a monitor, mouse, or keyboard, the Siemens Virtual Touchscreen (SIVIT) materializes almost out of thin air thanks to a ceiling-mounted "beamer" that projects images onto any flat surface. Users then wield their pointer finger as a mouse, selecting items by letting it rest on any image of their choosing. By 1999, Siemens had deployed several hundred touchscreen-based information kiosks around Germany and neighboring countries to demonstrate the concept—with a more dramatic unveiling scheduled for Expo 2000, the World's Fair set to open in Hannover.

Researcher Christoph Maggioni got onto his invention while lounging in the bathtub—as he raised two fingers to signal his girlfriend how many pieces of chocolate he wanted. That led him to wonder why computers couldn't be programmed to respond to similar gestures—and he began developing his vision thanks to a special innovation grant set up to support promising ideas that didn't fit into existing projects. Maggioni and a few colleagues first experimented with a head-tracking system that recognized where the user was looking and showed a corresponding view of the computer screen. But that turned out to have few applications, so they retreated to hand-tracking—a harder problem because hands provide less contrast to follow.

In the final system, a video beamer placed well above the user's head projects a large and bright image onto virtually any innocuous flat surface. Another small camera then tracks finger movements by capturing successive contours of the hand, enabling a cursor to follow those meanderings. Any time the finger comes to rest for more than a few hundred milliseconds, a tiny computer interprets the delay as a mouse click and changes the image accordingly.

The system is unique in that all electronics are placed in a small, hard-to-reach, and damage-resistant box—decreasing the opportunity for vandalism. SIVIT runs on an ordinary Pentium processor and draws a tenth the power of a typical personal computer. Even better for Maggioni, who has filed more than a dozen patents on the system, the touchscreen is so easy to use that "visitors intuitively know what to do."

Glowing returns from its field trial encouraged Siemens to push the virtual touchscreen as a new standard in information kiosks: one early adopter was Bayerische Vereinsbank, which is using SIVIT for its customer information stands. By early 1999 the Man-Machine Cooperation Lab was devising new applications that included multimedia payphones. The idea is to enable virtual dialing, faxing, e-mail, and even web surfing without easy-to-damage screens and phone books. In most instances, callers will just point to projected buttons to activate the function they desire, with a hands-free microphone and small loudspeaker enabling voice communications.

systems employ "static" pattern-matching algorithms that analyze the appearance of the completed signature—measuring such parameters as the contour and height of letters and how far names spread out in X and Y coordinates. But not only are these vulnerable to forgery, many people alter their signature slightly with each signing. So Wirtz designed a system attuned to harder-to-mimic dynamic characteristics—like the sequence of strokes and pressure of the pen—that remain essentially constant even though the outward appearance of a person's signature varies. By analyzing these underlying features in real time, she devised a unique technology that verifies a person's signature even as it is being written.

Wirtz demonstrates her technique with an electromagnetic writing pad that senses changes in hand position and pressure—both on the pad surface and directly above it—as a person signs his or her name. "This makes it possible to analyze not only the visible written lines but also the connecting strokes in the air," she explains. Users train the system by entering their signatures five or six times. After that, logging on to something—a corporate computer, say—is simply a matter of writing their names once. Authorized users are greeted by the notification: "original." Gate-crashers see: "forgery."

Static recognition remains important for things like verifying checks, which the bank doesn't receive until long after they've been signed. But Wirtz says Siemens's dynamic technique can bolster security on myriad other fronts. For instance, shoppers and restaurant diners typically pay for purchases and meals by handing cashiers a credit card. The employee, in turn, places the card in a machine that transmits key information to a verification center that simply checks whether the person has enough credit to pay for the bill. In the future, she says, "smart" cards will themselves contain all a person's necessary identification features. By placing these cards in a reader attached to a handwriting pad, stores can not only check credit—but verify the customer is the card's authorized owner.

Since the signature analysis takes place on the customer's own card—and not over a network—Wirtz figures few people will object, especially since it dramatically increases transaction security. Moreover, the system's error rate with a random forgery attempt is a mere tenth of one percent—meaning it will be fooled only once in every thousand tries. Even in the case of expert forgeries, where people spend days practicing someone's signature, the misfire rate rises to only about 3 percent. That's at least as good as other biometric methods—and in late 1998 Wirtz moved to the Siemens Semiconductor group to help turn the technology into products.

Siemens has even bigger hopes for its Biometric Sensor Fingertip™, a complete fingerprint analyzer on a microchip that was a finalist for the prestigious German Future Prize awarded in December 1998. By fall 1999, roughly two years into the project, the sensor had been incorporated into prototype cell phones—with other possible applications being studied for use with guns, computer keyboards, smart cards, and even electronic commerce.

In an age when spy-fi thriller concepts like iris recognition and DNA screening are gradually moving into consumer markets, checking fingerprints may seem like yesterday's news. But not only is fingerprint analysis more socially acceptable and less invasive, there's still no documented case of two people bearing identical swirls and grooves. Compared to pin numbers, passwords, and maybe even keys, fingerprints win hands down: there's nothing to memorize—and people never leave home without them. With credit card fraud surpassing $1.5 billion a year in the United States alone—not to mention other forms of theft—experts predict the market for chip-based biometric fingerprint analysis should top $1 billion by 2003. By that time, Siemens expects as many as 150 million fingertip sensors to have hit the market.

Just the kind of interdisciplinary effort Weyrich fantasizes about, the program traces back to late 1996, when a strategic study showed a huge potential market for low-cost fingerprint analysis. At the time, leading fingerprint analyzers employed optic fingerprint imaging, which cost upwards of $1,000 per unit. But the Siemens team—Christofer Hierold, head of silicon process technology for ZT's Microelectronics Department, researcher Thomas Scheiter, and Paul-Werner von Basse of the Man-Machine Cooperation group—realized a chip-based sensor built with the same CMOS technology used to produce integrated circuits could be manufactured for under $50. Initial support came from ZT's "white space" program, which subsidizes promising projects too nebulous to secure all their funding from business units. But within a few months, project leader Hierold secured half his budget from two interested Siemens business arms: Information and Communication Networks and Automotive Systems.

The official kickoff came the following March. By September 1997—in a record seven months—a prototype chip was ready. About 1.6 centimeters square, the chip contains over 65,000 sensor elements that rapidly register skin details at high resolution. On the software side, specialists in Siemens's Program and System Development group in Graz, Austria, adapted big system fingerprint image-processing algorithms—the kind used in police computers—for a standard Pentium. The result is that when someone rests a finger on the chip's coated surface, sensor data is transmitted to the processor, which then extracts up to two dozen characteristic points—line endings, whorls, bifurcations—and compares them with stored patterns. A person's identity is verified when a match is found with an authorized user who has previously entered fingerprints for recognition.

"This is the first microchip you can touch directly," says Scheiter. Lucent Technologies spinoff Veridicom and a handful of other rivals were working on similar chip-based technologies early in 1999. However, outside experts said the German giant's rapid prototype execution placed it at or near the head of the pack. Since cell phones, computers, and cars already possessed enough processor power to run the system, Siemens initially focused on those applications; that's why Hierold approached the communications and automotive groups for funding. The company hoped to deliver prototype cell phones—called handies in Germany— using the technology sometime in 1999. The sensor was to be embedded in a groove on the telephone face. A prototype keyboard with the fingerprint sensor—a first step toward dispensing with computer passwords— was due out later in the year.

Siemens was also designing fingerprint-activated car ignitions— dubbed "keyless-go"—although that technology lingered further off. Meanwhile, out to curtail accidental shootings, researchers were exploring the idea of building fingerprint sensors into handguns. Even more distant: fingerprint-activated smartcards for electronic commerce.

Hierold especially favors this last idea. Demonstrated in the lab in 1998, the sensor-equipped card plugs into a computer through a normal parallel interface. The computer then issues a "go" signal when an authorized user places a digit on the chip. As with Brigitte Wirtz's smartcard handwriting analysis idea, the key is that verification takes place on the card itself—or some other miniaturized, self-contained system. No fingerprint data is transmitted over a network. "With such a personal system, you can get access to your computer, to networks, to the Internet—or do electronic commerce on the Internet," Hierold explains.

ERLANGEN

TWO AND A HALF hours by train from Munich and a short car ride through this busy university town stands Siemens's Erlangen R&D facility. The complex has sprawled out considerably since its 1965 inauguration under Siemens-Schuckertwerke. Still boasting its fish-stocked ponds and shady streets named for famous scientists, it is far more scenic than Perlach. But it is also older and more downtrodden—with a distinct weather-beaten air. The ponds are marred by brown water, while inside the lab buildings once-gleaming linoleum floors are faded and worn.

Nearing the turn of the century, Erlangen harbors some 7,500 workers. Of these, more than 7,000 serve in the two business groups headquartered on campus: Power Generation and Power Transmission and Distribution. Besides their sales, marketing, and administrative staffs, each group maintains extensive research and development components that between them include work on conventional and nuclear reactors, automation, and a variety of other industrial processes.

The remaining 365 staffers belong to Corporate Technology. Nearly a third serve in administration and intellectual property. On the R&D side, the old semiconductor studies of people like Schottky and Welker have long since moved to Munich. What's left concentrates almost exclusively on energy. Indeed, the Energy Division led by Jürgen Vetter comprises about two hundred researchers divided into six centers of core competency. Like virtually everything else in ZT, the projects under his jurisdiction focus on discrete systems and components—from tiny fuel cells to gigantic turbines, transformers, and generators. But since complex power systems take years to design and develop, Vetter's group performs what is probably the most long-range research in all of Siemens. "Seven years from now," he explains, "you'll find the prototypes in, say, the first pilot applications."

Nowhere is this focus on long-range, heavily applied research more apparent than in high-temperature superconductivity (HTSC). The field fairly exploded in 1986, when K. Alex Müller and J. Georg Bednorz of IBM's Zurich Research Laboratory showed that ceramic copper oxides lost all electrical resistance and became superconducting at temperatures 60 degrees higher than any previously known materials. This allowed them to be cooled with relatively cheap liquid nitrogen instead of the prohibitively expensive liquid helium required for other superconductors—setting off predictions of such wonders as wide-scale electric transmission without losses and frictionless trains suspended on powerful magnetic fields.

Seeing superconductivity's immense ramifications for its vast energy and transportation businesses, ZT used special set-aside monies to jump into the field on the heels of the Müller-Bednorz discovery. As the years dragged on and costs rose, program managers were forced to find additional funding from Siemens business groups, the German government, and academic and industrial consortia. But finally, on the verge of the twenty-first century, the first real-world applications are becoming evident. With its superconductivity and cryogenics department swollen to twenty-five researchers, Erlangen's projects include superconducting railway transformers, power cables, switches, and high-efficiency rotors. The company expects that by 2020, sales associated with these and other HTSC areas will hit DM 15 billion, or around $9 billion.

One of the most promising early applications is being developed in a big second-floor room in Building 36. There, amidst tanks of liquid nitrogen, cryostatic vats, and sundry other equipment strewn across lab benches, Hans-Peter Krämer and colleagues design superconducting limiters that promise to save millions of dollars by helping utilities deal far more efficiently with the short circuits that periodically plague electric power networks.

Limiters are meant to contain—or limit—the effects of short circuits. Such events can cause switching station currents to rise dramatically. Left unchecked, these powerful currents would quickly overwhelm the system. But the surge triggers a switch that turns on the limiter and thereby breaks the current.

Conventional limiters—vacuum circuit breakers—work well enough. The one caveat lies in the 30 milliseconds or so it takes to trip the switch and trigger their action. During this delay, currents can surge to as high as 60,000 amperes—some thirty times their normal value. Therefore, all switches and other equipment down the line past the limiter must be designed to briefly handle such a surge.

Avoiding that costly step is where superconducting limiters come in. A key characteristic of superconductors is what's called their critical current. The exact value of the current depends on the material used. Simply put, though, materials remain in their superconducting state as long as the current stays below the critical level. However, if the critical level is exceeded—as it would be in a short circuit—they at once develop electrical resistance that limits the amount of current they can carry. In contrast to the 30-millisecond reaction time of conventional limiters, this self-triggering reaction takes only a single millisecond. In that brief instant, the current does not have time to rise much past 4,000 amperes— meaning downstream equipment can be built to handle far lower power levels. "All the remaining equipment can be normal," Krämer says. Even more, because the size of switches and other equipment depends on the short-circuit current they're designed to handle, the entire system can be scaled down—another potential cost savings.

The Siemens project, which began in 1994 under the direction of superconductivity and cryogenics department head Heinz-Werner Neumüller, uses yttrium-barium-copper oxide as its base material. One-quarter-micron-thick films of this black superconducting ceramic were deposited on a sapphire wafer and covered with a gold shunt layer designed to compensate for any deficiencies in the superconductor. Depending on the voltage and power required, as many as ten or twelve wafers were joined to form the limiter, then dowsed in a liquid nitrogen bath to cool them to their superconducting state.

It took until late 1997 to master the process well enough to build a prototype. The device, for a time the world's largest superconducting limiter of its type, consisted of ten plate conductors each about 10 centimeters square. A slow-motion video camera aimed into the chamber through a small viewing window captured the current flowing through the resistor as a short developed. Once the limiter's critical current value was exceeded, the device warmed up rapidly, kicking off the circuit breaker and evaporating heat into the liquid nitrogen chamber in a gush of gas-film bubbles. After only a second or two, the nitrogen cooled the limiter back down to its superconducting state—and the device resumed its guard.

While the prototype handily demonstrated the viability of the Siemens approach, Krämer says any real world applications will require limiters rated at around 10 megavolt-amperes (MVA)—a factor of one hundred higher than the lab's initial creation. Given the difficulty of preparing substrates, depositing the superconducting material, and controlling its properties, that dream was still far from realization in the fall of 1999. But the Erlangen crew was well on the way to building a 1-MVA device, which it planned to field test late in the year in cooperation with Hydro-Quebec, a Canadian utility. Krämer and department head Neumüller note that in addition to improving power quality and equipment performance and safety, superconducting limiters offer another great advantage. Traditionally, whenever a fuse blows somebody is dispatched to reset it—at no little expense or danger. But since superconducting limiters reset themselves automatically, that would be unnecessary.

Neumüller cautions that there's still a long way to go in reducing costs and demonstrating the long-term reliability and economic feasibility—not just of limiters but other superconducting applications. However, he adds, "Generally, the technical and economic progress in power engineering is characterized by the ongoing development of existing technologies. Innovative advances are rare. HTSC has the potential to significantly alter the design and use of electrical transmission and distribution equipment. The worldwide race to translate the basic innovation of high-temperature superconductivity into marketable products is on."

PRINCETON

SET IN A WOODSY office park just off the famous Route 1 corridor that gave birth to RCA's Sarnoff Laboratories and a traffic-jamming gauntlet of fellow R&D establishments, the two-story aluminum-sided laboratory known as Siemens Corporate Research, Inc., ranks as one of the most successful and longest-lived foreign-owned research outposts on American soil. From its 1977 inauguration in Cherry Hill, New Jersey, SCR has grown to around 140 researchers. And whereas energy is the hallmark of research in Erlangen, the lab has successfully harnessed the power of computers and information technology for a broad swath of Siemens businesses. SCR is a place where researchers don virtual reality goggles to "travel" through the body, fashion ways to combine handwritten notes with voice recordings and digital documents, and listen to the World Wide Web over car radios.

It has not been easy to blend a relatively freewheeling American lab with the more conservative style back home. Moreover, while the Perlach and Erlangen facilities are staffed almost entirely by German nationals, Princeton's melting pot employees hail from twenty-two countries—creating language difficulties that exacerbate the potential for culture clash. Still, SCR chief executive Thomas Grandke feels the lab's diversity and related ability to come at problems from many angles greatly enhances creativity. "I'm convinced that this is one of the special assets that we have," he asserts. During a company-wide ideas contest to commemorate Siemens's 150th anniversary in 1997, over half the submissions from Corporate Technology came out of the modest Princeton facility.

A friendly physicist with reddish hair and beard, Grandke spent some sixteen years in various R&D posts before taking over the Princeton lab in 1996. While SCR was established to create a window on U.S. technology, establish local corporate citizenship, and benefit Siemens's American operations, he notes that in the early 1990s push to establish core competencies the lab took global responsibility for four key areas: adaptive information and signal processing, imaging and visualization, software engineering, and multimedia/video technology. Anytime a business division needs special assistance in one of these four areas, they come to SCR. But Grandke emphasizes that since ZT represents only 6 percent of the overall Siemens R&D budget, the lab is not set up to handle routine requests. "We are really supposed to provide the top-notch things, the leading-edge things that they need. It's not a last resort, it's the ultimate resort."

By far the lab's best customer is the Erlangen-based Siemens Medical Engineering business group (the modern incarnation of Siemens-Reiniger-Werke), which accounts for about a third of all SCR activity. That's largely because Siemens's medical business—with $4.5 billion in sales in 1998, the world's largest medical equipment concern—has a strong focus on imaging equipment and derives great benefit from SCR's unique image processing expertise. Also, since "Med" derives the lion's share of its revenues from the huge U.S. health care market, the headquarters staff is adept at English and highly attuned to the American landscape—facilitating collaborations with Princeton researchers. Even more, the lab houses the Innovation Field Health, a special ZT group chartered to spur growth in medical fields (see box, p. 200).

The fruits of the collaboration with Med permeate the lab—and the market. Multimedia experts have contributed MMR, the MultiMedia Reporting system, which enables doctors to supplement digital X-rays and other records with dynamic annotations that combine drawings, gestures, and voice. Meanwhile, researchers in the 40-person Imaging and Visualization department have crafted a variety of tools for enhancing medical evaluation and diagnosis. One of the hottest projects, the Fly-Through, gives surgeons a *Fantastic Journey*–like ability to travel through the body—virtually.

The guru of virtual surgery is Ali Bani-Hashemi. In his team's systems, doctor's don 3-D goggles to navigate through precise simulations of a patient's body—evaluating medical options without the need for exploratory surgery or endoscopy, the insertion of a flexible tube outfitted with a miniature camera. "We're trying to basically replace some of the invasive procedures that patients go through—without bothering them—through simulation and through good visualization," the computer scientist explains. "Just like video games, where you have representations of objects in space, we have representations of human organs that we can visualize from any direction, any side—and interact with them."

It all started in 1995, when Bernhard Geiger arrived at the lab from the French National Research Institute for Computer Science and Automation, where he completed his Ph.D and did postdoctoral work. As part of his dissertation, Geiger had created an impressive childbirth simulation designed to show doctors whether a baby's head could pass through the pelvis successfully, or whether a Cesarean section was needed. From that seed sprang the virtual endoscope—a tool designed to ease diagnosis and treatment of bronchial tumors by giving surgeons a ringside seat inside the body.

Typically, whenever a bronchial tumor is suspected, physicians

□

Back from the Future: ZT's Innovation Fields

Shahram Hejazi can throw out some wonderful visions of the future. For one thing, he foresees a world where doctors and nurses make house calls. Not in person—but virtually. Thanks to the growing ability to digitize everything from thermometer readings to X-rays and MRI scans, and then transmit the data over phone lines, he believes medical personnel will check on patients, answer questions, and go over test results without the patients even visiting the doctor's office—saving everyone time and money. And it gets even better. This telemedicine represents just one aspect of the future of home health care, where diabetics and others requiring regular tests draw their own samples and place them in special analyzers that automatically recalculate their treatments. Such devices, in turn, might be part of the automated house, where all communications systems and appliances are networked for easy Web-based control, robots cook and clean, and, well, Hejazi could go on . . .

Every big company struggles to spark innovation and growth; some even employ futurists to help identify unseen long-term opportunities. But at Siemens, a small group of dynamic thinkers like Hejazi are more aptly termed Back from the Futurists. That's because brainstorming up scenarios that look a decade or more down the road is only part of their job. The really challenging task comes in then working backwards to find ways to address that vision today—through novel products that provide a foundation for entirely new businesses. A group examining energy, for example, might study trends in oil and gasoline consumption, laws requiring reduced carbon dioxide emissions, and European Economic Community growth as pointing to a future populated by new types of power plants and hydrogen-fueled cars. Working backward, these scenarios might justify increased fuel cell research.

The concept is called Innovation Fields—and it's proven so successful that it was slated in 1999 to become a separate department inside Corporate Technology. When unveiled in the early 1990s, the effort spanned ten areas. To increase simplicity and encourage interdisciplinary thought, however, Claus Weyrich cut the number to five: Energy, Transportation, Information and Communications, Industry and Environment, and Health. Each field is staffed by three to six energetic young people, usually Ph.D.s from engineering fields, tempered with a veteran or two with real business experience.

The mission of these innovation teams is to canvass experts, read literature,

order a computed tomography scan of the presumed site. This provides a series of three-dimensional, cross-sectional images. A surgeon can then use these pictures to visualize the tumor in relation to nearby structures

and generally survey a field to combine what's technologically possible with economic, social, environmental, and regulatory factors to produce a more realistic view of the future. On one level, the exercise influences the technology road maps used to plan core areas of focus. More important, perhaps, it also points to unexplored synergism between several Siemens businesses that might open new markets. In that case, team members examine potential revenues and growth and the prospects of Siemens grabbing a significant market share. If things still look good, they seek management approval to drum up business unit support—and launch a project designed to give the company a running start. Since business groups can't afford blue sky projects, the key is to identify products that can be developed quickly—but that lay the groundwork for the grander vision. "You try to initiate projects that have short-term return with a long-term outlook," explains Hejazi.

A doctorate in electrical engineering who spent two years studying biophysics, Hejazi heads one of these innovation fields: Health. This five-person effort took root in late 1994 inside Siemens Corporate Research in Princeton, and is the only such body outside Germany—a step taken both because the United States is the world's largest health care market, and because the Princeton lab specializes in medical technologies.

As of early 1999, Hejazi's initiatives spanned telemedicine, home health care, home networking and automation, medical informatics, and computer-aided diagnosis. The latest project, centered on genetic testing and associated analysis and imaging, got going in mid-1997, after his group's investigations pointed to the field's explosive growth. Among the most promising future technologies was the idea of placing DNA samples on specialized semiconductor chips that would test for genes or mutations that increased a person's risk of contracting cancer or other diseases. Such tests threatened to cut into Siemens's keystone medical diagnostic market by making traditional imaging-based screening less necessary. At the same time, however, they heralded a potential new line of business—not just for medical, but for the company's semiconductor arm, as well as its Automation and Drives group, which makes the process-intensive systems required to conduct genetic testing.

Between the medical group and Corporate Technology, Hejazi quickly secured enough funding to launch a project that might well have slipped through the cracks because it spans several businesses. He won't say exactly what Siemens is working on, but calls genetics a near perfect innovation field topic. Asserts he, "It's visionary, it's growing very fast with extremely large market potentials—and it's strategically very important for medical because it could increase our business or it could kill us."

□

and plan the best route for obtaining a tissue sample with an endoscope—a slender, camera-equipped tube that is inserted down the patient's throat and into the bronchial tubes that branch off the trachea towards the lungs. However, not only does the procedure require seda-tion, the bronchial tubes can narrow and twist so as to make it impossible to reach the desired area.

That's where the virtual endoscope comes in. Working with Geiger and Arun Krishnan, Bani-Hashemi developed a simulation system that uses CT scan data to provide a spatially accurate 3-D model of the wind-pipe, bronchial tubes, and lungs. Simply by maneuvering a computer mouse, surgeons can explore this virtual world. "If the virtual endoscope can be brought to within a few millimeters of the desired location," notes Bani-Hashemi, "then a real endoscope will also be able to reach the spot."

The virtual endoscope debuted in Europe in November 1997 under the product name Fly-Through, arriving in the U.S. market a year later. By then, Bani-Hashemi and colleagues were extending it to other forms of surgery. The first new application involves colonoscopy. Each year in the United States alone, an estimated 1.3 million people undergo this frightening screening procedure for colon cancer. The patient is sedated, and the camera-fitted colonoscope inserted up the rectum, on the hunt for large polyps at risk for turning malignant. But because the colon con-sists of a series of hard-to-navigate folds, there's a risk of perforation—and something like a 20 percent chance the scope will not be able to reach the end.

Bani-Hashemi believes more effective processing of the scores of to-mography images generated to help surgeons plan their strategy can obvi-ate the need for optical colonoscopy. "If you're getting the CT anyway, why not just use that," he explains. "But instead of going image-by-image through a hundred or so images, which is very labor intensive, virtual colonoscopy is basically combining all that information into something like a movie."

Siemens hopes to have the tool approved by the Food and Drug Ad-ministration early in the new century. But even that is likely only the be-ginning for virtual medicine. In the case of brain surgery, Bani-Hashemi notes, the minute a physician begins to operate, tissue swells and things shift around, "so what you have on the table no longer matches what you have in your computer as preoperative data." But suppose real-time imag-ing allowed the virtual model to be constantly updated, highlighting key brain centers so surgeons don't mistakenly damage a speech or motor function. Three-D images also could be combined with force-feedback systems to simulate the "feel" of different tissue types. In that case, sur-geons might practice on virtual renderings of their patients before at-tempting the real thing—just like airline pilots or astronauts. Exults Bani-Hashemi, "That would be the ultimate."

■

Even as the colonoscope drew nearer to market, other researchers strove to bring similar technology to everything from computer-aided design to the repair and maintenance of power plants. In the latter case, technicians could conduct inspections on-line, traveling virtually through an industrial plant or factory in the same way Bani-Hashemi's colonoscope lets doctors wend their way through the body. Imaging and Visualization department head Alok Gupta considers virtual rendering a potential "bread and butter area" for Siemens. "Just like everything is becoming digital," he notes, "things are becoming 3-D."

Everything is also going Internet. Across the building from the second-floor labs hosting Gupta's department, Multimedia/Video researchers delve into different aspects of the on-line society. For one thing, instead of bringing researchers into the virtual world, they're creating new ways to bring the virtual world to people.

Until late 1998, when he left to open a branch lab near the University of California's Berkeley campus, the department was headed by Arding Hsu. A veteran of some seventeen years in Siemens research, the Taiwan native likened the department's mission to a Chinese stir-fry chef. While every stir-fry recipe involves cooking at a set temperature, throwing in some standard ingredients, and stirring, Hsu explains, outstanding chefs add a few secret ingredients that make the dish even tastier. By the same token, he notes, SCR must meld off-the-shelf items with proprietary concoctions to fill specific needs others have overlooked. "Our job," Hsu asserts, "is to provide these secret ingredients for Siemens and give Siemens a competitive advantage."

The stir-fry approach to research is embodied in WIRE—for Web-Based Interactive Environment—an exciting technology for listening to electronic messages and audio-surfing the Internet over a telephone or car radio. The central trick involves generating audio renditions of Internet documents. Siemens's proprietary advance lies in a key algorithm developed by multimedia researchers Michael Wynblatt, Stuart Goose, and Dan Benson that analyzes and elicits electronic document structure—e-mail, HTML code, and so forth. From there, off-the-shelf synthesizers convert text to speech, allowing users to browse, search, and follow links without images or keyboard.

Although listening to the Web isn't the same as being there, Siemens thinks it has the next best thing. WIRE has spawned two main offshoots—one for the telephone, the other for the car radio. The phone version is called DICE, for Delivering Internet Content in a Cellular Environment. Designed in conjunction with the Munich-based Information and Communications Network group, it's actually intended for any handset, not just cellulars. To peruse e-mail, users will simply dial an access number—Internet providers must have the technology—and select the audio option.

A tinny voice might say: "You have four new messages." It then lists sender addresses and all the normal headers. By hitting #, plus the desired message number, dialers might hear:

"Message three. From researchers at scr.siemens.com. Subject: Modem. You won't need your modem again."

Surfing the Web works along the same lines. Users can preprogram things so that tapping certain numbers zaps them directly to favorite sites. Depending on the mode, full or partial page contents will be relayed. A bell chimes whenever links appear, with people able to follow links simply by punching another button. Bookmarking is as simple as one more tap on the pad.

Delivering the car version is trickier, since radio manufacturers and automakers must be along for the ride. Still, the concept—LIAISON, for Linked Interactive Audio Information System Over the Net—stands to help anyone far from home or with a long commute. One minute a driver might be tuned into a talk radio program, the next, he or she can log onto SunWorld for technology news, courtesy of some electronics and the car's antenna. Granted, uplink costs could get expensive, but one idea is to make hook-ups free in exchange for allowing advertising into the vehicle. That, say Siemens researchers, is not unlike the way radio works anyway.

Inside the car, everything runs essentially like the phone version, only instead of touching keypads, users push radio and tape player controls. Other schemes for car-based Internet access involve putting screens inside vehicles. But Hsu argues that 90 percent of drivers fly solo. "Our idea here," he explains, "is not to use the screen, because we target the driver, not the passenger."

With both DICE and LIAISON, documents speak out in English or German. Although speech synthesizers can be hard to understand, voices and intonations change frequently, livening things up.

As of September 1999, with prototypes of both technologies drawing raves from potential customers, Siemens would only say a product announcement was on the way—and that DICE will debut first. In the meantime, Wynblatt, Goose, Benson, and colleagues were hunting for a way to let people answer e-mail and surf the Web as easily as they leave voice messages—via spoken commands.

■

To veterans like Heinrich Schindler, who headed energy research at the Erlangen complex until his 1997 retirement, Siemens's focus on applied research has sharpened with each postwar decade. But that, he said just before stepping down, is what it takes to compete in modern times. "I think I had to change my culture in every decade—but why not?" Schindler stated. "I enjoyed the period of being free and working in the field of fusion, which was so far away—and now I enjoy our project-oriented period. I enjoy that, it's no problem for me."

The methodical drive to create a research organization closely at-tuned to business needs will continue into the twenty-first century. How-ever, while the basic mantra of connection to customers and markets is shared by every industrial research organization, Siemens has long taken a different approach to research than counterparts in Japan and the United States.

Since World War II, in concert with Japanese firms helping rebuild their own economy, it necessarily adopted a more applied focus than American corporations. In contrast to NEC, Hitachi, and others, however, Siemens also supported a viable program of fundamental re-search—most notably in solid-state physics. Not until the mid-1960s—when the heyday of science in U.S. firms still had another two decades to run—did it essentially eliminate basic studies and launch an all-out com-mitment to applied research.

The German giant renewed that commitment in the 1980s—first with Karl Beckurts and then Hans Danielmeyer—even as the Japanese established a host of basic research labs, including the free-wheeling NEC Research Institute just down the road from Siemens's own Princeton lab. And it clung steadfastly to this philosophy even throughout the prosperous late 1990s, when U.S. firms such as IBM, Lu-cent Technologies, and Hewlett-Packard carefully expanded their scien-tific horizons.

As he leads Siemens R&D into the twenty-first century, Claus Weyrich isn't fooling himself about ever getting it right. To continue serv-ing the company well in the future, Weyrich expects ZT will have to be-come even faster on its feet, with greater reliance on interdisciplinary teams that are rapidly formed—and disbanded. It must also do better at creating start-up businesses. Late in the twentieth century, Corporate Technology still offered too many individual solutions—as opposed to the integration of many solutions to open up verdant areas of opportu-nity, he says.

Of course, doing all that means asking researchers to simultane-ously think out of the box and inside the box—so that they can both find creative opportunities and maintain their focus on the reali-ties of costs, customers, and markets. Weyrich laughs at the dilemma. Programs and practices like ZT's Innovation Fields, road mapping, and handing more responsibility to researchers can help. However, he says, "Well, what I can tell you is it depends on my recruiting the right people. It's not a big secret, it's a simple secret. But it's hard to be done."

CHAPTER SEVEN

NEC: BALANCING EAST AND WEST

"The big difference for these ten years is our facilities have expanded over the island, over Japan, over the world."
—SATOSHI GOTO, GENERAL MANAGER, C&C MEDIA RESEARCH LABS

"The how to balance is the tricky thing."
—TATSUO ISHIGURO, ASSOCIATE VICE PRESIDENT, RESEARCH AND DEVELOPMENT GROUP

NEC Research and Development Group

Central research established: 1939
Company sales 1998: $37.4 billion
1998 company R&D expenditures: $2.9 billion
1998 R&D group budget: $290 million
Funding paradigm: 70% central funds, 30% business unit contracts
Total group staff: 1,850
Research technical and scientific staff: 1,400
Principal research labs: Kawasaki, Sagamihara, Ohtsu, Ikoma, Fuchu, Tsukuba, Japan; Princeton, New Jersey; Berlin, Bonn, Heidelberg, Germany

---

"*I* AM PLEASED *to see you all here today for our dedication ceremony. . . . We have representatives of a number of leading research institutions, both academic and industrial, and we expect in time to prove our worth to join their select ranks. We think that with the well-publicized cutback in basic research by many organizations in the United States, there is plenty of room for us. . . .*"

Dawon Khang's words carried out to the two hundred-odd guests gathered in an unheated tent next to a small pond in a wooded reserve outside Princeton, New Jersey. The date was May 2, 1990. Attendees had come to witness the dedication of a modern four-story scientific laboratory just behind the tent: the $25-million NEC Research Institute, an ambitious enterprise founded to promote basic research in the computer and communications sciences.

It was a champagne day for NEC. The Princeton lab was the first corporate enterprise devoted to fundamental research to open in the United States in decades—and marked a stark contrast to American concerns like Bell Labs, which were taking a scythe to basic science. Moreover, NEC was borrowing a page from Bell's book by recruiting top-flight researchers and letting them choose their own paths—trusting that good things would come. Khang, for one, the institute's first president, was a Bell Labs veteran who had pioneered in semiconductor memories.

Far more than a tribute to basic science, however, the lab's inauguration heralded the start of an impressive globalization of NEC's research organization—probably the most extensive by a Japanese company and among the most ambitious of any industrial concern. In the decade following the Princeton institute's creation—formal operations began in 1989—NEC would establish three small labs in Germany, and another pair in the United States. These were joined by five new facilities on its home soil, including a sister enterprise to the Princeton lab that was also dedicated to fundamental research.

Taken as a whole, these moves marked a bold attempt to break out of the strict Japanese mold and strike a better balance between East and West—to blend the team-oriented applied research practiced by Japanese companies with the more individual fundamental studies long the hallmark of U.S. firms. In choosing the Korean-born Khang to run an American-based lab for a Japanese company, NEC spoke volumes about where its research was heading.

Achieving this balance across oceans, languages, and cultures poses a significant challenge for any company—and over the first decade of its ambitious research expansion NEC has encountered its share of obstacles. For one thing, it *does* seem harder for a Japanese corporation to assimilate outsiders than for American companies more used to melting pot cultures. Then, too, like virtually every Japanese concern in the late 1990s, NEC is mired in a long-running economic slump. Meanwhile, the American firms it outmaneuvered in the 1970s and 1980s have transformed into fierce competitors—and the Koreans are rising. All these factors dampen research growth.

Nevertheless, the company remains a global leader in semiconductor and display technologies, and in bringing about the fusion between computers and communications. More to the point, unlike the way American firms slashed their research budgets amidst their own economic woes of the late 1980s and early 1990s, it has not let hard times dramatically affect research. This is the result of a uniquely articulated, three-tiered philosophy: research for "today," "tomorrow," and "the day after tomorrow." Under this approach, no one element should be forsaken for the others—no matter how difficult the times.

This perseverance has paid off in a dramatic inventive surge that's seen the company move from fourteenth place in the number of U.S. patents received when the 1990s opened to second place in 1999. Along the way, it roared past big names like Motorola, Philips, Hitachi, and Toshiba to trail only Canon and perennial top dog IBM. Even more exciting, the company may be on the brink of reaping the rewards of the basic science program kicked off by the NEC Research Institute. Between 1990 and 1997, NEC placed twenty-fifth in the world in a survey of high-impact physical science research papers—those cited at least 150 times in other articles. Its 3,132 citations paled against the staggering 18,840 received by AT&T-Lucent-Bell Labs—as well as second-place IBM's 13,020. But in between Big Blue and NEC came only universities and government labs: no other industrial organization made the list.

How well the company can capitalize on this growing patent portfolio and scientific base—with its forays into largely esoteric areas such as nanotechnology, bioinformation, and holographic data storage—remains to be seen. A significant part of the answer will likely depend on the success of its efforts to blend the best of East with West in a global research organization. Though it faces formidable obstacles in this quest, the company might get a leg up from its past—as Dawon Khang intimated in his inaugural speech.

"It should be noted that NEC was founded initially ninety years ago as a joint venture of some Japanese investors and the Western Electric Company, which many of us may remember as a part of AT&T," Khang recalled. "So, in a sense NEC's efforts in America merely reflect a return to its original roots. . . ."

NEC sprang to life from the wave of foreign capital that flowed into Japan at the end of the nineteenth century and jump-started the nation's modernization program. Until 1853, when American Admiral Matthew Perry's "black ships" appeared outside Tokyo Bay and forced open Japanese ports, the country had lived in virtual isolation from the West throughout nearly seven centuries of Shogunate rule. The U.S. fleet's arrival, "the spark that lit the powder," set the stage for Emperor Mutsu

Hito to seize power fifteen years later, inaugurating the Meiji era and the westernization of Japan. For about thirty years, western nations kept a strict rein on commerce. But widespread treaty revisions late in the century—coupled with legal changes that allowed foreigners to own stock in Japanese firms—paved the way for technological revolution.

On July 17, 1899, the day the new treaties took effect, the one-year-old Nippon Electric Limited Partnership was converted into the Nippon Electric Company, a Japanese-American concern controlled by the international branch of Western Electric Company, the manufacturing arm of AT&T. The new NEC bought out a failing Tokyo electrical equipment maker and began importing switchboards, telephones, gauges, meters, and testing equipment—almost all of it from Western Electric or General Electric. The company also adopted Western Electric's production and management systems, quickly evolving into a modern manufacturing concern. Its expansion rolled along through World War I, when it branched out into washing machines, ovens, fans, vacuum cleaners, and other appliances, and began exporting items to Asian nations previously dominated by the West.

In 1925, Western Electric sold its overseas holdings to a subsidiary of the International Telephone and Telegraph Corporation. Under ITT's auspices, NEC continued its rapid growth, pushing into radio, vacuum tube manufacturing, and other communications-related areas. During World War II, the company produced military electronics—primarily in communications. Manufacturing operations were nearly devastated by Allied bombing. But recovery came quickly amid Japan's dramatic postwar reconstruction. In 1952, when the country's main telephone and communications concern was reorganized as the government-run Nippon Telegraph and Telephone Company, NEC solidified its fortunes by becoming one of four main NTT suppliers—along with Hitachi, Fujitsu, and Oki Electric.

The company plowed through the 1960s, becoming the world's leading exporter of microwave communications systems and sallying forth into consumer electronics, TVs, and tape recorders. In 1954, NEC had timidly begun to offer small and medium-sized computers. But it took a 1962 agreement with Minneapolis-Honeywell Regulator Company, precursor of Honeywell, before the business mustered the technological cachet to take off. NEC based its 2200 series computer on a machine built by Honeywell—and with it dominated Japan's domestic computer market in the last half of the 1960s.

Transistor development started in 1950, as NEC moved to replace vacuum tubes with solid-state devices. That work led it into integrated circuits a decade later, with semiconductor manufacturing beginning in

1967. While other Japanese firms focused on gallium arsenide, NEC took an exclusive license for Fairchild's silicon manufacturing technique, specializing in memory chips and large-scaled integrated circuits. It opted to continue expanding the semiconductor business amid the economic havoc caused by the Arab Oil Embargo of 1973 and 1974. Then, when chips took off after the economy finally bottomed out, NEC rode the wave to international fortune.

The architect of NEC's emergence as a global technological power was Koji Kobayashi. An engineer who worked his way up through the ranks to become president in 1964—the first company head to reach that position from within—he would keep rising and eventually serve a dozen years as NEC's chairman and chief executive officer before retiring in 1988. A student of American business leaders like Alfred P. Sloan and Peter Drucker, Kobayashi embraced concepts such as the Zero Defects quality program, the business division system, and decentralized management and controls. At the same time, he consolidated marketing, engineering, and manufacturing—increasing ties between the technical and sales sides of the company. "I want to develop NEC Corporation into the best company in the world," he told employees to kick off the 1965 New Year.

Kobayashi also oversaw the company's emergence from under the veil of ITT—which sold its remaining stake in 1965—into an entity more valuable than its onetime majority stockholder. Ahead of other corporate leaders, he saw the coming convergence of computers and communications and pioneered the concept of C&C—Computers and Communications—that provided an umbrella theme for the company's strategic direction long after he stepped down. In 1998, revenues totaled nearly $40 billion. The company ranked as the world's fifth-biggest telecommunications equipment concern, fourth-largest computer maker, and second-leading semiconductor manufacturer, behind Intel.

Although NEC's research history reflects a unique corporate heritage, its general course through scientific and technological waters cannot be separated from Japan's. The nation's peculiar historic insularity from other cultures, the pre-1900s dampers put on economic growth by foreign powers, and defeat in two world wars all conspired to delay the development of a full-fledged system of industrial research by more than fifty years when compared to Europe and the United States.

The transformation to a true high-tech power—a necessary precursor to establishing viable central research organizations—began around 1950, with the importation of methods and machinery for turning out nylon, televisions, and transistors. For roughly a quarter century, a signifi-

cant proportion of all Japanese R&D spending went to absorbing and di-
gesting this technology. Then, around 1975, even as other Japanese firms
took center stage in steel and autos, electronics and telecommunications
concerns such as Hitachi, Fujitsu, and NEC shifted into full-scale development mode—focusing on refining and improving an array of technologies necessary to sustain economic growth. Over the next decade, a
wave of capital investments, coupled with major innovations in high-
quality development and manufacturing, enabled these companies and
others to emerge as world leaders in carbon fiber, liquid crystal displays,
and semiconductors. It was in 1979, as all the hard work was set to pay
off, that Ralph Gomory, Jim McGroddy, and the rest of IBM's technological entourage arrived in Japan for their shock therapy session.

Just behind this startling expansion—commencing with a few early
efforts in the mid-1960s but only hitting full stride the following decade—
firms also began to create Japan's first truly effective central research organizations. Research then entered a two-decade-long growth phase
nearly as impressive as that of the Japanese economy as a whole. Tech-
nology-oriented companies began supporting research in everything from
nuclear power to computers—backing them up with impressive programs
in materials science, chemistry, semiconductor physics, and robotics.
Labs sprang up around Japan, especially in the high-tech corridor southwest of Tokyo that runs through Kawasaki, Yokohama, and Atsugi—and to
the northeast in Tsukuba, the Japanese R&D enclave known as Science
City. The push soon extended beyond the home islands, as well. By 1985,
Japanese companies had acquired or built about two dozen research and
development enterprises in the United States: seven years later, the number had surpassed one hundred, including NEC's Princeton lab. Europe
came next. In the early 1990s, Hitachi inaugurated facilities in Ireland,
England, and Germany. In 1994, NEC opened the first of what would be
three German labs.

Japan's great research expansion reached its zenith between 1985
and 1992 and marked a fundamental transformation in Japanese technology policy and practice, according to Fumio Kodama, professor of science, technology, and policy at Tokyo University. Kodama argues that by
the early 1980s, as older business sectors such as steel, autos, computers,
and some aspects of semiconductor production matured, developing
new industries became critical to Japanese economic growth. Realizing
that most new businesses would spring from the fusion of several technologies—such as communications, computing, measurement, and software—companies considered it vital to strengthen their intellectual base.
That explains why some Japanese firms not only increased the breadth
of research in the late 1980s and early 1990s, but also launched forays

into basic science—kicked off by Hitachi's April 1985 opening of its Advanced Research Laboratory, a stunning facility of impeccably landscaped grounds, fountains, and open areas for conversation or contemplation.

In 1992, investments in research and development overtook capital expenditures as an aggregate of the Japanese economy. In Kodama's eyes, that event confirmed the nation's shift from a manufacturing society to one bent on knowledge creation—an essential step toward the establishment of new industries needed to spur growth. As late as 1999, however, Japanese firms were still struggling to make the transition to industry creation, a shortcoming that in his view loomed as a significant factor in the country's ongoing economic malaise.

Before established corporations can play a major role in launching new industries, the status quo invariably must be jolted out of its complacency, learn to embrace novel ideas—and then spring into action. Every big company resists such upheaval. But change has proven an even tougher sell for the Japanese. In Kodama's words: "That, we hate."

Heading into the twenty-first century, the struggle to accommodate a new order while also staying true to traditional business principles represents one of the greatest challenges facing Japanese companies. Perhaps no industrial research organization has worked harder at striking that new balance than that of the old Nippon Electric Company.

NEC opened its first small research unit in the early 1920s. Although descriptions of the lab's exact size and range of projects are not available, the driving force was Yasujiro Niwa, a talented engineer who joined the company in 1924 and three years later took charge of the Engineering Department. Hoping to curb NEC's reliance on western technology, Niwa pushed a program to recruit top engineering graduates from prominent Japanese universities.

The effort's first big payoff came in November 1928, as Emperor Hirohito ascended the throne and Japan's elite news organizations—*Asahi Shimbun* and *Mainichi* newspapers—raced to transmit photos of the event from Kyoto to Tokyo. *Asahi Shimbun* chose Siemens phototelegraphy equipment, while *Mainichi* employed an NEC-developed set. NEC's official history reports that the home-grown product beamed the photos in less time, and with better picture clarity than its German competitor—enabling the company to seize control of the national market.

Niwa spearheaded NEC's research and development efforts for nearly two decades. When a central lab was finally established at the Tamagawa plant in 1939, during the buildup to World War II, he was named the first head of research. Two years later, the main operations moved to Tokyo's Ikuta suburb, and several branch labs were created. Most of the work focused on radio and military electronics research. But NEC's early postwar production of microwave communications systems and TV broadcasting equipment also drew heavily on Ikuta and its sister laboratories.

The immediate postwar years brought hard times, as the company recouped from heavy bombing damage and a major labor strike. In 1949, all NEC research labs were closed, and the company concentrated exclusively on importing technology and refining it for mass production of TVs, radios, communications systems, and eventually transistors, integrated circuits, and computers. Even after central research got going again in July 1953, much of the R&D effort continued to take place in development-minded factory labs. Still, by 1955, NEC's engineers had invented a method of improving FM receivers, leading to advanced over-the-horizon communications systems that attracted many domestic and foreign customers.

By 1960, full recovery was in sight. Mass production of transistors had begun two years earlier, and with the company's mainstay telephony business booming, growth accelerated dramatically. The next year, research was reorganized into five labs: the Fundamental Research Laboratories; Communications Research Lab; Electronics and Mechanics Research Lab; Nuclear Energy Research Lab; and Manufacturing Techniques Lab. But although these operations were brought together in Tamagawa, they operated independently rather than as a central group.

That structure lasted until 1965, when new president Kobayashi engineered a company-wide reorganization that extended to his disjointed R&D operation. At the time, apart from the five Tamagawa-based labs, the three operating groups ran their own development labs—while each of the fourteen business divisions under the operating groups maintained engineering units. With the exact mission and responsibilities of these various labs far from clear, Kobayashi consolidated three Tamagawa enterprises—the Fundamental Labs, Communications Research Lab, and the Electronics and Mechanics Lab—into the new Central Research Laboratories: the Manufacturing Techniques Laboratory continued to operate independently, but the Nuclear Lab was soon abolished as NEC abandoned atomic energy.

The reconstituted central lab was placed on a par with the company's operating groups and given a loose rein. Its job was to focus on long-term problems associated with broad technological areas in communications and computers. Well-defined shorter-term projects were left to the three operating group development labs. Finally, the engineering units inside each business division concentrated on immediate problems relating to their specific product lines.

Thus was born the three-tiered philosophy—research for "today, tomorrow, and the day after tomorrow"—that guided NEC's research and development into the twenty-first century. In Kobayashi's words: "Because the division is most sensitive to current market needs, its engineering unit has the responsibility for developing technology for 'today.' At the next level, the development laboratories within each operating group have the responsibility for R&D into technology for 'tomorrow,' nurturing the seeds of what will eventually grow into new business ventures. Finally, primary responsibility for research and development for the 'day after tomorrow' belongs to our Research and Development Group, an organization that is not constrained by current market needs."

The first manager of the new central labs was board member Yujiro Degawa, who reminded his scientists to direct their activities toward corporate needs. But research was still greatly distanced from company goals two years later, when a dynamic Japanese-born engineer arrived at the central labs after a decade at Bell Laboratories. It was this newcomer who would eventually put the meat on Kobayashi's structure, emerging as a guiding light of Japanese research policy. The organization and mission of NEC research for at least the next thirty years would reflect his vision. His birthname was Michiyuki Uenohara. But friends called him by his nickname from the United States: Mickey.

Mickey Uenohara was born in Kagoshima Prefecture on the southern Japanese island of Kyushu, earned his undergraduate degree at Nihon University, and then ventured to the United States for graduate studies. In 1957, a year after receiving his doctorate in electrical engineering from Ohio State University, he joined the Bell Telephone Laboratories, where he pioneered in microwave and satellite communications. That ten-year stint at the world's premier industrial research organization—where scientific excellence coexisted with glaring shortcomings in business sense—proved critical in shaping his vision of how research should be managed (see box, p. 216).

When Uenohara arrived at NEC's central lab as manager of its Electron Device Research Laboratory in 1967, he found an operation that paid little attention to NEC's needs. "The Central Research Laboratory was like heaven—free to do anything," he recalls. "Business divisions did not rely anything [sic] on the Central Research Laboratory. They thought they could do anything they needed to do."

The new manager wasn't particularly surprised. With orders streaming in for existing products, the business divisions didn't have time to

"Mickey" Uenohara

After completing his doctorate at Ohio State, Michiyuki "Mickey" Uenohara spent a decade at Bell Labs, where he established himself as a gifted inventor. He then joined NEC—emerging as one of the country's most effective
research managers. He reorganized investigations along core competencies,
expanded research internationally, and pioneered basic research in Japanese
industry. Following is his perspective on some key aspects of research:

On Bell Labs and on research: "I am not a basic scientist, I am practical engineer. But still I did very basic work at Bell Laboratories—and I learned good
and bad at the Bell Labs. So I tried to adapt the good parts, and I avoided the
bad parts. You see, Bell Labs hired top scientists from key universities—and
they were so proud of themselves, very individualistic, very creative, but very difficult to coordinate. They had excellent, top, key technology. But they could
not integrate—so they could not make even 16-kilobit random access memory."

On his concept of Ko no dokusa to Gun no sozo: "Individual originality and
group creativity," which became a company-wide initiative in the 1980s: "We
have to promote individual originality, but in order to utilize the technology,
we have to promote group creativity. People have to integrate each small
idea—that makes a big building."

On the role of central research in today's decentralized world: "I asked the
general managers of SBU [small business unit] for many, many years—please
collaborate. One day, one came to me and said, 'I fully understand your point,
but in reality it is extremely difficult because that SBU is my enemy. We are
fighting for very similar markets.' So I said, 'Yes, that is the truth.' I then decided corporate lab had to act as catalyst—because SBUs themselves cannot
fully cooperate. But a corporate laboratory can act as a catalyst and the results
can be very similar as if the SBUs collaborate."

On basic research: "Most outcomes of basic research do not immediately
contribute to industrial development; they are the legacy to our descendants.
Therefore, basic research has to be open to everyone. The university should be
the center of basic research. However, it does not excuse industry from performing basic research. We have to have excellent basic research, otherwise
we cannot fully utilize universities' basic research. . . . Most company managements are shortsighted and cannot tolerate long-range R&D. In order to pursue long-range research, we have to have a network of key persons who
understand the importance of R&D and its social implications. I was able to
form a network of key persons in NEC. Our middle managers established networks of key persons in business units during the development of core technology program. Those networks helped to overcome many difficult problems
that I faced in order to pursue my technology management."

On research and gardening: "If I give good care to flowers, grass, trees, and
others, they respond accordingly. The R&D management is very similar. The
R&D organization consists mainly of researchers, very sensitive living creatures. . . . [No matter] how good R&D strategy and planning are, if you do not
give good care to individual members of laboratory, you cannot accomplish
your objective."

worry about technologies years down the road. But Uenohara had encountered similar thinking at AT&T and was convinced it spelled disaster. The lab would likely sail along while the good times rolled, publishing papers in academic journals and not producing much of value for NEC. But come the downturn, he figured, business managers zeroing in on eliminating waste would see the central facility as largely irrelevant and attempt to slash its budget—or eliminate it entirely.

Uenohara's idea was to head off that day by making research relevant before it really needed to be. That way, when the bad times arrived the business units would view the Central Research Laboratory as essential to their recovery—not a hindrance. "I intended to make CRL the organization that was most trusted and relied on by the business units," he notes. "So I insisted the Central Research Lab had to be tightly coupled to business divisions."

One way of achieving that synergy was through joint planning. Not long after he joined NEC, in order to better understand his new company, Uenohara began studying the major business units. He found they had a sophisticated agenda relating to markets, budgets, and manufacturing—but no long-term plans about how to evolve products to meet those goals. As far as the newcomer could tell, their technical strategy hinged simply on studying the American market to identify promising business lines, and then developing those as quickly as possible—a practice he felt devalued research. "For such a business strategy they did not need any activity of central research," he notes. "Business divisions could do it."

Alarmed, Uenohara successfully lobbied for the creation of a corporate strategy development office. The idea was to bring together key young aides—from the business divisions, marketing, manufacturing, engineering, and research—to help clarify future company directions. The group studied the strengths and weaknesses of each NEC business segment, and from this analysis sketched out a proposed general strategy that was presented to the corporate board. It was a controversial document that stirred vigorous debate. But Uenohara says the discussion sharpened company goals and helped integrate research into overall strategic thinking. Just as important, the office served as an excellent training ground for corporate up-and-comers. Many—himself included—eventually joined the board of directors. And later, when it came to a more dramatic makeover of research, he says, having so many friends in high places "helped me to overcome many difficult problems that I faced in order to pursue my management philosophy."

Not long after being elevated to the central lab helm in 1971, Uenohara crossed a major hurdle in the pursuit of that philosophy. Still looking for ways to increase ties between his researchers and their business unit colleagues, he resolved to make research more financially dependent on the SBUs. That way, he explains, the relevance of his people's work was almost assured: the business units wouldn't pay for things they didn't want.

Until this time, the central lab was funded entirely from general corporate monies. Uenohara wanted his managers to drum up 30 percent of their funds by accepting contracts from the operating groups for specific jobs. Theoretically, the divisions would pay a high percentage of the costs for short-term, low-risk projects. The farther out or more chancy the work, the greater the share covered by central funds. Uenohara insisted the exact percentages—along with timetables and other pertinent details—be spelled out in a formal contract.

Here the research boss was evolving a core element of his philosophy that he later termed RAP, for receiver-activated paradigm. "The key to these technology transfers is they are largely dependent on the motivation of receiver," he explains. "The one who tries to utilize the technology has to be very active, very motivated." Therefore, an SBU's willingness to enter into such agreements gave him a valuable clue as to how serious it was about a project. But even that was not enough. Uenohara also asked the business groups to send key engineers to participate in the projects. "And if they agreed," he relates, "we are sure they are very committed."

Uenohara unveiled his plan in late 1973, when the first Arab oil embargo sparked high inflation that among its other effects squeezed R&D spending. Until that time, research managers had resisted any efforts to forge closer ties. But he told his managers if they didn't come around now, their staffs would be cut. "I took the chance of business crisis in order to increase the contact with business units," he explains.

Already by then, Uenohara was laying the groundwork for reorganizing research around core technologies. "In order to be fully trusted and relied on by business units, CRL has to create and establish necessary basic technologies—generic technology—ahead of the needs of business units," he notes. "To do so CRL has to identify the core technologies for the growth of company business in the future. This is the R&D strategy."

Beginning in 1973, research middle managers spent one day a week surveying the business units and identifying core technologies. The two-year preparation phase included three extensive studies: one concentrating on market trends, another evaluating the direction of science and technology, and the last analyzing small business unit performance, especially each group's technological strengths and weaknesses.

From these results, Uenohara's staff fashioned a huge matrix. The vertical axis showed key NEC markets. Core technologies—for present and presumed future products—ran across the top. That made it easy to see what technologies were linked to a given SBU—and whether they concerned several business units or just one. While not ignoring technologies that served only a few sectors, Uenohara focused on those such as multimedia that spanned the biggest swath of NEC businesses. "From the Central Research Lab management point of view, [for] the core technologies which have broad market demand, the risk is the least—very small," he explains. The research itself may be risky, he notes. But the decision to concentrate on these areas is not. In fact, failure to invest heavily in these critical technologies could spell disaster.

The core technology program went into effect in 1975—about the time NEC's new central lab opened on a 14-acre site in Miyazakidai, in the Tokyo suburb of Kawasaki. At first, the business divisions didn't really understand the concept. Later on, though, they began cultivating their own areas of expertise—making it easier to cross-correlate programs with central research. Ideally, the central lab would create generic technologies. Then the business units would tailor them to distinct end products—the way ASIC memory chips are modified for use in mobile phones, personal computers, and other devices. NEC might have ten thousand products, Uenohara notes. "But if you analyze these products, they utilize very common technology."

With these changes, Uenohara modernized NEC research. Soon after being elevated to the corporate board in 1976, he was named head of the entire R&D group—making it easier to pursue his agenda. To complement the core technology program, he launched an aggressive patenting campaign—asking researchers to file disclosures on even seemingly mundane advances. It's extremely difficult to establish core technologies and protect them without a strong patent portfolio, he explains. And because it is hard to identify key patents right away, Uenohara liked to blanket a given field of innovation with lots of filings. "Research managers might think this is just cheap ideas—but such patents make big money," he says. Besides, patents provide a good bargaining chip for entering into strategic alliances that bring in additional technologies.

Uenohara spurred many other changes before his 1989 retirement to become an NEC executive adviser. But perhaps his greatest contributions came in establishing a global presence for NEC research—and advancing the cause of basic research in industry. Uenohara felt strongly that Japanese firms had for too long forsaken both these responsibilities. Erecting labs on foreign shores, he argued, would bring the company closer to international markets, open windows on outside technology, and establish NEC as a good corporate citizen—making it easier to recruit personnel. Meanwhile, never forgetting his Bell Labs days, he felt that a strong basic research program was critical to building ties to universities — a key source of new ideas and people. These concepts led directly to the creation of the Tsukuba lab and the Princeton Research Institute.

The challenge of safeguarding this research legacy into the twenty-first century goes to Tatsuo Ishiguro, who took over the R&D Group in 1994 after the death of Uenohara's immediate successor, Yasuo Kato. Under Ishiguro, and Kato before him, the globalization of research has gathered momentum throughout the decade — as NEC opened a cadre of new labs and expanded to a third continent: Europe. At the turn of the century, Ishiguro presided over seven home country labs, three in the United States, and another trio in Germany.

Sitting in an executive floor conference room in NEC's rocket-shaped supertower, Ishiguro acknowledges the difficulties of managing research across these vast geographic and cultural distances — and against nonstop competition from corporate behemoths and start-ups alike. On one hand, researchers must continue to defend NEC's bedrock businesses in such areas as high- and low-end computers, communications systems, semiconductors, and flat panel displays. On the other, as growth in more mature lines slows or enters downturns, it's becoming critical to find major innovations that spawn additional opportunities.

Ishiguro's solution to these dilemmas lies in working even harder to strike the right balance between West and East. Following the lead of American firms, the R&D group has moved vigorously to spin out promising technologies, entered into more alliances like NEC's ongoing partnership with Lucent Technologies to make next-generation logic chips, consolidated individual labs, and abandoned its often-moribund departmental organization in favor of an elastic group structure that makes it easier to assign researchers to different projects as new opportunities or needs arise. Meanwhile, a corporate-wide initiative called *senmon shoku sedo* breaks with tradition and grants managers and researchers alike more flexibility and decision-making powers (see box, p. 222).

At the same time, in true Japanese fashion the overall direction of research has held remarkably steady through the company's travails; the slashing and burning found in American companies is not evident. Although the growth in total R&D expenditures ran under inflation for most of the 1990s, the budget inched up annually. For fiscal year 1999, NEC's research and development outlays stood at nearly $3 billion, a higher-than-average 8.6 percent of company sales. The R&D group under Ishiguro got about a tenth of this figure, or nearly $300 million.

As it has since Uenohara's early days, Ishiguro's group devotes about

30 percent of these resources to the needs of "tomorrow"—typically through business division contracts. Twice as much activity—60 percent of the overall effort—is focused on the "day after tomorrow," projects whose payoffs lay some three to five years down the road. The final tenth—a sublayer to the three-tiered research philosophy—is allocated to fundamental research for the "day after the day after tomorrow."

"As a management issue, we have to keep some good balance between future technological development and technology transfer," Ishiguro explains. "This ratio is almost constant for these ten years, and still I intend to keep this ratio—even though the environment is changing." That philosophy extends to the company's relatively new thrust into basic research, as well. "It takes a long time to cultivate new technologies—so we think the first five years were spent for the buildup of the organization," notes Ishiguro. "We are now in the second five-year period, and in this period we want to get some output in the basic ideas which may have great possibilities in the future."

Starting inside the Kawasaki central labs and broadening to the fundamental explorations in Tsukuba and Princeton, at some potentially path-breaking projects show the promise of this long-term approach to research.

CENTRAL RESEARCH LAB

LOCATED IN A tranquil residential area in the low-lying Tama Hills about a half-hour southwest of downtown Tokyo, NEC's trailblazing Central Research Laboratory in Kawasaki served as the company's only dedicated research arm for a dozen years after its 1975 inauguration.

A source of pride and joy when it opened, "CRL" now looks tired and weatherbeaten. Architecturally uninspired, the flat white building stands three stories tall in the front—the administration, or "A" block—and rises to five floors in the two parallel research wings in the back ("B" and "C" blocks). A scenic garden—large by Japanese standards—stretches from the main gate around the sides of the complex: signs denote the species of each tree or plant, even what types of birds might be spotted. In addition to garden tennis courts, the lab harbors a rooftop nine-pin bowling alley and mini-golf course. But perhaps the most notable amenity is its pub—One Shot—created because the area's residential zoning prohibits bars where employees might congregate after work.

Despite its lack of outward glamour, however, the lab remains the hub of R&D group activities, with close to a thousand administrative and technical workers. Besides the R&D group's big planning office and a

□

Senmon Shoku Sedo

Like many Japanese companies, NEC is big on tradition. Among its ongoing practices: promoting workers into management based on seniority. But while ensuring managers are long on experience, this policy handcuffed the company during its late 1990s economic crunch—especially in R&D. With a hiring freeze curbing the influx of fresh talent, and NEC's most experienced technical people "trapped" in management, innovation was squeezed. That reality prompted one semiconductor lab to break with tradition by allowing managers to return to the bench with no loss in status—simultaneously handing individual engineers greater decision-making powers. The novel program raised lab productivity so profoundly that it was soon extended corporate-wide.

The program is called *senmon shoku sedo*—specialist work system. The central figures are Nobuhiro Endo, assistant general manager of NEC's ULSI (ultra-large-scale integrated circuit) Device Development Laboratories in Sagamihara, and Nahomi Aoto, one of NEC's few women engineers. Endo implemented the program. Aoto showed its power by shedding her white-collar role and launching a bold campaign to raise the cleanliness of NEC's semiconductor plants.

The effort started in 1997, when sluggish growth in chip sales sparked a mini-crisis for managers like Endo. His three-hundred-strong operation, part of NEC's Sagamihara Research Laboratories, develops ultra-large-scale integrated circuits three to five years—or two generations—down the road. But while the crunch made increasing productivity vital, the seniority-based promotion system kept getting in the way. Typically, Endo explains, engineers move into management after fifteen years with the company. First comes *kacho*, or section team manager. Then comes *bucho*—group team manager. In theory, managers wear two hats: one technical, one administrative. But administrative chores often dominate a manager's time. In light of the hiring freeze, this meant Endo's most experienced engineers were chained to their desks just when he most needed them back in the lab.

Early in 1997, Endo convinced NEC to change the system—for his lab only—by effectively making two separate jobs out of some management positions. One job was for an engineer who focused on technology, the other for a project manager concerned about meeting schedules, filing patents, and so forth. Specifically, the rank of expert engineer became equivalent to *kacho*, while that of senior expert engineer corresponded with the *bucho* level. The idea was to avoid adding new jobs by having those opting for the engineering side take up the technical responsibilities of two managers, and vice versa—hopefully raising productivity by allowing people to concentrate on a single type of work.

The move's simplicity belies its profound break with tradition. Historically, notes Endo, employees gain respect based on the number of their subordinates. Therefore, asking his scientists and engineers to forsake management would not work unless he found other ways for them to gain respect. His novel solution involved granting technical staff more power to choose their own projects or tackle specific problems—things that normally would have to be cleared with a *kacho* or *bucho*. He also afforded expert engineers and senior expert engineers greater freedom to attend technical conferences and write scientific papers—things that elevate their status inside and outside the laboratory. While some other Japanese companies have implemented similar specialist systems, Endo says, "the role and job of the specialists cannot be clearly discriminated with those of the managers." NEC's clear distinction between managerial and technical specialists, he asserts, makes its system "unique."

Senmon shoku sedo went into effect in July 1997. Of the ULSI lab's fifty-odd managers, seventeen switched back to the technical side—a godsend for Endo. Says he, "That is the equivalent of hiring at least that many freshman, probably more because of their greater experience." In its first year, the added work force and reduced red tape helped slash the time needed to develop a new generation of memory cells by 50 percent—from two years to one. The lab's role in computer-aided chip design also broadened. Before the program got going, lab-developed CAD systems were simply shipped off to NEC's device designers. But *senmon shoku sedo*'s flexibility allowed Endo to send his people to provide technical support for their creations. Other parts of NEC began reporting much greater satisfaction with new chip designs, he says.

Then there's Aoto. A fifteen-year NEC veteran, she made *kacho* in 1994— but jumped at the chance to get back to full-time engineering. In particular, she felt that as semiconductors grew smaller, with a corresponding shrinking in the size of what constituted a contaminant, NEC's clean room atmosphere had to be rethought. Her increased authority under *senmon shoku sedo* allowed Aoto to introduce new technologies and practices for circuit layout, polishing, layering, and a host of other aspects of chip production. When her efforts paid off by dramatically improving the "cleanliness" of the ULSI lab's clean rooms, she spread the word to plants worldwide, becoming NEC's first "specialist cleaning engineer."

In July 1998, an impressed NEC extended *senmon shoku sedo* corporate-wide. The jury was still out early the next year on whether the program could be adapted to nontechnical fields such as sales and public relations, where lines between management and workers are fuzzy. But inside the R&D Group, returns were more glowing. Kohroh Kobayashi, general manager of NEC's Fundamental Research Lab in Tsukuba, praises the system for allowing him far greater flexibility in how research projects are assigned and staffed—a flexibility he feels is necessary to compete in today's fast-moving world.

In Endo's view, if there's a catch to *senmon shoku sedo*, it comes back to Aoto. Her efforts proved so successful that in order to better coordinate all the changes she was making in NEC's clean rooms, the company moved her back into management.

few development operations, CRL harbors four main research labs. These four branches, each with a series of sub-laboratories, specialize in functional devices like liquid crystal displays; optoelectronics; computers and communications (C&C); and resources and environmental protection. Inside their linoleum-lined confines one can find many of the same general activities common to other computer and communications labs around the world, from IBM to Xerox PARC and Siemens: futuristic TV screens and multimedia concepts, semiconductor lasers, "expert" software guides, new approaches to Internet search and retrieval, or storage disks with double the capacity of digital versatile disks (DVDs).

One of CRL's biggest areas of focus lies in the display realm. The Display Device Research Laboratory (a branch of the Functional Devices Lab) led by senior manager Hiroshi Hayama conducts investigations in three broad areas. For the over 100-inch diagonal market of image-rich data presentation and high-end home theaters, work centers on extremely high-resolution projector displays. In the middle range—40- to 100-inch screens—NEC is betting big on plasma display panels (PDPs). This type of screen, which should soon be thin enough to hang on a wall, offers a wider range of colors than conventional television and holds the potential to become a large-screen multimedia display for personal computers, DVDs, and a variety of other applications. Finally, on the small end of the spectrum—surfaces of 30 inches and under—come various flat panel technologies, most notably the liquid crystal displays used in everything from watch faces to radio dials and laptop computer screens.

Roughly half the Display Device Lab activity centers on liquid crystal displays. It's no wonder. With some $1 billion in annual LCD sales, NEC holds 30 percent of the worldwide market and ranks behind only Sharp among the world's largest liquid crystal display makers. In 1999, Hayama's lab was exploring an array of technologies that offer hope for all those frustrated by their laptop's diminished brightness at certain angles, the way portable computers eat up battery power to illuminate their displays, or who just have tired eyes from making out the fine print on their screens. Thinner panels, better detail, more colors, and lower power is Hayama's mantra.

An especially intriguing project underway in 1999 involved so-called ultra-fine-view displays that promised to sharpen up computer screens to hawk-like clarity. Conventional cathode ray tubes—like those in most computer screens—have a top resolution of about 80 pixels per inch. That might be fine for reading English text, Hayama notes. But when it comes to displaying complicated Japanese and Chinese characters—especially in small font sizes—these screens fall flat on their faces.

"We complain always these characters cannot be read with normal resolution display," he asserts.

Hayama sees ultra-fine-view displays as the answer. Central lab researchers have created an 11.3-inch laptop screen that achieves a resolution of 177 pixels per inch, more than double the value of a standard screen. The CRL group entered 1999 trying to interest business units in the technology, but faced a number of problems. The most fundamental: a severely depressed LCD market that made any new technology a hard sell, especially a high-end one like ultra-fine-view displays. Nevertheless, relates Hayama, "I still think high-resolution LCDs are the best to break this bad situation and to make money. So we continue the development." One promising early application lies in displaying medical data such as digital X-rays or magnetic resonance imaging pictures that contain many details hard to see on normal computer screens. Japanese newspaper production marks another potential market. Hayama explains that papers are printed in small type sizes that cannot be easily proofed on a computer screen. Either they must be enlarged, making it difficult to view layouts, or proofreaders must print out each page in order to do simple copy editing. High-resolution LCDs could end the trauma.

Like most CRL activities, liquid crystal display investigations almost always involve small-scale advances in the current state of the art. But a different class of research aims at achieving breakthroughs that either generate completely new business opportunities or advance conventional technology by leaps and bounds. Historically, research managers admit, Japanese companies have done better at incremental advances than hitting research home runs. But a concerted push to foster individual initiative and help novel products get out the door may be changing the game. "I think the Japanese are very good at doing evolutions, but not so good at revolutions," sums up Satoshi Goto, general manager of the C&C Media Research Laboratories, one of the four main CRL labs. "I think that might be true—up to now."

A concrete example of what Goto's talking about can be found in the work of Masatoshi Iji, one of CRL's few chemists, who doggedly pursued a hunch that led to his discovery of a uniquely fire-resistant and easily recyclable plastic. The material, called NuCycle, holds the potential to spur a new generation of environmentally friendly plastics for computer and other electronics housings, battery covers, and even building materials—and is already proving a welcome source of additional revenue for NEC.

Iji is a group manager in the Resources and Environment Protection Research Laboratories, where part of his job involves trying to improve the general environmental "friendliness" of the tons of plastic-based materials used each year in NEC's computers, monitors, circuit boards, and other electronics. For example, many plastics contain a flame-retarding additive—usually a halogen such as bromine. But while such compounds reduce the risk of fire, they possess two major downsides. One is that during combustion—as things heat up—they can generate toxins such as dioxin. Another is that the standard recycling process weakens the overall structure and reduces the flame-retardant capabilities of these plastics, limiting their future uses.

These shortcomings drove Iji, among others, to hunt for alternatives. Phosphorous compounds offered some potential. Their flame-retardant properties rival those of the halogens—and they are distinctly less toxic during combustion. However, they do not generally provide good heat resistance—a different property from fire-retardancy that involves the ability to prevent melting. Moreover, they can leak from waste plastics and contaminate soil and water, and also suffer from a lack of strength and flexibility—so in the end leave much to be desired. That left the silicones, polymers with inherently high heat resistance. In 1995—when he was considering the body of research into silicone—an insight came to Iji.

Other researchers, he says, focused on ways to bolster silicone's heat resistance even further. However, Iji realized that this strategy served to limit the substance's fire-retardant capabilities. In particular, the silicones with the highest heat resistance were coarsely dispersed in plastics—and tended to be extremely viscous at high temperatures. Both factors worked against their ability to migrate to the surface during combustion and form a uniform flame-retarding barrier. "In general silicone researchers think high-heat-resistance silicone is very good," he explains. "But I don't think so."

Armed with this realization, Iji got to work designing a chemically altered silicone that dispersed finely in plastics. The silicone he devised enjoyed low viscosity, making it easier to remove to the surface during combustion and set up a kind of organic firewall. But in order to prove its effectiveness, he needed to incorporate the additive into a new type of polycarbonate plastic.

Since NEC was not a plastics manufacturer, there was no business unit to help sponsor a project. Therefore, Iji couldn't even be certain his superiors would be willing to test the idea. But when he approached Kouichi Yoshimi, then head of the environmental studies lab, the young researcher got the green light to continue his effort with central funds. The project got going in mid-1996. By March of 1997, Iji's six-person team had succeeded in creating a silicone-polycarbonate that seemed to fit the bill. They announced their results to the press and the Japan Chemical Society, publishing some details in Japanese academic journals. This led Sumitomo Dow Limited, one of Japan's largest plastics manufacturers, to contact the group about a collaboration to commercialize the new material. NEC jumped at the opportunity. Since Iji's small team had limited resources, having Sumitomo Dow's technical expertise was critical in readying the process for production. The plastics giant was also a perfect marketing and distribution partner.

Only a year later, NuCycle was pronounced market-ready. Tests showed it offered the same flame and heat resistance as polycarbonates containing bromine. It also matched the flame-retarding capabilities of those containing phosphorus, while outperforming them in heat resistance. It produced no toxic substances, possessed at least double the impact strength of both its main rivals, and kept its strength and flame retardant properties after recycling. If that weren't enough, NuCycle plastics could be produced at the same cost as polycarbonate resins containing bromine or phosphorus.

In the summer of 1998, NEC began using NuCycle in the casings for its Thin Film Transistor, or TFT, computer monitors, and some laptop batteries. Later that year, Sumitomo Dow started marketing the material worldwide. Despite its price parity with polycarbonates doped with flame retardant additives, however, NuCycle still cost more than the "ordinary" plastics used in low-end personal computers. With PC prices tumbling—putting pressure on all components—it proved unfeasible to incorporate the novel offering in low-end casings. That's why the initial introduction was confined to high-end products like the TFT, which already employed flame-retardant plastics.

Iji hoped production prices would drop as demand increased, opening new markets for NuCycle as a casing material. In the meantime, a variety of international chemical concerns were studying its potential in plastic building materials and other areas. On a related front, the NEC team was hard at work on an environmentally friendly replacement for the epoxy resins used as insulating materials in a variety of electronic parts. While effective insulators, these compounds are flammable and so require the addition of flame retardants such as halogen. NuCycle's limited heat resistance and insulating properties rendered it insufficient for such applications. But at a technical conference in February 1999, Iji announced the creation of an epoxy resin with inherent fire-retardant properties, obviating the need for toxic additives. To develop Iji's resin for commercial applications, NEC entered into a collaboration with Sumitomo Bakelite, Ltd., a leading Japanese producer of insulating materials.

BASIC RESEARCH: TSUKUBA

IN A DIZZYING business climate where relevance, profits, and growth are the watchwords of industrial research, it's hard enough for any company to justify even a small basic research program. When NEC opened fundamental research centers in Tsukuba and Princeton in the late 1980s, it joined IBM as the only company to build labs that conduct basic studies on two continents. The two NEC labs probe the frontiers of quantum computing, race to fashion the world's smallest transistors, and explore storing computer data on holograms—while also investigating the fundamentals of more immediately practical problems in semiconductor lithography and computer security.

On the home front, basic science is conducted in Tsukuba, a city carved out of flatlands some 60 kilometers northeast of Tokyo in an effort to create a scientific and technological paradise. "Science City" has evolved into a high-tech Brasilia, isolated and dreary despite a slew of impressive R&D facilities. But the NEC lab, inaugurated just after Uenohara's 1989 retirement, continues to thrive.

The facility consists of two main buildings—a three-story administrative side with conference rooms, cafeteria, and library, and a connected five-story edifice where most of the research takes place. In addition to its Fundamental Research Laboratories, Tsukuba contains branches of the Silicon System Research Laboratories and the Optoelectronics & High Frequency Device Research Labs based in Sagamihara and Kawasaki, respectively. For the most part, these more applied organizations occupy the north wing of the five-story research building, while the fundamental group sticks to the south side. All told, close to three hundred people work in the building; just under a third belong to the Fundamental Lab.

General manager Kohroh Kobayashi—no relation to former company chairman Koji Kobayashi—has directed the Fundamental Lab since July 1997. In describing his job, he points to a balancing act of his own. "The Fundamental Research Lab is still inside an industrial organization," he relates. "So we should think more or less about . . . how to make our results practical." At the same time, the lab is chartered to develop a basic scientific understanding of biological and physical processes that might not bear fruit for decades. That means he must resist pushing too hard to find near-term applications for its work. "For near future, many other laboratories are working very hard," Kobayashi explains. "I'm afraid if we come to think more about the business aspect, maybe the topic targets will become smaller."

That doesn't mean thinking big won't produce smaller, more immediate payoffs. The lab's early 1990s examination of the free dynamics of molten silicon and other fundamental processes led to an important technique for improving the uniformity of semiconductor crystals that early in the twenty-first century will become essential to producing 1-gigabit memory chips. But the seven research groups—each consisting of ten to twenty scientists and support personnel—aim to create knowledge, not products.

One major area of interest is nanometer technology, or nano-tech. Here, from a first-floor clean room equipped with a scanning tunneling microscope and e-beam lithography equipment, researchers strive to keep pace with the worldwide push to put ever more transistors onto ever smaller devices. In September 1997, scientists announced they had fabricated the world's smallest operational metal oxide semiconductor transistor. With its key operational unit—or gate—just 14 nanometers long, the device marked an early step in the development of memory devices capable of holding 10 trillion bits of data, some hundred thousand times the capacity of the 64-megabit chips coming into production in the late 1990s. Early in 1999, Kobayashi's researchers were set to announce the gate size had been cut to 8 nanometers. "At the moment, we are top—smallest," he related.

Almost hand-in-hand with the nanotechnology work goes "advanced characterization," where scientists probe atomic or electronic structures. Armed with a powerful NEC SX-4 super-parallel computer, for instance, the lab's strong computational physics team tries to model such mysteries as what happens when light hits a single hydrogen atom adhered to a silicon surface. Yet another important effort covers "new information," where the electrons harboring the ones and zeroes of digital information are replaced by "qubits"—binary entities such as nuclear spins, or atoms jumping between two energy states, that enable computers to operate at lightning speed and on an extremely small scale.

"New information" also extends to bioelectronics—studying biological processes in order to apply them to computing. Here, researchers seek to develop algorithms for machine vision—and collaborate with food and drug companies to model how proteins work. New information is also the home of biophysicist Johji Miwa, who has spent a quarter-century studying a microscopic worm called *C. elegans*. By 1999, scientists will have sequenced this creature's entire genome—and Miwa foresees a potential revolution in biology. Until the mid-1990s, he explains, studies concentrated on *C. elegans*'s simple nervous system—in the hopes analogies would be found for wiring electronics. "I think that's still right," he says. But the outpouring of genetic information is adding a second dimension to the worm's possible payoffs. For instance, researchers have long studied how chemicals like dioxin affect living organisms by examining its action on a few genes. Once scientists lay out an entire genome, Miwa says, "We can analyze the whole network and pattern of effect, and from that can theorize how the network functions and how genes interact with each other. . . . You just open up a completely new world."

While many of the fundamental group's projects have similar power to fire the imagination, none has put the lab on the map like the work of Sumio Iijima. The NEC research Fellow employed a beguiling combination of insight, serendipity, and dogged determination to discover needle-like sphericals of graphite called carbon nanotubes. These strange shards of pencil-stuff—barely 10 nanometers long and 2 wide—turned out to have a host of intriguing structural and electrical properties, kicking off wide-ranging scientific investigations that have led researchers to theorize they may pave the way to such wonders as microscale transistors, a viable fuel storage medium for hydrogen-powered cars, and super-long-life laptop computer batteries. Indeed, by 1997 Iijima's landmark 1991 paper in *Nature* announcing his discovery had received 790 citations. According to the Philadelphia-based ScienceWatch, which conducted the survey, that established it as the world's fifteenth most-cited paper in the physical sciences for the years examined. All nine of the lab's highly cited core papers sprang from Iijima's group.

Friendly and down-to-earth, Iijima grew up in a Tokyo suburb, played flute in his college orchestra, and by his own admission "didn't study hard" until he arrived at Tohoku University to pursue his physics doctorate. He completed that work in 1970, came to the United States as a postdoctoral fellow and ended up staying a dozen years at Arizona State University, where he took up microscopy and began to study materials at the atomic level. In 1982, he returned to Japan and spent five years in a government lab before joining NEC. Initially assigned to another R&D arm, Iijima brought his keen scientific acumen—tempered by an American-style casualness—to the Tsukuba lab in 1989 as a charter employee.

Iijima works out of a cavernous first-floor room that holds his electron microscope—a monstrosity of a machine with weird metallic arms wrapped in glittery aluminum foil. He got onto nanotubes while rushing to study fullerenes—or buckyballs—so named because their shape mimics naturally the soccer ball-like geodesic spheres that R. Buckminster Fuller showed offer maximum structural stability. The great promise—still unrealized—of fullerenes is that they'll one day lead to a new generation of lightweight, durable materials. Iijima and others think nanotubes just might go them one better—as he says, "much, much better."

Carbon comes in three main forms: diamonds, charcoal, and

graphite. Like fullerenes, nanotubes are graphite-based. Iijima's initial samples spanned several nanometers in width and several hundred nanometers in length. At such relatively large dimensions, their properties resemble those of ordinary graphite crystals. But in smaller configurations it turns out that unlike fullerenes, nanotubes possess characteristics resembling both a molecule and a solid—and with the right tweaking can act as either semiconductors or metals. This startling versatility has caused the field to explode in recent years, with Harvard, IBM, and a host of leading institutions launching nanotube studies.

Behind this flurry of activity, hard-wired reality is catching up with theory. In 1998 a group at Holland's Delft University built a working transistor from a single carbon nanotube—an important step toward nanoprocessors. In a different configuration, nanotubes also offer great promise for hydrogen storage. Scientists have long looked to nature's most abundant element to power clean-burning cars, but they've been unable to find a safe and efficient container. A metal such as titanium can contain hydrogen but weighs too much to be practical given the volumes that must be stored. However, nanotube-like materials absorb hydrogen so efficiently that they offer the potential to compactly store enough fuel for a 5,000-kilometer jaunt—say from Los Angeles to New York City, or up and down Japan several times—without replenishing. "The people in the automotive companies love it," laughs Iijima. "All of them are studying this now."

Iijima hesitates to reveal NEC's exact plans for nanotubes. But one big possibility is as a field emission source for electrons—meaning nanotubes might be used to provide electrons without the need for bulky, power-sucking heated filaments. That could lead to super-thin TV screens—or pave the way for a new generation of lightweight, long-lived lithium batteries.

Any such payoffs may be decades away. But Iijima isn't worried. "Maybe it will take a while," he says. "But we still have dream."

BASIC RESEARCH: PRINCETON

NESTLED BESIDE its pleasant pond and surrounded by woods, Mickey Uenohara's attempt to create a small facility with the creative flavor he found at Bell Labs continues to offer a near-idyllic setting for a basic research laboratory. As the trailblazing facility enters its second decade, however, the blush is oh-so-slightly off the rose. Computer scientist C. William "Bill" Gear, who took over in 1992 after the death of founding president Khang, notes that in the institute's heady youth NEC cared little about what was produced. Not only was the company minting money, the Princeton lab provided a public relations gold mine. By the late 1990s, though, a subtle feeling of change hung in the air. "Although they don't say anything to us, I'm listening very hard to what's between the lines," Gear reports. What he's picked up has spurred the director to move some technologies out the door—an action that hasn't come easily: witness the painful 1998 spinoff of digital watermarking technology (see box, p. 234).

Even with a slightly more market-oriented focus, however, the institute remains a special place. With an annual budget of around $22 million, the lab has grown to some eighty researchers and thirty support personnel—triple its 1990 size. What's more, it's *people* who are funded, not projects. Each researcher gets a set amount to spend as he or she will, with additional expenditures subject to peer review. With no grants to write, and with senior scientists spending time in the lab instead of supervising students, there still seems to be nothing like the institute—at least outside of Microsoft. "This is paradise," proclaims Fellow Boris Altshuler, a mathematical theoretician from St. Petersburg, Russia, concerned with quantum chaos.

Touring the lab, in fact, is a bit like being a kid in a candy store— only one selling future candies. In addition to Altshuler's largely theoretical work, the smorgasbord of projects runs from hard-wired physics to creating novel software algorithms. Here, researchers seek to co-opt DNA molecules for computing, examine fly vision in hopes of unearthing clues to processing external stimuli, or study chess to try to divine new programming tricks and explain consciousness. Two wildly different approaches to storing computer data—one hands-on physics, the other mainly a software vision almost out of a science fiction novel—illustrate somewhat the range of investigations at hand.

On the hardware end of these two extremes comes the work of Richard A. Linke. The senior research scientist's lab is a mass of optics circuitry and other equipment. When he flips a switch, the green or blue hue of argon laser light dashes across a small countertop, bounces off mirrors, and splits into two beams that end at right angles to each other in a crystal cube smaller than a die.

It's not the laser, but the cube that fascinates Linke. The crystal is made of cadmium fluoride, on its surface a standard optical material. But it's what lies beneath the surface that counts—for with the right tweaking the little crystal exhibits unique properties that might hold the key to a revolutionary data storage medium, one that preserves the "on" and "off" signals of digital bits not in electrons but as aspects of light: in the form of holograms. If everything pans out, one cube about an inch on each side could store 30 terabytes of information—a library's worth of data. "I see it as a replacement for hard disk drives, but with maybe a thousand times more capacity," Linke notes. "I think that people are going to want something like that kind of storage in every PC."

Linke joined the institute at its 1989 inception after spending seventeen years at Bell Labs, where he explored radio astronomy with future research director Arno Penzias—and then moved into optical communications. He left the hallowed lab largely because he had risen into management and saw the NEC institute as a way to get back to full-time research. At Bell Labs, Linke explains, senior people who stayed in research were often viewed as failures. "What I liked about this place, and it continues to be true, is that you can be a success and retire as a researcher."

The erstwhile Bell Labs physicist also wanted to apply his optics expertise to computing—never exactly an AT&T strongpoint—and was excited about working alongside the computer gurus being recruited to the upstart lab. In Princeton, he dived into the emerging field of optical computing. But within a few years, for a variety of technical reasons, the once white-hot arena was in a state of collapse. It was just about then—sometime in 1991—that institute solid-state physicist Jim Chadi approached Linke with an important theoretical explanation for a poorly understood semiconductor defect known as the DX center, setting Linke on the road to holographic storage.

The defect had been discovered several decades earlier but remained somewhat esoteric. Its chief physical manifestation was a phenomenon called persistent photoconductivity: PPC. This referred to the fact that if a semiconductor carrying the defect was cooled to a low enough temperature, it became an electrical insulator. However, the sample would turn into a conductor if a light was then shone on it—and remain in this conductive state long after the light was extinguished.

It turned out that Chadi had shown the defect to be the result of the "dopant" atoms routinely added to semiconductors to enhance their electrical properties. He reasoned that these atoms could exist in either of two states. One—the DX state—was basically an electron trap that sprang into being to prevent conduction when the material was cooled. However, he deduced, when an outside stimulus like light was applied, the dopant atoms switched to the donor state, releasing their electrons so that the material again became a conductor.

Chadi and Linke held many discussions about how to capitalize on the phenomenon. Still, it wasn't until 1993 that Linke realized the change in conductivity probably also involved changes in the material's optical properties—specifically, its index of refraction. If true, he saw, a

---

□

Watermarking: How (Not?) to Launch a Spinoff

Ever since it flew against the grain of corporate cutbacks in science and opened its basic research institute in Princeton, NEC has drawn raves for its visionary attitude toward research. Privately, however, competitors noted the institute's Ivory Tower approach, its isolation from development and manufacturing, and the vast geographic and cultural separation from company headquarters in Japan—questioning whether the scientific outpost could ever produce anything meaningful to its mother corporation.

On the verge of the twenty-first century, some of those questions were being answered. Although research findings had long been shared with Japan, the institute had sought its first outright commercial transfer by creating a subsidiary company to develop its watermarking technology for digital copyright protection. Funded by NEC's own venture capital arm, the 1998 spinoff—Signafy Inc.—apparently marked the first such enterprise launched by a major Japanese corporation in the United States. But while optimistic about Signafy's chances, its Princeton lab architects acknowledge suffering through a painful process that highlights important lessons in managing research.

A big problem in today's increasingly digital times is that perfect copies of photos or movies can be easily distributed without anyone knowing whether a person has a legal right to use the images. Watermarking involves hiding data—in this case invisibly—within the image in a way that is easily scanned for copyright protection but makes it hard to remove without degrading the picture. The NEC institute got into the field in a big way late in 1995, after three researchers co-authored a paper on the subject with a Massachusetts Institute of Technology professor. The concept quickly mushroomed into the institute's first attempt to construct a prototype; and as the rapidly growing team—including several university collaborators—moved toward that goal, members began investigating commercial opportunities. That's when the headaches began.

Since watermarking did not fit well into NEC's business, and direct commercial work violated the Princeton lab's charter, the initial challenge hinged simply on getting the company to agree to launch a spinoff. James Philbin, the institute computer scientist tapped to head the effort until a chief executive could be hired, found the idea ran counter to the standard Japanese business model. Top people understood, he notes, but the middle managers who execute decisions proved another story. Beginning in January 1997 and continuing for several months, he traveled regularly to Japan to explain the rationale.

---

precisely positioned laser beam could bring about the DX shift—and with it a decrease in the index of refraction—in an area as small as a cubic wavelength, about half a micron in every dimension. The refraction index in that tiny spot would stay low for some time after the laser

The in-person meetings, supplemented by informal discussions over dinner and drinks, proved critical. "Everything was theoretical until then," notes Philbin. "Almost everything I closed was at a face-to-face meeting."

Amidst this process, three team members left the lab, largely over the institute's choice of projects to be pursued and the decision to place the endeavor under a single head like Philbin, with no watermarking experience. One of those leaving was Robert E. Tarjan, a winner of the prestigious Turing Award in computer science. In his resignation speech, Tarjan referred to "the litany of ugly episodes" the watermarking project had spurred and railed against an institute he felt had become corrupted by NEC's corporate mentality.

"I see an Institute that has lost its way, an Institute whose original vision, the pursuit of excellence, has become lost in a sea of mediocrity, an Institute where the appearance of excellence, rather than excellence itself is rewarded," he said. "I see an Institute weighed down under a bureaucracy that becomes more stifling every day, an Institute where the objective of the administration is to prevent failure or embarrassment at any cost rather than to take risks, to innovate, to be flexible."

Tarjan's departure stung the lab. Philbin says the incident illustrates the need to set expectations early on. Had the institute clearly spelled out such issues as leadership, the company's stake in the venture, and its researchers' likely equity, he feels the worst might have been avoided.

The next stages moved forward more smoothly—but the headaches persisted. To fund the enterprise, the institute hooked up with NEC's recently started venture capital arm, based in Cupertino, California; and Signafy was founded on May 2, 1997. That October, the company announced its new chief executive would be Frank Richardson, who had previously headed Internet software start-up Firefox, Inc. But within about eighteen months, Richardson was gone and plans to base the company in Silicon Valley were shelved; instead, Signafy found space inside the Princeton institute building.

Watermarking promises to be a highly competitive field. But Signafy got a big boost in July 1998, when NEC entered into an agreement with IBM to develop a watermarking standard for digital versatile disks. The basic concept is to license technology that will enable DVD players to scan for watermarks and refuse to run bootleg copies—with Signafy handling the main NEC end of the deal. Even if things go awry, however, the endeavor heralds a new era for the Princeton research haven, which is already planning other expeditions beyond its Ivory Tower.

□

was turned off—while the half-micron next door would be left in its normal high state.

The existence of two optical states—one highly refractive, the other not—began to sound to Linke like the basis for a storage medium. After all, typical disk drives rely on two magnetic states, or orientations, to represent the ones or zeros of binary data. The big difference with the DX center was that these bits theoretically could be written, stored, and read with lasers by selectively changing or detecting the refractive index associated with dopant atoms. More important, since photons don't interact with each other, data could be distributed throughout the crystal—increasing geometrically the volume of information a semiconductor could hold. As Linke puts it, "Other storage media are all planar. Everything has to be written on the surface—so you waste a lot of volume." If that weren't incentive enough, he adds, putting so much information in a small space makes for extremely rapid access.

In October 1993, Linke conducted an experiment that proved his hunches were right on target. Encouraged, he began studying optical storage more intensely. It was not a new concept. But apparently no one had explored a DX-based approach. Instead, most existing studies focused on so-called photorefractive materials like lithium niobate. While offering some promise, however, these compounds were plagued by shortcomings Linke felt limited their capabilities.

Linke, though, had his own materials problem. In order to exhibit the DX effect, his samples had to be super-cooled to around 50 degrees Kelvin. Since that represented an extremely expensive undertaking, the NEC team went hunting for alternatives. Members explored a variety of materials—with limited success—until around 1996 Chadi met up with Alex Ryskin, a solid-state physicist at the Vasilov State Optical Institute in St. Petersburg. Ryskin had spent his professional life studying cadmium fluoride. He told Chadi that when doped with indium or gallium it exhibited a DX center that was stable at around 180 K. While still a far cry from room temperature at 290 K, that represented the breakthrough Linke needed—sparking a fruitful collaboration with the Russian scientist and paving the way for practical experiments later that year.

Under Linke's scheme, laser light is split into two paths: the signal beam carrying in data and the reference beam needed to "write" the information into the crystal. Before entering the crystal, the signal beam passes through a small glass mask—much like those used for photolithographic etching of integrated circuits—that holds a mock-up of the data to be stored. This mask is laced with intricate patterns of metal. Therefore, while the laser beam passes unencumbered through the clear portions of the glass, it is blocked by those containing metal. The result is a complicated tapestry of light and dark—ones and zeroes—that rides the signal beam into the crystal.

At this point, classic holography takes over. The interference between the signal and reference beams forms a stationary pattern of bright and dark regions throughout the crystal's volume. Since the refractive index is lowered only in regions struck by the light—and stays low even after the signal beam is turned off—what remains is essentially a recording of the mask data. Moreover, if the reference beam later enters the crystal in its original form—at the same wavelength, angle, and polarization—it will interact with the refractive index patterns written into the crystal and recreate the signal beam, allowing the stored data to be "read."

Data from any given mask—or page of information—is distributed all through the crystal: hence, the term "hologram." The number of independent holograms that can be written determines the crystal's capacity. In theory, a fresh page can be written by rotating either the crystal or the laser beam as little as 50 arc seconds. Linke says as many as ten thousand holograms—each carrying millions of bits—can be stored in a 7-millimeter cube of cadmium fluoride.

By the first months of 1999, Linke and colleague Ian R. Redmond had managed to store 11 million bits—around 1.4 megabytes of data—on a single holographic page. More important, they had been able to implant a handful of these holograms onto the same crystal, and were on their way toward raising that figure by the hundreds or even thousands. Even the higher figure would only bring the system up to around 1 percent of a crystal's theoretical 30-terabyte capacity. However, Linke notes, "That we would consider a real success, because in this business nobody has gotten anywhere near that."

Even if Linke and Redmond can pack a *lot* more holograms into a single crystal, some formidable obstacles remain to be overcome before holographic storage is ready for commercialization. One involves "decay time"—how long it takes the excited DX center to decay back to its ground state, wiping out data. At around 180 K, the phenomenon persists for about a year. But at room temperature, things go haywire in 10 seconds. Another problem is that there is no way to selectively erase data. The entire slate must be wiped clean in order to eliminate even one bit.

Undeterred, Linke explains that even with its short decay time, holographic storage could be used at room temperature if the data were automatically read out and rewritten every five seconds or so. In fact, he and colleague Warren Smith have patented a "dynamic refresh" technology for doing just that. "In that way," Linke asserts, "we can have essentially an infinite lifetime with the present material." What's more, he notes, such a refresh scheme also solves the problem of erasing data—simply because users can choose not to rewrite any portions they don't want.

While wrestling with such problems, Linke's team is also exploring applications for DX materials beyond data storage. One promising idea lies in wavelength division multiplexing—enabling lasers to operate reliably at closely bunched frequencies in order to pack more communications channels onto a given fiber optic cable. Hirohito Yamada of NEC's Kansai Electronics Lab spent a year at the Princeton institute developing ways to exploit the DX center to achieve the necessary reliability. And although the company wouldn't confirm anything, a new laser based on holography might be forthcoming.

Still, Linke favors storage as the ultimate payoff of his studies. Given the problem of decay time, and the somewhat unreliable nature of a dynamic refresh system, he says likely uses might be confined initially to downloading movie rentals or temporarily capturing images from security cameras—both areas where data is not meant to be stored forever, anyway. It might also work for normal personal computer use—as long as a cheap and reliable backup system were available to guard against power failures. However, he admits, "It's not the system you would use for archival storage."

But that's okay. It turns out that some other institute researchers have some thoughts about that.

The story Andrew Goldberg and Peter N. Yianilos tell dates back to August 1799. While strengthening a fort near the Egyptian city of Rosetta, a French engineering officer chanced upon an ancient basalt slab. Carved into the stone—in Egyptian, Greek, demotic, and hieroglyphic—was a priestly decree dating back to around 196 B.C., in the reign of Ptolemy V. This Rosetta Stone proved remarkable not for its content, but because its multilingual message, which had been ordered distributed to temples around Egypt, provided the key to deciphering hieroglyphics dating back thousands of years.

Nearly two centuries later, the historic message, recorded in a durable medium and dispersed to multiple locations, has inspired the two NEC computer scientists to attempt a similar preservation of modern society's mushrooming body of electronic data by scattering bits so widely they're virtually invulnerable to natural or man-made disaster. In their view, essential corporate documents, family trees, or even people's ideas can be protected for eternity against earthquakes, war, pestilence, sabotage—even spilled coffee.

The concept is called the archival intermemory. And while physicist Linke focuses on future data storage hardware, his colleagues are all about software. Their central idea is to ask computer users around the globe to designate a portion of their hard drives as a safe haven for a sliver of the world's knowledge. In return, contributors will be allowed to deposit some of their own data into the intermemory. Hard disks, floppies, and other backup strategies, the NEC pair argue, are all vulnerable to local disasters like theft and fire—let alone cataclysmic events. But the intermemory should outlast almost anything short of global annihilation. "Your ideas," says Goldberg, "will be there in a thousand years, even if there's earthquakes or wars."

Say a person had a 10-gigabyte hard drive. A fraction—maybe 500 megabytes—would be donated to the intermemory for a set period of time, perhaps two years. The person would then leave the drive on-line, configured in such a way that subscribers to the intermemory could access that portion—and only that portion. In exchange, the donor would get the eternal right to deposit a smaller amount of data (maybe 50 megabytes) into some other corner of the intermemory. If he chose to close down access to his own hard drive when the subscription period ends, the data he deposited would live on.

Goldberg and Yianilos contend that as long as the number of subscribers—or the storage space donated—grows faster than the dropout ranks, it should work. Even if something went wrong for a time, new memory allocations could be suitably regulated. Moreover, what works for a public intermemory should also work for smaller-scale *intramemories* set up inside multinational corporations seeking to protect vital records—or among a consortium of libraries out to safeguard electronic collections. To lose the data, the NEC researchers argue, hundreds of offices or libraries would have to be compromised. Yet the documents don't have to be stored hundreds of times.

As for the data, various algorithms and cryptographic tricks come into play. Essentially, though, the mathematically transformed files would be broken into many smaller parts—32, 1,024, or even more—with each piece assigned to a different system node. The way the math shakes out, the intermemory could typically reassemble the complete document with the data from any half of the nodes. For example, if a file had been dispersed to 1,024 hard drives, it would remain safe as long as at least 512 stayed healthy. "Even if a nation for political reasons decides to withdraw from the Internet, that would not be enough to destroy the memory," says Yianilos.

In late 1998, the researchers tested a prototype at the NEC institute that simulated an intermemory with thirty users. "We don't feel it's in any way ready for prime time, but it works," sums up Yianilos. For 1999, the biggest challenges were to develop the cryptographic and security protocols needed to make the system robust against hackers—and to continue drumming up interest in the computing community.

Even while wrestling with such practicalities, Yianilos notes that whenever he talks about the intermemory, folks bombard him with queries. While he's not even certain how NEC might profit from the venture, that gut-level interest makes him optimistic about its prospects. "When a line of research starts generating really interesting questions, you know you're onto something," he says. Besides, he adds, "It's tantalizing. It makes you smile."

■

For the corporation as a whole, there wasn't a lot to smile about closing in on the millennium. The first few months of 1999 looked grim. Margins were tight in its bedrock liquid crystal display and semiconductor memories businesses. Packard Bell NEC Inc., the world's fourth-largest personal computer maker, was hemorrhaging money—having lost nearly $1 billion over the previous two years. A late 1998 defense contract scandal had led to the resignation of Tadahiro Sekimoto, although the former chairman was not personally linked to the scandal.

Research represented a rare bright spot. Tatsuo Ishiguro still had his hands full trying to blend East and West, short-term research, and basic studies. He was putting it mildly when he said, "The how to balance is the tricky thing." At the same time, patents continued to explode out of the labs, and more than any other Japanese corporation NEC was tapping the skills and talents of people around the world—raising its odds of a brighter future.

Notably, while becoming increasingly aggressive in pushing innovations out the door, the company remained committed to nurturing the Tsukuba and Princeton labs—and letting these flowers of basic research have time to blossom. The closing words of Dawon Khang's inaugural speech still held true a decade later: "How well we succeed, time will tell . . . [But] our parent is patiently waiting."

CHAPTER EIGHT

THE PIONEERS: GENERAL ELECTRIC AND BELL LABS

I N LATE 1900, under Willis Whitney, General Electric created the first great central corporate research laboratory in the United States. AT&T organized a formal research department in 1912, then dramatically expanded the role of research with the 1925 creation of the Bell Telephone Laboratories. With the single exception of DuPont in the chemicals arena, these two enterprises dominated industrial research through the first half of the twentieth century. Not only did they define standards of engineering excellence, their halls became hallowed scientific breeding grounds, pushing the boundaries of solid-state physics, nuclear studies, and even astronomy. Two General Electric researchers have won Nobel Prizes, while Bell Labs boasts an unrivaled eleven Nobelists.

Through much of the early 1900s, the two labs competed fiercely in the development of the triode and other vacuum tubes—and especially in radio, with their patent wars proving instrumental in the creation and evolution of the Radio Corporation of America. Since World War II, however, the two have increasingly gone their separate ways. Bell Labs, now under the auspices of Lucent Technologies, specializes in communications systems and technology, while GE has become a financial services and medical imaging giant that also owns NBC.

Despite their divergent paths, however, these two pioneers of industrial research share similar legacies: their establishment early in the century, their embrace of science, and the subsequent need to reinvent themselves. Indeed, both have been reborn in recent times, emerging as guiding lights in a new era of corporate research.

GENERAL ELECTRIC—FROM HOUSE OF MAGIC TO HOUSE OF HARD WORK

"The truth be known, not much has changed in twenty-five years, but everything has changed in twenty-five years."
—WALTER BERNINGER, TECHNICAL DIRECTOR, GE RESEARCH AND DEVELOPMENT CENTER

"My boss, I talk to him five or six times a year. But the guy in the business unit, I talk to him every day."
—TOM ANTHONY, GE'S LEADING INVENTOR

GE Research and Development Center

Central research established: 1900
Company sales 1998: $100.4 billion
1998 company R&D expenditures: $1.9 billion
1998 R&D Center budget: $190 million (est.)
Funding paradigm: 52% business unit contracts, 30% corporate funds, 18% external contracts (government and strategic customers, chiefly Lockheed Martin)
Research technical and scientific staff: 1,130
Principal research labs: Schenectady, New York

ONE OF THOMAS EDISON'S desks guards the entrance to the General Electric Research and Development Center. Across the wide lobby, a huge glass wall provides a striking view of the meandering Mohawk River. For nearly a half-century, this has been the home of GE's pioneering research organization—a scenic hilltop on The Knolls, a 525-acre estate once owned by a star salesman of Dr. Williams' Pink Pills for Pale People. It is a sprawling modern facility with two million square feet of floor space spread over perhaps a dozen buildings—and a far cry from a carriage barn on the Erie Canal. Still, the atmosphere harkens back to more leisurely days. The main building's long hallways are covered with linoleum. Outside each laboratory door hang metal "in" and "out" trays, like those in a hospital. The effect is a mood of quiet investigation and discovery.

The House of Magic, though, is no more. The days of entertainment and taking the show on the road are long gone. Despite its two Nobel Prizes and a legacy of innovations in lighting, radio, X-ray equipment, and synthetic diamonds, GE research has all but fallen off the map of the world's premier labs—at least in the public perception. There is no attempt to blaze scientific trails. The company no longer ranks in the top ten U.S. patent winners. The lab is not part of a worldwide central research organization: there's not even a U.S. satellite. Meanwhile, its 1,130 technical and scientific workers total 500 fewer than those in rival Siemens's central R&D organization. And the lab's relatively small size, coupled with the very diversity of GE's operations—covering everything from Mickey Mouse lights to jet engines, credit cards, and nuclear reactors—means Schenectady researchers are spread far more thinly than counterparts at more tightly focused organizations such as Intel, Microsoft, or even NEC and IBM.

Yet while it may lack outward pizzazz, the pioneering lab has quietly reasserted its standing as a highly effective central research facility that can indeed claim a place among the world's best. It's unique among central labs in taking a major role in its parent's quality initiative. And through hard work and a dramatic reinvention accomplished without the acute pain and budget-slashing that accompanied more tumultuous turnarounds at Bell Labs and IBM, it is delivering value to every aspect of GE's far-flung businesses—churning out better industrial diamonds and other superhard materials, computed tomography scanners that cut imaging times from minutes to seconds, remote monitoring systems that track aircraft engines in flight, and even more efficient ways to handle financial transactions. While the pressures of these endeavors have caused it to reduce exploratory projects and narrow time horizons, through its novel Game Changers program General Electric's vintage lab has even managed to find a way to back risky projects that may take a decade to pay off.

"I would say I've gone from science to more engineering and development," admits researcher Tom Anthony, who with more than 160 inventions may be the United States' leading active inventor (see box, p. 244). But that's what it takes to compete today in GE's universe. The corporate world doesn't look to labs for science, it wants ways to grow the bottom line—both in the near term and for the further-off future. Asserts Anthony, "I think we've done the best job of all the labs I know."

Despite its storied past, GE research has only survived into its sec-

---
□

Tom Anthony and Harvey Cline: Chasing Steinmetz

Located at the end of one of the R&D Center's three long wings, the adjoining ground-floor labs occupied by Tom Anthony and Harvey Cline don't look much out of the ordinary. Anthony's side is filled with the stuff of the physical chemist: lasers, a chemical vapor deposition apparatus, annealing furnaces, spectrometers, computers. Cline's area is packed with an oscilloscope, voltmeters, various tools, bookcases, and more computers. But while similar equipment might be found in a host of industrial laboratories, what comes out of them isn't. For more than two decades, the two colleagues have been turning out patents at a rate that may be unequaled anywhere in the world.

GE has always been an inventor's company. It was formed around corporations created by all-time U.S. patent leader Thomas Edison (1,093) and third-place Elihu Thomson (696). Its ranks have also included radio and TV pioneer Ernst F. W. Alexanderson, who garnered 320 patents as a GE employee—and the legendary Charles Steinmetz, who amassed 201 patents during his company career.

In the modern era, though, none of its slew of talented staff members have shown the inventive prowess of Anthony and Cline. On the eve of the research center's hundredth anniversary, the little-known GE scientists and Coolidge Fellows—the company's highest technical honor—were both closing in on lab founding father Steinmetz. Front-runner Anthony, fifty-eight, had amassed 161 patents, with six more pending. Cline, also fifty-eight years old, was right behind—with 155 patents to his name and five more in the pipeline. Not only that, both had apparently steamed silently onto the list of the top twenty inventors in American history—quite possibly ranking as the nation's leading active patent holders.

Anthony, who joined GE research in 1967 after earning his doctorate in applied physics at Harvard University, is an unassuming blue-jeans-and-tennis-shoes type who rides his bike four miles to work each day and clears his mind with beekeeping. When asked about the secret of his success, he quips: "Good lawyers." But his father directed research at the Mellon Institute, before it became part of Carnegie Mellon University—and Anthony confesses to an inherent passion for solving problems. "When I get onto a technical problem, I can't get rid of it," he notes. "I keep thinking about it day and night. It never goes away until I solve it."

Cline came to Schenectady two years before his colleague, fresh from com-

ond century by periodically reinventing itself to serve the evolving needs of its parent. The pioneering era made famous by Whitney, Coolidge, and Langmuir encompassed an ever-shifting mixture of basic and applied research. The lab became far more heavily applied during the Depression and throughout World War II. Then, in the postwar years, it swung toward basic research.

pleting his doctorate in metallurgy at M.I.T. A snow skier and general out-door lover, he, too, downplays his inventive skills, noting: "I have been fortu-nate in being able to work on several new technical areas ripe for invention."

Both men took a while to find their inventive feet. Anthony's first patent— "Semiconductor Device Production"—didn't come until 1975, some eight years after he joined the lab. Then the flood gates opened. For the next twenty-three years, he plowed away at the rate of nearly seven patents annually. Cline took four years to break through the patent barrier. His initial patent—and the next fifteen—came in the field of exotic alloys designed to withstand the ex-tremely high temperatures inside gas turbine engines. Since then, he has clipped along at the rate of just over five patents annually.

In 1970, Anthony and Cline began a long collaboration that produced fifty-plus patents relating to their invention of the semiconductor fabrication tech-nique of thermomigration, which enables the ultrafast production of certain types of microchips. Since the early 1980s, however, the two have followed di-vergent paths. Anthony went on to tackle laser processing of materials and dia-mond synthesis: his latest work is for GE's Ohio-based abrasives business, a leading manufacturer of industrial diamonds and other superhard materials. Cline also delved into lasers. But since 1985, he has concentrated on medical imaging, where he holds nearly forty patents. Among his important recent inventions—in collaboration with GE graphics engineer William E. Lorensen—was the so-called "marching cubes" algorithm, which enables computers to rapidly transform raw computed tomography and MRI data into accurate 3-D images. The algorithm was recognized as one of the fifty most important contributions to computer graphics.

As the leader in GE's undeclared patent race, Anthony is on track to pass Steinmetz in 2005, the year he turns sixty-five. If he's anxious to catch the leg-endary inventor, it doesn't show. In fact, he talks most readily about simply doing what he loves, comparing his passion for invention to the joys of bee-keeping. "Both beekeeping and inventing require pleasurable periods of in-tense concentration that cause you to temporarily forget everything else," he notes. "In the case of beekeeping, you open the hive partly out of curiosity and partly to examine and improve the situation in the hive. You proceed quickly, carefully, and economically so that you do not disturb the queen bee and you do not disrupt honey production and do not get stung by the workers. Invent-ing . . . is not that dissimilar."

□

After twenty years, in 1965, the enterprise merged with the company's Advanced Technology Lab to form the GE Research and De-velopment Center. More familiarly known as CRD, for Corporate R&D, the very name signaled a return to applied activities—and the message to Guy Suits's successor, Art Bueche, had been to get things better con-nected to the businesses. Afterward, under both Bueche and Roland

Schmitt, the lab had continued to pursue exploratory research—but increasingly that work was motivated less by scientific curiosity than by commercial interests. As the publication rate went down in physics journals and up in engineering venues, recalls Schmitt, who directed the lab from 1978 to 1986, "We received a lot of criticism from academic circles, and, quite frankly, from other corporate labs, that we were drifting away from basic research. . . . Our own attitude was that we were learning to be more effective as a relevant research center for GE—maybe 'relearning' is a better word, considering that we were beginning to do again what Langmuir, Coolidge, and colleagues did early in the century."

This modern era of research picked up steam when Walter Robb took over in 1986. Robb was the former bench scientist who scaled back unfocused longer-term research even more and upended the lab's funding structure to make it largely dependent on business unit contracts. A dynamic individual who just in advance of his seventieth birthday in early 1998 would climb Mount Kilimanjaro with his three sons, he had cultivated his research philosophy during fifteen years as an executive at GE's plastics and Medical Systems businesses.

In particular, Robb had been running the Wisconsin-based Medical Systems division in 1974 when competitor EMI began usurping GE's X-ray customers with its pioneering CT scanners. He turned immediately to his former Schenectady research colleagues for help—and the result was a novel fan-beam CT machine that dramatically reduced scan times over the competition. The project sped through the labs, into testing and then production in the unprecedented time of about two years—drawing $50 million in orders during its first year of commercialization. A year after that—just the fourth year from inception—it was number one in U.S. market share. In the early 1980s, when competitors began introducing magnetic resonance imaging scanners, GE pursued more advanced technology through another research project. Although it delayed GE's entry into the emerging business, that effort led to an advanced high-field MRI machine that catapulted the company to the market lead; GE remains number one in the field.

These two dramatic successes helped make Robb chief executive Jack Welch's choice to take over the research lab in September 1986 as then-director Roland Schmitt prepared for retirement. Robb assumed the CRD post on September 1, 1986. The next month, at a corporate officer's meeting, he got a rude shock. "I found out that our relationship between the lab and Medical Systems was unique," he relates. "No one else had coupled that closely. And in some cases the businesses were so unhappy with their relationship with the lab, which they were funding, that they wanted to shut the place down."

The depressing situation led Robb to invite business group vice presidents to the lab for a two-day bull session. As he recalls: "They all uniformly said, 'Hey, you got great scientists here, but there are too many hobby shops—and we've got to have more control over what they're working on. And I said, 'Yeah, but you can't control scientists.' And they said, 'Well, someplace in between.'"

So arose the revamping of the lab's funding structure. Until that point, the central lab had received about two-thirds of its funding through the proverbial "tax" on business units. Robb devised a new structure whereby three-quarters would come from contracts for specific jobs— primarily from the business divisions, but also from the government and strategic customers. To help land the contracts, he beefed up the liaison program established by Guy Suits. Although the system had been strengthened at least once since Suits's days, Robb says, these emissaries served primarily as tour directors showing off the lab to business group contacts. He made the liaison position a management responsibility, assigned at least one "ambassador" to each of GE's dozen or so business divisions, and told them to start bringing in deals.

The changes went into effect in mid-1987. Because the current year budgets had essentially been set already, the lab wasn't really on its own until 1988. Still, that summer marked a crisis of confidence for research. Notes Robb, "All the scientists thought I had just sold the lab down the river. . . . They all thought the lab would simply close down in '88 because we wouldn't get the orders." But Jack Welch also backed the changes and pushed his business executives to start funding projects. By late 1987, notes Robb, the orders started coming in "and we began to see we would be okay."

Not all the gears clicked immediately into place. But on the other hand, GE research managed its dramatic reorientation without the bloodletting that accompanied similar shifts at other great research houses. That was largely due to the fact that CRD had never really gotten disconnected from the business groups to the extent some other research organizations did. By the time Robb retired in January 1993, the lab had a host of big things—either on the market or well into the development pipeline—to show for its redirected efforts. Among its chief contributions: all-digital ultrasound, two major diamond-making advances, and the reinforced plastic fan blades for the superquiet and efficient GE90 jet engine that powers the Boeing 777 aircraft. "And so," relates Robb, "we destroyed the hobby shops."

Lonnie Edelheit is looking relaxed in a red polo shirt emblazoned with a Design for Six Sigma logo. "DFSS" (see box, p. 248) is a quality

---

Design for Six Sigma

One day in early 1996, employees strolling into GE's R&D center were amused to find Lonnie Edelheit and his top managers dropping paper helicopters in the big three-story atrium. The senior staffers had cut the choppers themselves, experimenting with paper weights, wing widths, and other variables—then timing each craft's descent. "He [Lonnie] looked like a little kid," one observer recalls.

It wasn't all about fun, however. The exercise is part of the Design of Experiments (DOE) training that's included in the company's Six Sigma quality program. The aim is to keep the helicopter aloft as long as possible. But instead of changing one variable at a time to identify the best specifications for each individual parameter, multiple factors are altered with every trial. Then, powerful statistical software crunches the results to determine the optimal combination—an outcome that might well be different when variables are considered en masse as opposed to separately. Furthermore, the program predicts the *distribution* of expected flight times. After all, one design might be perfect in ideal conditions—no wind, say. But an entirely different configuration might prove steadiest throughout all conditions, and ultimately be more reliable.

Edelheit's delight in fashioning paper helicopters marked a turning point for his research organization. Originally, attention to design practices formed only a small part of GE's quality training that applied equally to engineers and business executives. But with the R&D head aggressively backing a far more dynamic approach, it soon blossomed into a second program called Design for Six Sigma (DFSS) targeted chiefly at the company's ten thousand engineers. Although other corporations have implemented similar efforts, GE's may be the only quality initiative led by a central research organization. From that unique base, it seems to be sparking a measurable improvement in everything from ranges to X-ray machines to locomotive engines. "It's really amazing what they've done," says Maurice Berryman, the Texas consultant who helped bring DFSS to General Electric. "While many companies have experimented with the concept of a DFSS methodology, none have deployed it as efficiently and rapidly as GE through their research center. He [Lonnie] is the one who's really leading and pushing it out to the businesses."

GE has made quality a hallmark ever since chairman Jack Welch embraced Six Sigma in late 1995. Arising out of Motorola in the 1980s, the program seeks to take manufacturing quality far out on the Gaussian curve—to six standard deviations from the specification limit. While a Four Sigma operation would generate 6,210 defects for every million items produced (and Five Sigma reduces that to 233), getting to Six Sigma means lowering the error rate to just 3.4 per million.

Welch wanted all processes and products at Six Sigma by 2000. To that end, every salaried employee started getting the basic training—ten days of classes spread over several months. But while the program markedly improved overall manufacturing quality, it soon became apparent that moving much beyond

Four Sigma would be impossible unless all products were designed with "Six" in mind. Late in 1996, GE hired Berryman, based in the Dallas suburb of Flower Mound, to introduce the design program to its Medical Systems business. Its success stirred the passion of Edelheit, who argued that CRD was the ideal hub for spreading the practice company-wide. In early 1997, a massive training effort was launched in the R&D center, while a cadre of missionaries began carrying the word to the business units. They were soon joined by CRD "tiger teams," experts from different fields assigned to identify or invent design tools that enhance the program. Blue DFSS banners went up around the Knolls lab.

Design for Six Sigma exemplifies the old adage of getting things right the first time. Utilizing a variety of engineering and statistical tools that include elements of Design of Experiments, it takes a rigorous systems engineering approach to product design that assimilates expert opinion, historical performance data, materials quality, and more to match customer and product requirements to production capabilities. Making heavy use of computer modeling and simulation, it then seeks to predict a planned product's real-world behavior—feeding results back into the design to improve performance, reliability, and even serviceability (for instance, by ensuring key parts are easily replaced) when the costs of such changes are relatively low.

GE's first DFSS product was the Performix X-ray tube, which debuted in early 1998. Not only does the average tube last 400,000 scans, as opposed to its predecessor's 75,000, the early failure rate has gone from 15 percent to nearly zero—driving support costs down from $7 million annually to $1 million. On its heels came the LightSpeed CT scanner, which utilizes Performix's high-speed tube to take four patient "swaths" in one pass, dramatically reducing exam times: GE received a stunning $60 million in orders within the first ninety days of its introduction. What's more, CRD quality program manager Mark Sneeringer says DFSS virtually eliminated the need to shake post-installation kinks out of the scanners—freeing engineers who typically would have stayed with LightSpeed after launch to work on its successor. "So the next generation is going to come that much faster, that much better—and so this thing gets to snowball," he says. The latest DFSS beneficiary in mid-1999 was GE's Spectra Electric Range, where more exact heat distribution modeling led to a revamped broiler element design the company claims finally overcomes the age-old problem of food cooking faster in a broiler's center than on its sides. GE bills the range as the world's "most consistent oven."

GE employees are not required to learn DFSS. Still, by July 1999 some nine thousand of GE's ten thousand engineers had gone through its extra three-day training period. Beyond reduced service calls, more evenly broiled food, and improved engineering and process models, the program is credited with bringing a common approach to design that facilitates sharing best practices throughout the company. "For the first time . . . across GE we've got an engineering community," Sneeringer says. "It's really heady stuff in terms of the power it gives us to move things from one business to another."

program that Edelheit has made research's special domain—with the lab spearheading its practice throughout General Electric. There's no Nobel Prize-caliber work going on at CRD anymore—no basic science to speak of. Instead, the lab is using DFSS for things like making better jet engines, X-ray machines, and washer-dryers. And Edelheit is very happy.

Edelheit is the man destined to take CRD into the millennium. A physicist, he had been the first project manager of the research center's CT program. He then moved to the Milwaukee area to work under Robb as the project's engineering manager—later rising to general manager for the entire CT business. He eventually left GE to become chief executive of Quantum Medical Systems, a Seattle-based maker of ultrasound equipment. But Robb and Welch wooed him back to GE in 1991, about a year after Quantum was sold to Siemens. Edelheit rejoined the Schenectady lab as head of its electronics research center—taking over as senior vice president of corporate R&D in late 1992, just before his predecessor's retirement. He also serves as GE's chief technology officer.

Like his predecessor, Edelheit was profoundly influenced by the CRD-Medical Systems collaboration that produced the fan-beam CT scanner—the CT/T7800. People made such a huge deal of the machine's being delivered on time and budget, he recalls. "And that's when I realized how unique that project was. So my motto was that the whole lab has to be vital like that."

He's got a lot on his plate. In 1998, GE became a $100 billion concern whose main businesses included Aircraft Engines; Appliances; Capital Services, Industrial Systems; Lighting; Medical Systems; Plastics; Power Systems; Transportation Systems; and the NBC broadcasting network. Chairman Jack Welch plans to drive the company's growth into the millennium through the soaring services businesses associated with many of these operations—as well as the globalization of GE's finance and manufacturing endeavors.

The astonishing diversity of General Electric's businesses has changed the way research operates. Edelheit says the lab remains a leader in steam hydrodynamics, polycarbonates, and high-temperature materials critical to its turbine and engine businesses—and perhaps industrial statistics and the mathematics of medical imaging. But fundamental science is a thing of the past: in fact, CRD conducts somewhat less research of any kind than it once did. General Electric's scientists and engineers increasingly distinguish themselves not by inventing new technologies but by finding novel applications for existing technologies. "Now we only want to do research if it's vital to the company and we can't get it from someplace else cheap," Edelheit notes.

That being said, the CRD director has some well thought out views on how research can stay vital in the twenty-first century (see box,

Lonnie Edelheit: Being Vital

In Lonnie Edelheit's eyes, technology is changing so rapidly it has become almost a commodity—dramatically shifting the role of a central lab. When only a few firms dominated a given field, researchers could work more leisurely, almost assured that their creations would succeed financially. But this 'If we build it, they will come' strategy no longer wins. Success, says Edelheit, springs from rapidly combining outside technology with in-house advances to create low-cost, high-quality products. His keys to getting there:

Eliminate boundaries: "GE is a very nonbureaucratic culture," explains Edelheit. "We don't want boundaries, we really want to move fast." The goal is to meld research, development, marketing, manufacturing, service, and even vendors and customers into an innovation team. When the company developed its screw-in fluorescent lamp in the mid-1990s, the R&D center formed the glue for a worldwide collaboration that saw design managed out of England, electronic ballast work done in Ohio, and manufacturing centered in Hungary. Team members kept in touch through e-mail, video conferencing, and personal visits—with group photographs distributed to everyone involved.

Tackle several product generations simultaneously: In addition to its Design for Six Sigma initiative, the lab relies heavily on multigenerational product plans (MGPPs) that identify the key technology components, processes, and features needed to make new offerings a success—and lay out a timeline for bringing everything together. MGPPs are hardly unique to General Electric. But the central lab routinely takes the lead in helping shape the plans. Critically, says Edelheit, work proceeds concurrently on two or three generations. The farther out one looks, the more risks are involved. Tackling these risks from the start increases the odds for each subsequent generation. Says he, "It's a way of moving faster toward best practices."

Exploit synergies: Products are developed more rapidly and economically when core technologies are combined to address multiple business opportunities. Edelheit's researchers helped improve locomotive traction through software that identifies conditions likely to cause wheels to slip and orders appropriate counteractions. These same control technologies have been applied to aircraft engines, gas turbines, even refrigerators and other appliances. What's more, cultivating such synergies fosters better design and simulation practices that help an even wider array of efforts.

Be vital: "Being vital means being necessary for maintenance of the life of the business. Technical work is not vital because it is interesting or fun. . . . Technical work that is vital keeps customers alive by enabling them to survive and win in the marketplace. . . . To the extent that central research labs have different goals that aren't tied to the company's goals, they're very vulnerable."

p. 251). And while most lab projects involve incremental advances to existing products, his CT success convinced Edelheit of the need to plant longer-term seeds. Following Robb's lead, he therefore uses the 25 percent of lab funding that still comes from central coffers to build expertise in core areas and take a few risks outside the sphere of the contract work that dominates his agenda.

These central funds represent about $75 million annually, with the money going into three big pots. Perhaps 40 percent—or $30 million—is spent on technology development. This money is parceled out to the thirteen labs in roughly equal amounts to enable them to maintain their skills in key areas of competency. Another roughly equal amount is doled out to the labs—one year's worth of funding in blocks of between $300,000 and $500,000—to finance seed projects that lay the groundwork for meeting key technological milestones associated with its longer-range product plans. Typically, this work isn't expected to bear fruit for three to five years, and so is too risky and lies too far down the road for business groups to fund directly.

The final portion—maybe $15 million a year—goes to the novel Game Changers program. Its intent is to provide funding for long-shot projects that might take five to ten years to come to fruition—but hold the potential of adding at least $20 million annually to GE's bottom line. Unlike the rest of the central funds, which are distributed to all the various labs, the entire center competes for this money, which is awarded based on the merits of individual proposals. Instead of receiving a few hundred thousand dollars for one year's funding, winners are typically backed by $6 million in seed money. Even this figure can be augmented if the opportunity seems promising enough.

Rather than existing apart from the lab's highly applied and short-term endeavors, Game Changers is designed to feed on it. GE officials explain that CRD's extremely close association with the business groups leads researchers to understand these operations better—making it easier for them to conceive truly valuable longer-range projects that really will change the game. Notes Edelheit, "We don't use the Game Changers program to, as Edison used to say, 'study the fuzz on a bee.' We use it to find the products and services that are going to significantly grow GE's revenue."

Some particulars of three hot areas of research show how the lab's mix of projects—from small product improvements to Game Changers—is supposed to work.

INFORMATION TECHNOLOGY—THE APPLIED STATISTICS LAB

PEOPLE KNOW GE for its jet engines, appliances, light bulbs, medical equipment, financial services, and maybe even NBC—not as an information technology company. Nevertheless, collecting, managing, and processing information lies at the core of just about everything the diverse giant does—from monitoring the performance of its engines and medical machines to tracking loans and figuring out where to place TV advertisements. Fittingly, CRD's Information Technology lab has become a cornerstone of center activities in the 1990s. Home to only a few dozen researchers when the decade opened, it was closing in on 150 nine years later—making it both the largest and fastest-growing segment of Lonnie Edelheit's domain.

IT lab offices line the fifth and uppermost floor of the main research building, taking up the bulk of B and C wings. From this base, manager Nancy R. Martin oversees eight roughly equally sized programs. These run from e-commerce to control systems, where activities include developing algorithms for smart dishwashers and other appliances, to a software integration group that helps link previously incompatible systems—for instance, by finding ways to provide NBC digital access to its analog tape archives.

The biggest push comes in e-commerce, which chairman Welch proclaimed in early 1999 was "one, two, three, and four" on GE's priority list. But a far more unique IT lab effort involves its powerhouse Applied Statistics Program. This collection of sixteen mathematically minded analysts, whose offices span a C-wing hallway one floor below the main IT body, forms only a small supplement to the software-oriented lab. But with three of its current members elected fellows of the American Statistical Association, the group is rivaled in industry perhaps only by a similar body at AT&T Labs. Whenever NBC puts on TV shows, GE Capital provides a loan, or Medical Systems builds a new X-ray scanner, chances are the statistics team has had a hand in it. "All of these problems are heavily data driven, and our role is to help the businesses identify what data they should be taking so that we can translate that data into information," sums up program director Gerald Hahn. "It's very exciting, because we literally work with every component of the company."

GE Research has maintained a statistics group since around 1952, three years before Hahn joined the lab. Growing out of World War II operations research, it initially dealt with tasks such as optimizing manufacturing control processes to reduce scrap. That remains part of its mission. But another branch of activities covers areas undreamed of before the data crunching power once confined to mainframes arrived on every desktop, enabling statisticians to calculate in seconds what used to take weeks by hand. In recent times, Hahn's group has worked with NBC to determine the best time to advertise for certain television programs. Researchers also help GE Capital Services determine what line of credit is appropriate for an individual—and how best to collect overdue accounts given a person's purchasing habits, income, and past payment history.

Statistics also figures in tracking telephone call response times to watch for patterns of slowdowns—and identify their root causes—as a means of continually improving the quality of customer service.

In many of these endeavors, the lab's statisticians act almost as pollsters gathering statistically relevant samples aimed at ensuring product reliability. Whether it's for jet engines or appliances, the goal is not just to make things work, but to make them last in the real world. To that end, a new washing machine, for example, is tested not only in the lab prior to its introduction, but in employees' homes and Laundromats, where six months of use might equate to five years of normal wear and tear. Statistics is essential to determining what kind of testing to undertake in order to spot potential failure modes, how many tests to run, and how to translate information from an accelerated environment like a laundry to a real world setting. Statisticians also help identify ways to sample the manufacturing process—a safeguard against hidden production problems. In all these ways, researchers can verify quantitatively whether GE's reliability standards are being met. "Our job is to go beyond the opinions," asserts Hahn.

When applied correctly, statistics is integrated with other test and control measures to provide GE with a critical quality edge. Typically, the benefits of statistical analysis are hidden deep in the design—making something that was good just a little bit better. But in the case of the TrueTemp electric range introduced in early 1999, the payoff was more evident. Part of achieving a statistically relevant sample involved testing the unit in various parts of the country to see how different environmental factors might affect performance. It turned out that in the hot and humid south cockroaches crawled through the range and short-circuited the burner elements. That led to a revamped design that protected against such an intrusion. Quips Hahn, "Not only did we improve our range, but we also improved the lifetime of the roaches."

ENGINE OF GROWTH—SERVICES

THE TRAUMA PATIENT came into the San Francisco hospital's emergency room around 9:00 one night suffering from a head injury. Doctors wanted to perform a CT scan, but their lone machine was down and no on-site technical help was available. No matter. A quick call to the General Electric Medical Systems Service On-Line Center in Milwaukee was routed to Paris, where an engineer set down her morning "café" and had the system back up and running in fifteen minutes.

It was all about service. It's a fact of manufacturing life that even the best products eventually wear out or break down—no matter how well they withstand roaches and other hazards. Therefore, it's not surprising that much like IBM, General Electric is parlaying services into a growth juggernaut. In 1998, services-related revenues surpassed $12 billion—on their way to an estimated $17 billion in 2000. But the days of merely being what Jack Welch calls "expensive wrench turners" are long gone. Instead of billing hourly to fix things, companies charge up front to keep things *running* and *competitive*. GE's service operations now manage and monitor entire railroad and power systems—and hospital imaging machines like the one in San Francisco. And instead of simply replacing parts or refitting this high-value equipment, the goal is to improve them. As one mission statement explains: "By adding higher and higher technology to the customers' installed base of machines, we will have the capability of returning them to operation not just 'overhauled' but with better fuel burn rates in engines, higher efficiency in turbines, better resolutions in CT scanners, and the like."

The R&D Center has long played a vital role in creating or assimilating the technology driving this initiative. All told in 1998, the center spent more than $40 million on services-related research—nearly a fifth of its budget—supporting nearly three hundred projects across the thirteen labs. Since every service operation has similar basic needs and functions, the goal is to leverage these efforts across a variety of fields. "The same solution, tailored for each business," is how Dan Smith, leader of the R&D Center's Services Initiative, sums up the vast body of work he helps coordinate.

Take the case of repair and inspection. As GE enters into ever more maintenance agreements, it is often prudent to fix existing components rather than swap them for new ones. But this can be a complicated process that involves cleaning and tempering materials to ensure viability. It happens that metallurgical process technologies originally developed for manufacturing—say for combining two materials to fashion a new blade—can help salvage parts previously considered unrepairable. Similarly, researchers have pioneered in nondestructive methods for "qualifying" manufacturing materials that can also be applied to services. This actually represents a case of double leverage. Many of these techniques arose from the adaptation of ultrasound, CT, and other medical imaging technologies to manufacturing. Lately, the same equipment has been redirected again for service applications. One project in 1999, for example, centered on utilizing GE's recently announced Revolution® digital X-ray detector for industrial inspection. Smith says the machine's superior contrast and imaging range should enable a rapid assessment of whether a component can or should be repaired.

Perhaps the hottest area of services research lies in remote diagnostics and monitoring—the practice that enabled a French service engineer to fix a machine in San Francisco. The effort started in the 1980s, when scientists who had established modem links to hospital CT scanners to conduct research got the idea of keeping them on line for troubleshooting. Today, imaging equipment at more than 15,000 hospitals worldwide is linked to GE support centers in or near Milwaukee, Singapore, Sydney, Paris, and Tokyo—with calls routed to whatever center is open and available. Edelheit is hard pressed to think of a better example of the synergy he is seeking. "Now that's expanding into a whole service strategy," he enthuses. "We took that technology and started monitoring turbines around the world—and that is spinning off into a business in jet engines. We can now monitor jet engines that are flying."

To facilitate such efforts, in early 1998 the lab formed the Service Center of Excellence, a team of about thirty scientists and systems engineers who scour the facility for technologies and best practices—and to launch new projects—that can give GE a further services edge. The latest stage in the evolution involves "proactive" troubleshooting, where service personnel not only respond to existing problems but also try to detect the harbingers of failures and make sure the worst calamities never happen. This is another area where Gerald Hahn's statisticians are paying real dividends. In recent years, through sensors placed in key components, the ability to collect operational data about engines, imaging machines, and other equipment has exploded. It's possible to cull through this wealth of data and identify statistically relevant indications of an impending breakdown—such as subtle changes in engine vibration and performance over time. Researchers then devise algorithms that analyze data from GE's remote diagnostics network and trigger an alarm when these thresholds are exceeded.

Smith tells how a sensor placed in one hospital's MRI machine picked up a temperature rise in the water chiller unit that kept the magnet operating properly. An alarm went off at the Milwaukee service center, prompting a technician to track down the hospital's maintenance engineer—who was on the building's roof, where he had shut down the chiller to perform some unrelated maintenance. The grateful engineer turned the system back on in time to avoid a magnet breakdown called a quench that would have taken up to a week to correct and resulted in thousands of dollars of lost revenues.

Although proactive servicing began in GE's Medical Systems business, the company is rushing to bring the practice to its railroad and aircraft services operations. In most cases, there should be time to take care of any problems during the next scheduled maintenance session. But sometimes a more pressing need will arise, and officials say proactive servicing could prevent needless delays, and maybe even save lives. In the case of trains, Hahn envisions a sensor picking up indications that a key part needs replacement. The alarm sounds in the company's service center in Erie, Pennsylvania, which alerts mechanics at the train's next stop and dispatches the part there if necessary. With mechanics now knowing exactly what to do, the problem can be fixed during the normal layover. "You as the passenger on that train, you're unaware of the fact that there may have been a breakdown," enthuses Hahn. "And your train arrives on time, when it might have arrived three hours late."

GAME CHANGERS—THE DIGITAL X-RAY DETECTOR

EVER SINCE William Coolidge invented his pioneering X-ray tube at the blossoming House of Magic in the early 1900s, General Electric has remained a worldwide leader in the technology. But while dosages have been lowered and image quality dramatically improved, X-rays remain a clunky analog technology that in the age of digital photography still requires film to be processed. For the past dozen years or so, GE has supported a massive project to break that paradigm. Finally, by early 1999, even as Dan Smith's services group labored to bring digital X-rays to industrial inspection, the technology was making its medical debut.

It marked a Game Changers effort if ever there was one. Of the dozen or so projects launched under the program since its inception, none has approached the scale of the digital X-ray detector effort. Spanning a dozen years, involving more than one hundred scientists and engineers, and costing upwards of $140 million, the project grew so big that it outstripped the program entirely—and had to be bolstered by special corporate funds, as well as support from the Medical Systems business. Still, GE's success in overcoming technical and institutional barriers to bring the detector to market marks a triumph for the central lab—and its determined effort to provide multiple ways to pursue a few well-calculated gambles. Lonnie Edelheit feels this one Game Changers alone—a project few other companies could even contemplate undertaking—could justify the entire program. "That's going to be a big thing," he proclaims.

Code-named Apollo, the effort goes back to 1984, well before the Game Changers program was even implemented. Veteran researcher Jack Kingsley, who in the early 1960s helped build one of the world's first laser diodes, was working to develop a superbright active matrix display for aircraft cockpits. Suddenly, he realized that the basic amorphous silicon technology for these displays could be combined with photo diodes and other devices and used to detect X-rays.

The idea was on the verge of its time. Other major imaging technologies—CT, ultrasound, MRI—were inherently digital. But X-rays still chugged away in the old analog world. Not only did the precision of digital data promise higher resolutions, unlike conventional X-rays digital images shot from different angles could be computer-reconstructed to provide a three-dimensional picture of a bone, say—or even a patient's vascular structure. If that weren't enough, the X-ray business consisted of a multitude of systems for different tasks—angiograms, chest X-rays, gastro-intestinal tract pictures—each with a host of expensive attachments. A digital detector could do all those jobs without the claptrap of associated devices. And, of course, digital records could be easily stored, transferred, or shared.

An initial proposal to Medical Systems was rejected as too long-term and expensive. In June 1987, however, the Milwaukee-area business learned that several competitors were pursuing a similar technology. That led to a joint Medical Systems–CRD project to test the idea's feasibility— and by January 1988, Kingsley had created a working prototype. A plethora of obstacles remained to be overcome before the postage-stamp display could be geared up to the standard X-ray film size of 16-by-16 inches— and then made robust enough for production. Still, as Kingsley's manager Bruce Griffing recalls, "We were convinced that we had a real deal."

So began the long saga that ultimately turned into a Game Changers project extraordinaire—teaming CRD researchers with Medical Systems counterparts and, ultimately, production engineers from the California firm of EG&G. But while the fast-growing group made a lot of progress over the next five years, it continued to be hampered by technical difficulties. In 1993, with the price tag running several million dollars annually, Medical Systems pulled the plug. It was at this point that the determined Griffing sought out new boss Edelheit for help. The R&D director, with his intimate knowledge of the medical equipment business, quickly put Apollo into the Game Changers category. When even that wasn't enough, he appealed directly to GE chairman Welch. Quips Griffing, since risen to head the Industrial Electronics Lab: "We went into the 'Game Changers–plus' program. We went to the 'See Jack Welch' program."

The end result was dubbed the Revolution® digital X-ray detector. As opposed to other digital X-ray approaches that piece together a large mosaic out of images captured on small detector "tiles," GE's creation takes in everything in a single panorama. The company claims that it not only provides better picture quality, but enables the same basic technology to serve a variety of X-ray applications—from mammography and angiography to real-time fluoroscopy.

Incoming X-rays strike a "scintillator" that transforms the waves into visible light and channels them to the detector—a layer of amorphous silicon photodiodes fabricated on a large glass plate. Each pixel in the detector—one million in the case of a small cardiac machine, four million for a larger mammography unit—consists of a photo diode that converts the light into ones and zeroes, plus a transistor switch that enables the digital data to be "read" out. Since digitized signals precisely describe the X-ray intensity striking each element, they can be combined to instantly produce a higher-resolution image than a conventional image—with no film or chemical processing. For the cardiac machine, the entire array is read out and pieced together at a rate of thirty times a second—fast enough to provide a video image of the heart in action.

GE introduced its first Revolution-based product—the Full Field Digital Mammography system—in Europe in 1998. As of mid-1999, the company was still waiting for Food and Drug Administration approval to sell the unit in the United States. Meanwhile, a digital chest X-ray machine—which does have FDA approval—was being marketed worldwide, with a cardiac imager coming down the pike.

GE estimates that digital technology will capture most of the current $5.6 billion X-ray market by around 2010. But the company's "game changing" doesn't end with simply replacing analog technology with digital machines. By early 1999, Griffing's lab had succeeded in using the Revolution detector in prototype systems that capture high-resolution, CT-like volumetric data of organs and other tissue—blurring the very lines between X-rays and computed tomography.

"The idea there would be you could look at a whole segment of the body quickly," Griffing explains. With X-ray machines common in hospital emergency rooms, that could be vital in assessing internal injuries in a trauma case, where physicians might otherwise have to wheel a patient to a special CT facility for several minutes of imaging. It might also revolutionize something like lung cancer screening. Conventional X-ray units cannot generally pick up tumors until they're fairly large, at which point survival rates are poor. Although CT machines can spot cancerous growths at a much earlier stage, it is not feasible to test everyone with the expensive units—so a combination of computed tomography and X-rays might hold the answer. Says Griffing, "You need something you can walk up to, put your hands over your head, it goes 'zoot'—and you're done."

■

Perched on its hilltop, surrounded by peaceful green fields and the lazy Mohawk River, General Electric's pioneering research lab lies far from the madding crowd. With no major city nearby (not counting Albany), no great universities, no booming high-tech industry, and no world-famous scientists, the lab is hardly a watering hole for twenty-first century digerati. Outside of a few new facilities and some fresh coats of paint, the place hardly even looks much different from the day it opened back in 1955. But that's just the surface view. Walter Berninger, one of the lab's five technical directors, came to the historic facility in the 1970s. As he puts it, "Not much has changed in twenty-five years, but everything has changed in twenty-five years."

That sentiment is echoed throughout the sprawling facility. In the past, while always striving to help General Electric compete in its far-ranging industries, the lab found room to push the bounds of science—winning two Nobel Prizes and contributing to fundamental knowledge in a wide range of chemical and physical sciences fields. All that, though, is no more. Even as places like IBM and fellow research pioneer Bell Labs cautiously expand basic studies in the late 1990s, the old House of Magic isn't wavering from its exclusively applied mission of helping the business—typically in the here and now. "The intensity is getting very more sharply focused on doing that job more quickly and efficiently than we ever did in the past—and the past includes last week," relates Berninger.

So is it good—or bad? Well, it's both. The lab's leading inventor, Tom Anthony, spells out the tradeoffs. The grueling pace of helping the businesses stay competitive means researchers like himself have lost a lot of flexibility. In the past, when all the lab's money came from corporate headquarters, it was much easier to follow a good idea—even if it meant dropping a project or changing fields. Now, with costs more rigorously controlled and researchers typically working under business division contracts, that's much more difficult. In the old days, too, if a project was killed, it was relatively easy to keep it going off the books. Now, such salvage jobs are much trickier. Perhaps most troubling of all, there never seems to be enough time to simply read the scientific literature and think about it—and thought is often the precursor of big things. "The downside," Anthony therefore sums up, "is that there isn't a new transistor coming out."

On the other hand, CRD is a place that knows what it's about. During the heyday of science, researchers were judged chiefly by the quality of their scientific papers and achievements. As the more applied era began in the late 1960s, the yardstick for a time shifted to patents. In the 1990s, the watchwords are how much money a researcher has saved or produced for the General Electric Company. And that, says Anthony, is reality. Labs have to be relevant, or they die. While that might mean science no longer holds the place it once did, the new age does hold its own joys. "In the old era, if you did something the chance of a business using it was zilch, so you couldn't get that satisfaction," Anthony notes. "Now, if something works, there's a really good chance it will be used."

LUCENT–BELL LABS—A CHEMISTRY OF PEOPLE AND PLACE

"There's a chemistry of both the people and the place that encourages interaction."
—RESEARCHER ALAN GELPERIN

Lucent Technologies—Bell Laboratories

Labs established: 1925
Company sales 1998: $30.1 billion (restated as $31.8 billion in 1999 to reflect Ascend Communications acquisition)
1998 R&D expenditures: $3.7 billion ($3.9 billion)
1998 Bell Labs research budget: $310 million
Research funding paradigm: 97% central funding (1% of Lucent revenues), 3% government contracts
Research technical and scientific staff: 1,200
Principal research labs: Murray Hill and Holmdel, New Jersey

ALL SEEMS SERENE at the legendary Bell Labs headquarters in Murray Hill, New Jersey. Broad green lawns highlight copper roofs aging into an eye-pleasing aqua-green. A small Japanese-style garden graces an interior courtyard. A row of chess tables await players in a glass-walled lounge area.

But behind this tranquillity lies a poorly understood odyssey of upheaval, transformation—and renaissance. The lab's glorious history—eleven Nobel laureates and a tsunami of world-changing inventions from the transistor to information theory—once led many to consider it a national asset. Almost as well documented is the supposed decline, spurred by a much-lamented and highly criticized 1990s make-over that has seen the lab's legendary research arm scale back some areas of fundamental science and emphasize applied projects and meeting business objectives.

What's missing from the picture, though, is the genesis behind the changes—and the story of Bell Labs' remarkable resurgence. Its transformation rocked the lab to its soul over the decade's first half. But on the verge of the millennium—and its seventy-fifth anniversary—the venerable establishment has reasserted its place at the forefront of industrial research. Today's Bell Labs is hungrier, faster on its feet, and smarter about business than at any time since the Cold War began, playing a vital role in the soaring success of its upstart parent, Lucent Technologies. At the same time, though, scores of scientists continue to pursue dreams that may not pay off for decades, if ever—be it wiring up slug brains to find clues to biological data processing, advancing mathematical theories, or mapping the universe's dark matter.

Science for science's sake doesn't hold the place it once did: in that the critics are right. Still, by creating novel ways to balance business realities with far-off explorations better targeted on areas likely to benefit the company, Bell Labs may be trailblazing a new era in corporate research—just as it defined the old science-driven model. Former research vice president Arno Penzias, himself a lab Nobelist and a leader of the upheaval, calls the transformation a carefully conceived renaissance. "There's been a rebirth of research," he asserts. "But in a well-aimed way."

Bell Labs opened on the western edge of Greenwich Village on New Year's Day 1925, abandoned Manhattan's grime and noise for pastoral Murray Hill amidst World War II, and no matter what its locale has left an almost unceasing stream of creation and discovery in its wake. Beyond its Nobel Prizes and historic inventions, the lab even has an Emmy—for contributions to high definition television.

In light of its pathbreaking achievements, the labs is almost universally viewed as a research haven. But Bell Labs has always been primarily a *development* organization. In fact, of its approximately 25,000 workers, nearly 24,000 labor on the "D" side of R&D. The research organization, by contrast, numbers only about 1,200 personnel—mainly physicists, computer scientists, and engineers. Big changes have enveloped the entire organization in recent years. But it's really the research organization, the historic fountainhead of science, that has attracted the public attention—as it moves beyond its old purview of invention and embraces the concept of innovation. As Penzias puts it: "Years ago, research used to focus just on invention. But invention is not the same as innovation. Invention is thinking of something new. Innovation is required to change the world in which people live."

Mastering the innovation game preoccupies virtually every corporate lab today. But as the standard-bearer of the old ways, the transition seems to have come more painfully to Bell Labs' esteemed research arm. The roots of its difficulties go back at least a half-century, to the immediate post–World War II years, when the enterprise virtually defined the model for corporate research as a tower of scientific excellence. A newly

initiated focus in solid-state studies led directly to the transistor's invention in December 1947. Then, as the Cold War heated up in the 1950s,
amid the widespread perception that science held the key to national security, science took over as the linchpin for much of the lab's work. For
decades, insiders note, research emphasized being the first or the best—
publishing papers, setting transmission records, building the most powerful laser diode.

Bell Labs could afford this modus operandi in large part because
AT&T was a regulated monopoly—shielded from competition and with
an R&D tax credit built into every phone call and telephone sale. But
even as Penzias took over as vice president of research in 1981, the litigation was already underway that would lead to the Bell System's formal
breakup into seven regional operating companies on January 1, 1984—
opening the era of full-fledged competition with low-cost challengers like
MCI Communications Corp. and Sprint. It didn't take a crystal ball to
see that wholesale changes were at hand.

Penzias was born in Munich in 1933, a Jew in Hitler's Germany.
The family fled Europe early in the war, when he was just seven, crossing
the Atlantic on an ocean liner and settling in the Bronx. He then became
the quintessential New Yorker—fast-moving and fast-talking, without a
trace of a German accent, and extremely confident. Penzias took his undergraduate degree at City College, earned a doctorate in physics at Columbia University, then joined Bell Labs in 1961. Four years later,
working with colleague Robert Wilson, he detected the cosmic background radiation. The discovery earned the pair the 1978 Nobel Prize;
long before then, Penzias began moving up the management ranks.

In those days, the upstart New Yorker adhered to the long-standing
philosophy that for research to be effective, it had to be shielded from
market demands and fluctuations—and from the short-term needs of the
rest of the company. He stuck to those convictions through the breakup,
which ultimately saw 7,000 of 25,000 Bell Labs employees funneled
away to other enterprises—almost half to Bell Communications Research
(the former Bellcore, now known as Telcordia Technologies Inc.), the
R&D arm of the new Baby Bells. It was a stressful time, recalls Bill
Brinkman, vice president of Physical Sciences and Engineering Research. "Washington was so concerned about doing damage to Bell Labs,
it became a key issue how to split it up. Arno would say, 'My job is to protect the crown jewel.' "

At the same time, Penzias could not shake the feeling that real trouble lay ahead. "When I first took the job, every department head would
impress upon me that they had my job in miniature," he recalls. Each
would stress the uniqueness of their department—how it worked in all

sorts of areas, from devices to theoretical physics. "They figured I would be pleased," says Penzias. Instead, the image that sprang to his mind was that department managers were like old-time trading post owners out in the woods. Only instead of outfitting prospectors with skinning knives, hardtack, and gunpowder in return for a piece of the action should they strike gold, Bell Labs' department heads backed different scientific or technological ventures. "This is basically how these guys worked," Penzias relates. "By having as broad a range as possible, the owner of the department or post hedged on bets—if one got famous, that was the one he took credit for." The trouble was, that kind of shotgun approach didn't work well in a competitive business world. "Everybody worked on a different thing, so there was no coherence. All they were doing was spending stockholders' money."

Penzias admittedly stepped lightly in trying to get his managers to focus projects better on the business. Little changed for several years after divestiture. But a central assumption of the breakup was that once freed from the constraints of its operating companies, AT&T would turn out a wave of innovations—thanks especially to Bell Labs. The lack of those innovations drove Penzias to confront the fact that things weren't working, with the dam finally breaking in Tanglewood to the strains of Beethoven's Third. That was when he realized his job was no longer to shield research from the company's business interests but to embrace them. "I finally figured out I couldn't put my body in the way anymore," he relates.

As a company, AT&T had been slow to take advantage of the transformation sweeping the communications, computing, and electronics worlds. In particular, managers had failed to grasp the power of the personal computing revolution, clinging to AT&T's old strengths in hardware while dramatically underestimating the driving role of software.

Many of these shortcomings were reflected in research. Penzias's internal review showed many of his fifteen or sixteen laboratories had lost a lot of their focus. Two thought their main job lay in condensed matter physics; four built lasers; silicon technology and software research was widely scattered. Suddenly, what he had previously seen as freedom and independence looked like isolation. Setting records and writing great papers was not unworthy. However, Penzias explains, that kind of internal focus on excellence just did not adequately take into account market realities. "While colleagues in our Lightwave Business Unit sought more powerful lasers," he notes, "they might have preferred to trade some of that device's performance for compatibility with their existing fabrication methods." The realization that research had to reorient itself, work more closely with business units and AT&T's customers, and focus attention on studies relevant to the firm, marked a profound change in its historical role that was not easy to accept—even for the one having the revelation. " 'Eighty-nine was a crucial year for me," Penzias sums up. "I had to change my way of thinking from what it was to this new paradigm."

Bell Labs was not a dictatorship. Even if Penzias knew exactly what steps to take, he would get nowhere unless key personnel were on board. As a first measure, he polled the heads of his four key divisions to see where they stood. One was already in tune with—or ahead of—Penzias's thinking. One resisted change adamantly. The third expressed willingness to change if others did, but would not take a leadership position. The last division head sat on the fence. He had no problems leading change but did question the full extent of what Penzias wanted to do.

That fourth man was Bill Brinkman, whose Physical Sciences and Engineering Research division served as home to virtually all of Bell Labs' fundamental physics research. As guardian of the spawning ground for Nobel laureates, Brinkman might have been expected to resist any altering of the status quo. But back in 1986, as part of divestiture, he had already reorganized his division to be better focused on business needs—among other steps beefing up silicon research and opening two photonics labs that concentrated on fulfilling AT&T's ambitions in light-wave communications. Now, three years later, as he considered frankly how Bell Labs was doing vis-à-vis its parent, he, too, concluded that things were still "too researchy."

For Penzias, the respected Brinkman was the key to building a consensus. It was not as if no one else in the Lab saw the need for change. Still, there was a fine line to walk—and Brinkman provided a much-needed sounding board for how to proceed. "His support was crucial," Penzias says. "Bill Brinkman, ultimately he became the foundation of all this change." Indeed, throughout the research organization, the general perception seems to be that Brinkman was at least as important as Penzias in charting the new course.

Together, the men began a long, hard struggle against the ingrained ways of the past. Organization theory maintains that middle managers fight change the most fiercely, because they personify the status quo. "In my case," says Penzias, "it was the department heads." These were the first-tier managers inside the individual laboratories—each overseeing ten to twenty people—who seemed so proud of their broad-based attack on science. Many would not be happy to rein in their labs and focus more on applied projects.

For help figuring out the best way to bring about the changes he sought, Penzias turned to Mike Hammer, a consultant who had previously advised AT&T on other matters. The research director didn't always agree with Hammer's specific ideas, but found his insights into general behavioral patterns dead on target. Right up front, Hammer told Penzias to get ready for two truths of dramatic change: leaders have to explain things three times before people grasp the new concept—and some good people always leave. "I didn't believe it," Penzias says. But sure enough, in the months following his Tanglewood epiphany in August 1989, he held three key meetings with department heads. During the first, people were screaming and protesting. But by the final get-together, most of the group had come around to the need to right the ship.

Penzias revealed his changes in the lab auditorium in July 1990. In addition to his overview statement vowing to "reverse the trend toward university-like research," the director sketched out some specifics. The restructuring would reorganize research, called Area 11, at the laboratory level, he told staffers. Each lab would maintain core competency in a specific area, such as silicon devices. That way, the organization could broadly serve the business units without creating separate mini-research departments for each one. Not only that, he stressed, Bell Labs would support fewer labs after restructuring, with more people in software studies and less work in the physical sciences.

At least initially, Penzias had feared that Bell Labs president Ian Ross would resist the dramatic reorganization. After all, with 90 percent of the enterprise already focused on development, the remaining 10 percent that constituted research held tremendous public relations and recruiting value to AT&T. He figured that Ross would be reluctant to assign it a more applied role—and foresaw the creation of study groups and meetings that solicited all points of view and would paralyze or severely water down any attempt at restructuring. It turned out, though, that Ross was wholeheartedly behind the change. Indeed, he implemented a similar reorientation on the development side. That didn't mean the transition came easily; veteran researchers knew all sorts of ways to stymie rules and directives. But Hammer had advised him early on to "communicate relentlessly." So Penzias tried to keep the lines of communication open, even while revving his plan into gear.

The central idea was to run research the way Patton fought a war: control the choke points. "And there are choke points," Penzias relates. In the case of communications research, they took the form of technology platforms in which Bell Labs had to excel. Lasers, which transmit communications signals, were an easy platform to identify—although Penzias confined the effort to two labs instead of the previous four.

Silicon, the building block of most semiconductors, marked another absolutely crucial area. Brinkman had considerably strengthened efforts in this field back in 1986—but Penzias went a step further, elevating it to the same status as mathematics, physics, and materials science by creating a formal Silicon Laboratory and entering into a partnership with NEC, a world leader in silicon technology. On other fronts, he cut robotics and most superconductivity and human interface work. But he maintained strong programs in ceramics, statistics, mathematics, billing technology, optical amplifiers, data switching, all kinds of networking, object-oriented programming, and speech recognition technology. Bell Labs already worked in these areas, but the mix was changed. Overall, hard-core physical sciences work was cut back to about half the total research effort, down from 80 percent. Software and networking research ramped up to fill in the other half. Finally, to go along with these program changes, managers were told that beyond maintaining high standards of excellence they would be held responsible for meeting the company's technological needs in their particular areas.

To some scientists Penzias was almost a traitor, forsaking Bell Labs' greatness. He sympathized: "Since the researchers saw themselves as guardians of traditional excellence, they naturally regarded new criteria as a lowering of standards." Still, he held firm—telling people to get on board or leave, almost encouraging Mike Hammer's predicted exodus. And many did depart. In fact, so many researchers took jobs at the University of California at Santa Barbara that folks back in Murray Hill began calling the school Bell Labs West.

Penzias hated to see them go. However, he stresses, people leave all the time anyway: the technological and scientific landscapes are littered with Bell Labs alumni. "Not just to justify," he adds, "but what I have done is for the good of research."

Lucent Technologies, insiders like to say, is the best thing that's happened to Bell Labs in recent memory. It may seem less than earthshaking to outsiders, but chairman Henry B. Schacht's decision to place his headquarters inside the lab and feature the R&D arm in the company slogan provided a ringing endorsement absent in the old AT&T days. Looking relaxed and content in his expansive office, executive vice president of research Arun Netravali* seems to embody that pride by wearing a polo shirt emblazoned with the message: "Lucent Technologies. Bell Labs Innovations."

Netravali heads the resurgent team. As the successor to Arno Penzias, who stayed on several years as chief scientist before retiring to become a Bell Labs consultant in May 1998, Netravali assumed daily

* In October 1999, Netravali was promoted to Bell Labs president. Bill Brinkman was later named vice president of research.

control of research a few months before Lucent's 1996 formation. But the native of India joined the Labs in 1972, not long after completing his doctorate in electrical engineering and computer science at Rice University. As a Bell Labs engineer and computer scientist, he pioneered digital image and video compression technology—work which in 1997 earned him the prestigious Computers & Communications prize awarded by NEC Corporation. His only other full-time professional job came right after graduation, when he wrote guidance programs for NASA and the space shuttle. "I had nothing to do with Apollo 13," he quips.

An almost palpable optimism courses through Lucent, and it extends to the research boss. A far cry from the situation only a few years past, when the lab was thrown for a loop by fast-shifting global competition, Netravali now sees incredible opportunity in the mayhem. Research is aligned into three main divisions that cover a gamut of hardware and software aspects of the communications industry: Communications Sciences, Computing and Mathematical Sciences, and Physical Sciences. Each consists of several laboratories, with each lab in turn harboring a host of departments. Boundaries between these groups have blurred markedly in recent years. And in stark contrast to the old days, when Bell Labs did almost everything internally, innovations can come from almost anywhere. Smaller companies and start-ups may move faster and excel in a few narrow areas, Netravali asserts. But Bell Labs' strength lies in the ability to make sense of and shape the bigger picture—by assimilating technologies from inside and outside the lab, and fitting them into systems.

To fulfill this promise, Netravali pursues his three fundamental tenets of speed, conquering complexity, and cannibalization (see Chapter 1). None could be adequately met under the old research model. To break down the entrenched ways, his administration has been far more aggressive in conducting systematic personnel reviews and weeding out poor performers. Research has also dramatically stepped up benchmarking—continually assessing itself against other companies in technology, costs, speed, and internal processes. One of its most distinct moves was the early 1990s establishment of a department of research effectiveness, charged with managing intellectual property, evaluating how well research contributes to the company, and acting as a bridge to the academic community. Netravali has underscored these changes by beefing up his reward-and-incentive structure to encourage greater innovation.

On a broader level, Netravali is also looking to expand the reach of research. As of fall 1999, Bell Labs boasted some fifty separate operations in twenty countries, counting the United States. Most serve in a development capacity inside Lucent's business units. But research has long maintained a strong presence in an optoelectronics R&D center in Breinigsville, Pennsylvania, as well as a wireless systems group with some twenty-two researchers in Swindon, England. Early in 1999, Netravali opened the first dedicated research laboratory outside New Jersey—Bell Labs–Silicon Valley, a twenty-five-person operation focused mainly on software issues and networking. "R&D talent, especially the top talent, which is what we are after, is also being sought by other aggressive competitors," he says. "So a lot of times it makes a lot more sense to put facilities in different parts of the world."

Market awareness is central to the new Bell Labs. Scientists and business colleagues interact more regularly with customers and know a lot more about how their customers operate. Top research managers serve as liaisons to specific business units. At the same time, representatives from research, development, and the business units jointly create technology road maps, outlining where developments are heading and trying to divine new opportunities for Lucent. Such communication has helped both ends of Lucent understand the need for a constant flow of new technologies. In theory, that makes business units more receptive to technology push—where researchers usher their creations into development and then to market. It also renders researchers more attuned to market pull, the idea that the customers and business needs can dictate what will be created. In the old days, says Netravali, there was some push—and almost no pull. "In the new environment there is both push and pull."

Since the early 1990s, about half of Bell Labs' researchers have worked in partnership with the business units on specific "joint projects." Almost alone in the research world, Bell Labs does not depend on contracts with business units for its funding. Outside of about $20 million it receives in government contracts, virtually its entire budget comes from central coffers—an amount fixed at 1 percent of Lucent's revenues. Still, either side can propose an endeavor to create a specific product—a new switch or a networking operating system, for example—that would be jointly developed and staffed by research and the business units. Up to fifty staff members may be assigned to the project, though most run much smaller. Specific milestones and timetables are created, and researchers sometimes transfer temporarily to the business unit to help introduce the completed product to the marketplace.

Around 1996, Netravali and then Bell Labs president Dan Stanzione took this a step further by creating a "breakthrough projects" category. The reasoning was that while joint projects always had backing from both a business unit and research, some were more vital than others and needed to be speeded through the pipeline. To qualify for breakthrough status, both Netravali and a business unit president have to agree that an innovative technology has strong potential to dramatically reduce costs, improve functionality, or create new markets. Then it's off to the races: typical breakthrough projects get increased staffing over a normal joint project and seek to slash the usual eighteen- to twenty-four-month time to market in half. Since the idea's creation, more than fifty projects have qualified as breakthroughs, including the PacketStar Internet Protocol (IP) switch for routing data over the Internet and Softswitch, a highly acclaimed software switching technology released in June 1999 that allows high-quality voice communications over these same IP networks. "Some of these have moved at a speed that has sort of dazzled us," says Mel Cohen, Bell Labs' vice president for research effectiveness.

The new climate has hardly come easily. Some researchers and managers understood fully the need to shift their orientation to reflect business realities. But Arno Penzias's 1990 reorganization nevertheless shook research to its core by questioning the long-standing role of the physical sciences at the lab. Just as things seemed to have calmed down, along came AT&T's 1995–96 trivestiture. While the decision ultimately proved a godsend, the voluntary breakup saw about 350 researchers—about a quarter of Bell Labs' research staff—follow their work to the new AT&T. Even though the rest went with Lucent (none joined the third spinoff, NCR Corporation), this third major shakeup since the original 1984 divestiture temporarily put the pinch on research all over again. "It was really pretty dire during trivestiture," says one veteran Bell Labs scientist. "A lot of people were looking around for opportunities elsewhere, and it seemed like there was going to be zero opportunity to do basic research of any kind."

This time, though, physical sciences skated by largely untouched—while the software and networking end took the big hit of having to replace the researchers who went to the new AT&T. Lucent was set up as a systems and technology company, while AT&T is chiefly a service concern providing voice and data communications. That mission virtually dictated that it keep a strong core of software experts as part of its neophyte AT&T Labs. After lengthy negotiations, it ended up taking some 300 such researchers, nearly half of those engaged in such studies. "It could not have been done well, but it was very painful," says Dennis Ritchie, the legendary creator of the Unix operating system and C language. After a lot of soul searching, Ritchie stuck with Lucent and the Bell Labs core. Still, he lost many valued computer networking colleagues.

In Penzias's mind, the mere fact AT&T bargained hard about divvying up research meant his organization had turned the corner and begun to offer real business value. "They wanted more than what Lucent could afford to give them," he says. "I think that is a real testimony to the fact that we've adapted."

Indeed, even though trivestiture caused the loss of many software researchers, the creation of Lucent—a far more nimble and better focused enterprise than the old AT&T—set the stage for the various initiatives put into place under Penzias and, later, Netravali, to blossom. "Lucent has been very good to us," says Horst Stormer, who won the 1998 Nobel Prize in physics for his work characterizing the fractional quantum Hall effect and until June of that year directed the Physical Research Laboratory, home to most of Bell Labs' far-ranging physical sciences studies. "We just feel much more in charge of our fate." He tells how new chairman Schacht and chief executive officer Richard McGinn brought a genuine enthusiasm for technology—to the point that people chased them down for demos. "This was something that with AT&T we never had."

Despite that passion to produce, the fledgling Lucent still had a tall order as it left AT&T's loins, however. Competitors included a formidable block of global communications and networking powerhouses: Cisco, Motorola, Nortel, and Ericsson among them. But within the first few years following trivestiture, Bell Labs' targeted approach to research had produced a slew of lucrative innovations that helped the company establish itself clearly among the first tier. These run from numerous fiber optics advances to a low-power digital signal processor (DSP) chip, a hot-selling dense wavelength multiplex system, and various Internet Protocol switches designed to route data with unprecedented speed and quality.

Since its inception Lucent has also operated a New Ventures group that helps spin off inventions outside its core areas of focus. "If we do our research work right we're going to create lots of pleasant surprises that are technologically exciting and that make more business sense to commercialize outside of our normal business interests," explains Mel Cohen. Under its old style, such products might well have withered on a lab bench. But as of late 1999, the group had funded fourteen start-ups based on Bell Labs innovations.

Even as these initiatives continue to pay dividends for Lucent—its stock soared more than 800 percent in the three years following its April 1996 initial public offering—research managers professed wariness about going too far to the applied side. The great challenge, notes Bill Brinkman, lies in becoming better attuned to corporate needs, but not to "overdo it so badly that you have no science." It's a fine line to walk. To directly help the company, staffers have to focus on applications. However, researchers can't do first-rate science if they're worrying too much about applying their work. What managers don't want is for researchers to lose their clarity of focus and end up with some sort of muddied hybrid. "At the in-between is a piece of junk," asserts Brinkman. "It's very difficult to keep a balance because you don't want the stuff in the middle."

It's true that "basic" science projects are less numerous than in the past and have been scaled back in scope to more closely match areas of core competency: lasers, optical communications, and materials research among them. That means Bell Labs cannot cultivate the comprehensive expertise across all areas of solid-state physics or materials sciences, say, that it once did. But wherever its impressive talents are trained, the Labs remains a world leader. Arno Penzias even argues the whole concept of basic and applied research should be recast. One common notion is that an increasing emphasis on applications always comes at the expense of basic research. Penzias calls this the Vodka model of research, meaning people think the strength of a research organization is determined the same way as the proof of an alcoholic beverage: adding water (applied research) means less alcohol (basic/strategic research).

But the former Bell Labs research director insists applied and basic research are not antonyms. "The opposite of applied is unapplied," he says. He calls this unapplied work academic research, because its practitioners are typically university scientists who worry only about advancing science. Their investigations are very basic—but that hardly means basic research is confined to academe. In fact, basic studies can be very applied, as when theoretical physicists are called in to investigate an anomaly in metal oxide semiconductor behavior at low voltages—thereby tackling a pressing production problem but also contributing to science.

The Vodka Model of Research

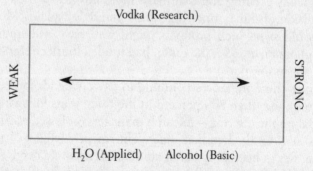

When making vodka, substituting alcohol for water increases strength. Too many people, says Penzias, view applied and basic research in the same way: they think if companies increase applied research, it must come at the expense of basic studies.

In truth, Penzias says, applied and basic research exist in different dimensions—leading him to classify work as academic or applied, short-term or long-term. The further out in time one tries to look, the more uncertain a project's outcome. Therefore, the longer the time horizon, the more academic the approach should be, with heavily applied and targeted work confined largely to the shorter term.

The goal at Bell Labs is to create a portfolio of projects that spans all quadrants of Penzias's box—with one exception. "The one thing you don't want is short-term academic," he explains. Such pursuits entail spending precious resources on esoteric questions with little chance of benefiting the company. Penzias maintains that the greatest percentage of a corporate lab's work should fall in the short-term applied category, because these efforts mark the surest bets to help the company. Far less should reside in the long-term academic arena, where outcomes are most uncertain. This goes against the grain of the old Bell Labs style. But why, Penzias asks, should Bell Labs research continue to compete heavily with academicians? "So there we were trying to scoop Harvard," he says. "In some cases, okay. But at some point, why not let the guys at Harvard scoop us? In fact, why not give them a donation—donate the equipment?"

That point made, however, managers stress that a key benefit of Bell Labs' fixed research funding model is that it allows room for a small but formidable core of scientists to pursue well-chosen, academic-like, long-term investigations without being sucked into too many applied projects. That freedom has also helped it to maintain a unique environment where people from different disciplines mingle in the halls and share ideas through seminars, forums, and lectures—and where new collabora-

A Better Way of Looking at Research

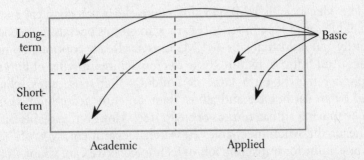

Applied and basic research inhabit different dimensions. When considering corporate research, therefore, draw a box that labels work as either academic or applied, short-term or long-term—any of which can include basic studies.

tions always seem to be forming, stirring up the pot, and keeping ideas flowing.

The result is a nearly unrivaled program of over-the-horizon scientific investigations. The ScienceWatch study of high-impact research papers showed that in the physical sciences Bell Labs led the world from 1990 to 1997. With its papers cited almost 19,000 times, the labs easily outpaced the 13,020 of runner-up IBM, as well as all the world's top academic institutions.

All three research divisions maintain some well-chosen fundamental work. When it comes to the labs' hallmark studies in areas such as solid-state physics, however, most long-range fundamental investigations take place inside the Physical Research Laboratory run by Cherry A. Murray (see box, p. 275), part of Bill Brinkman's Physical Sciences and Engineering division.

Staffed by about 140 researchers, the lab's activities span physics, materials science, wireless and optics studies, computer science, biophysics, and astrophysics. Half the projects involve science more than twenty years out, with almost all the rest focused on a five- to ten-year time frame. Even within this framework, there is a striking variety in how closely related the research is to business objectives. In the meantime, however, it's expected under the new climate that researchers should be ready, willing, and able to bring their expertise to bear on more pressing problems that might arise—as a trio of examples show.

THE OPTICAL SWITCH: THE $64 BILLION QUESTION

DAVID J. BISHOP LABORS in a Lilliputian land. Wander the halls around his office and the evidence is readily seen in blown-up pictures of his group's creations: miniature trampolines, microphones, and mirrors— each riddled with moving parts, yet so small hundreds fit on a pinhead.

The idea is called MEMS, for micro-electro-mechanical systems. The field has exploded in recent years, with groups operating at scores of university and industrial labs. MEMS sensors already control automotive air bags. And futurists picture these micromachines driving button-sized cell phones that fit on a lapel or buildings that sense stress changes caused by an earthquake and adjust their structure accordingly. Lucent won't be making air bag sensors or smart steel. However, explains Bishop, who heads the Micromechanics Research Department, MEMS "has a huge possibility for impacting lots of technologies we care about—particularly optics, acoustics, and wireless."

The Holy Grail of his powerhouse group—now more than twenty

☐

Cherry Murray

As a diplomat's daughter, Cherry Murray spent most of her formative years in Pakistan, Japan, and Korea. She's lost her Urdu, comprehends only a little Japanese, and never did learn much Korean. But along the way, she picked up a more universal language: physics. She taught herself as a high school junior because the U.S. embassy school in Seoul didn't offer the subject. "I was valedictorian," she admits. "But the embassy school in Seoul is not quite the same as the Bronx School of Science."

Even if it wasn't Bronx Science, it proved a good starting point for a long journey in research. Murray enrolled at M.I.T. to study physics, staying on to earn a Ph.D. This career path surprised her entire family. Murray's parents are both talented painters (they met in a Taos art colony), who assumed their daughter would become an artist. The main reason she chose M.I.T., Murray confesses, "was that my brother went to M.I.T. and studied physics, and he said no way would I ever be able to do that."

She did it—and has barely picked up a paint brush since. Her brother, John, became chief scientist of the National Ignition Facility at Lawrence Livermore National Laboratory. Murray found her star in industry. She joined Bell's research staff after receiving her doctorate in 1978, rising to manage departments in low-temperature and solid-state physics, condensed matter research, and semiconductor physics—and winning distinction for her studies of colloids. These are microscopic particles that can be crystallized to form structures with unique properties she hopes will lead to new and better optical materials that can be used in communications systems.

In June 1997, Murray was named director of Bell Labs' Physical Research Laboratory, a renowned group of about 140 scientists whose mission is to explore scientific and technological frontiers a decade or more down the road. Murray is the first woman in the storied history of Bell research to reach the director level, the organization's third-highest management rank. With Research known as Area 11 and Physical Sciences and Engineering as Division 111, Murray's group is called 1111. Research boss Arun Netravali calls her "the director at the outer edge."

After two decades in the Physical Research Lab, Murray finds herself with a far different mandate than most predecessors: not only to produce excellent research but to directly impact the bottom line. To that end, fundamental investigations have been better targeted on Lucent's core interests, and researchers occasionally work with business unit colleagues on shorter-range projects. "Now we're not there yet with the perfect synergy between research and the businesses, and we never will be—but we're certainly far better connected," says Murray. "We actually know what the stock price is, and who our competitors and customers are, whereas at AT&T ten years ago nobody had a clue. If we went to a training session, people would say, 'You're from research? You're from outer space.' Now, it's, 'You're from research? Oh, I've interacted with So-and-so.'"

☐

strong—is the optical switch, a means of keeping fiber optics signals humming along without the need to convert photons back to electronic form until they reach their final destinations. If it pans out, such a switch might improve transmission capacities by one thousand times over what's typical today. And forget about the traffic bottlenecks that lead conventional circuits to overload. Bishop sees the optical switch enabling a "data mesh"—a vast adaptive network offering ways around virtually any problem. With huge spoils awaiting the first out of the blocks—in gaining markets and setting standards—the race has drawn a host of entries, with almost as many technical approaches as contestants. However, Bishop believes Lucent has the inside track. "Our horse in this race," he proclaims, "is MEMS."

Bell Labs' run for the optical switch underscores an important research lesson: that focusing on the right areas is often more important than hitting a specific target. Until around early 1998, Bishop's top priority was to bring the power of fiber to the home. Given optical fiber's near-infinite bandwidth, such a plan promised to trump the efforts of broadband cable outfits—if it proved cost-competitive. MEMS was the key. Rather than expensive and power-hungry laser transmitters installed in each house, movable micromechanical mirrors would convey signals through patterns of reflected laser light beamed from neighborhood nodes, simulating lasers in every household for a fraction of the price. But Lucent misjudged the convenience of piggybacking transmissions on cable TV lines. When AT&T, once counted on to blaze these fiber trails, began buying cable companies, notes Bishop, "they very well may have decided the issue."

All of which might have spelled trouble, if not for the fact that Bell Labs had already begun shifting its MEMS horse to the optical switch race—where the payoff was looking even bigger. Fiber lines typically come together at major switching stations—dubbed NFL cities, because they correspond roughly to metropolitan areas with football franchises—where signals are rerouted toward their final destinations. In order for the electronic switches performing this task to function, however, optical signals must be converted to everyday bits. They're then assigned a new fiber and reconverted to optical ones and zeroes for the next phase of their journey. The big problem: traffic volume is growing far faster than the capabilities of electronic switches to handle it, vastly limiting data flow and leading to potential bottlenecks.

An optical switch could avoid most, if not all, of the intermediate conversions from photons to bits and back, conceivably freeing networks to handle as much as a petabit of data per second, one thousand times the typical terabit capacity of late 1999. If each phone call costs ten cents a minute, Bishop says, "that amount of data is worth about a billion dollars a minute." Therefore, he calls the switch "the sixty-four-billion-dollar question." And hence the horse race. Several technical approaches are being pursued, but Bishop says MEMS has the special advantage of being "cheap, fast, small, and robust." Computer-controlled micromirrors could route photons to their destinations simply by tilting on their axes. Moreover, MEMS switches can be made almost exactly like an integrated circuit using inexpensive "last-generation" equipment—potentially bringing the cost to pennies apiece. "This is a disruptive technology," he enthuses.

Lucent isn't saying how much progress it's made. However, Bishop notes that if the optical switch proves viable, it will mark a vindication for none other than Alexander Graham Bell. In 1880, the telephone pioneer patented an optical connection scheme that employed large mirrors to relay light signals. Bell didn't have lasers, optical fibers, MEMS, or even flashlights, so his dream went unfulfilled. Says Bishop, "We definitely see ourselves as individuals who are going to bail Alexander Graham Bell out."

MAKING SCENTS

WHILE IT'S EASY to see how MEMS relates directly to Lucent's business goals, other Physical Research Lab work has a more tangential relation to the bottom line and may take many more years to pay off. Witness Alan Gelperin, proprietor of the Slug Emporium, a triad of refrigerators crammed with trays holding several thousand chilled garden slugs: big slugs, little slugs, cigar-fat adults, and pencil-thin babies. Nearby, in Gelperin's cluttered lab, more of the creatures lay sprawled under microscopes. Or, rather, their tiny brains and sense organs do—kept alive in a nutrient bath and hooked up to various electrodes and devices that gauge their reactions to such stimuli as a nice cucumber mist.

A seventeen-year veteran of Bell Labs, Gelperin is a computational neurobiologist and neuroethologist, meaning he studies the algorithms nerve cells use to produce behavior. He concentrates on slugs—snails without shells—because the creatures possess an intriguing ability to rapidly and reliably learn about odors, with this "learning" able to continue even after their brains have been removed from the body for experimentation. Their ability in this regard, he quips, is comparable to that in "rats, pigeons, and undergraduates." The big difference is slugs have several orders of magnitude fewer neurons—and Gelperin aims to use these relatively simple nervous systems to derive basic biological design principles that might some day be applied to artificial neural networks.

Gelperin works primarily with *Limax maximus*, the spotted garden slug. The key to devising models that can be simulated in software or even wired into a machine lies in physiological experiments designed to get at how slugs store and access their odor memories, then take action based on their experience with certain scents. In collaboration with colleague Winfried Denk, a biophysicist and expert in optical recording, Gelperin studies dyed slug neurons through two-photon scanning, a microscopy technique that allows him an unprecedented view of the activity inside the processes of single nerve cells.

Similarly, by applying dyes that change their fluorescence if the voltage across the cell membrane changes, he and David Tank, head of the Biological Computation Research Department, have detected electrical waves and oscillations that originate at one end of the procerebral lobe—a slug's odor analyzer circuit—and propagate along it, starting over again as the previous activity dies out. One hypothesis is that the wave acts as a kind of time stamp for storing data. That is, with the detection of an odor and an associated stimulus—a shock, for instance—the memory of that odor is stored in a specific band of cells that run perpendicular to the wave. "Where the wave is determines where the memory storage is going to happen," Gelperin suggests. The next time the slug is exposed to the odor, it accesses the cells at the same point along the wave—and orders an appropriate response, like running away from an odor previously paired with shock.

Many experiments remain to be performed before this hypothesis can be confirmed—and possibly incorporated into tomorrow's neural networks. Even while pursuing such long-range studies, however, Gelperin tries to turn an eye toward more directly applied projects. Working with AT&T's NCR unit before it spun off as a separate company under trivestiture, he used his expertise in neural networks to develop an electronic nose for automated checkout machines. Electronic checkers have little trouble reading bar codes but run into trouble quickly distinguishing produce. Gelperin teamed with Bell Labs researcher Sebastian Seung, a neural network and machine learning theorist, to create a system that emits a vacuum pulse to pull odors over special sensors that tell broccoli from lettuce—or even distinguish a Macintosh apple from a Pippen. In November 1997, Gelperin received a patent on the device.

Gelperin expresses delight in being able to apply his knowledge of neurobiology to solve real-world problems. However, he acknowledges, accepting the need for such efforts has been difficult for some scientists. "Some folks just didn't want to think that way," he says. "They had their pure science, and pure was with a capital 'P.' And they just didn't want to be bothered."

IN SEARCH OF THE UNIVERSE'S MISSING MASS

IF GELPERIN'S RESEARCH is a fruitful mixture of the basic and applied, Tony Tyson's seems to be purely fundamental. Tyson is one of the world's most eminent astrophysicists. When his name comes up, Physical Research Lab director Cherry Murray deadpans: "He's discovered 90 percent of the universe—what can you say?"

Her statement is only somewhat glib, since what the Bell Labs researcher has actually done is find a way to image cosmic dark matter, the invisible "missing mass" thought to make up some 90 percent of the universe's total mass. Tyson has made a start on filling in the details. But, he figures, "At the rate we're currently going it will take me another fifty years."

The idea that invisible dark matter exists has been around since the 1930s. However, the theory attracted only a fringe following until the late 1970s, when modern techniques proved that the visible universe doesn't contain nearly enough mass to explain the movements of galactic gas and dust—a sure indication something else is out there exerting a strong gravitational effect. Early theories tapped neutrinos for the missing mass; but these particles have since been ruled out as major players. Tyson's bet is for a combination of unfamiliar objects and events, including weakly interacting massive particles, or WIMPs, magnetic entities called axions, cosmic strings, and breakdowns in the uniformity of the space-time continuum.

The thirty-year veteran of Bell Labs research has been hunting cosmic dark matter since 1977. "I'm a prospector," Tyson says. "I should have a donkey, a hat, a canteen, and a pickax." His work makes use of what are called gravitational lenses to map this invisible dark matter. Any mass exerts a gravitational pull that bends or deflects the light from something behind it with respect to an observer. It's a very imperfect lens—like looking through a Coke bottle. So, if something lies between the Earth and some distant galaxy, for example, astronomers equipped with the right camera sensitivity and processing software will detect multiple images of that galaxy. The distribution of those images makes it possible to figure out how much mass is out there affecting the light.

In one of Tyson's experiments, the Hubble Space Telescope was trained on a well-known cluster of several hundred galaxies known as CL0024+1654, some two billion light-years from Earth in the constellation Pisces. Previous observations had indicated the cluster was a good bet for a gravitational lens. Sure enough, at least eight images or partial images of the same background galaxy were detected. In addition, while the cluster acted as a gigantic lens, individual galaxies inside the cluster served as smaller lenses, revealing fine details of their masses. In May 1998, Tyson and collaborators Greg Kochanski and Ian Dell'Antonio published a map showing the distribution of cosmic dark matter at unprecedented resolution.

To complement the Hubble work, Tyson and University of Michigan astronomer Gary Bernstein built the Big Throughput Camera and installed it on a telescope in northern Chile. Offering two hundred times Hubble's field of view, the BTC will enable Tyson to probe a larger mystery. While dark matter seems to concentrate around clusters of galaxies, he postulates, maybe there's a bit of mass everywhere that taken as a whole forms the lion's share of what's out there. Tyson compares the clusters to whitecaps out at sea—anomalies attracting attention because nature has piled things up on each other. "What I want to do is measure the rest of the ocean."

■

Tony Tyson might seem a throwback to the old ways, pursuing a fascination with no apparent relationship to Lucent's business. But while practicing his basic science, the astrophysicist has worked on several applied projects. What's more, while hunting cosmic dark matter he pushed the development of charge-coupled devices for image detection and helped create novel image-processing software—advances incorporated into an automated fingerprint detection technology designed to replace locks and a valuable failure analysis tool that maps the surface temperatures of semiconductors while they're still in production.

Tyson's work—like that of Alan Gelperin's—can be taken to illustrate how Lucent's attention to applications can pay off. Conversely, it can be used to show that companies should support unfettered science—because far-ranging studies have a way of paying dividends where they're not always expected.

Indeed, the chief complaint from critics of the new Bell Labs is that the drive for relevance has overly constrained scientific inquiries—a strategy that will ultimately cause it to miss the kind of breakthroughs that brought the lab to glory.

Former Bell Labs researcher Charles Townes, the Nobel laureate inventor of the maser and one of Arno Penzias's instructors at Columbia University, understands the reason behind the transformation and doesn't know what could have been done differently. Nevertheless, he feels that a good deal of Bell Labs' pioneering spirit is evaporating. "Bell Labs was a rather unusual and exceptional place," notes Townes. "For a long time it could be different from other companies because it was a monopoly." Now that it functions like any other company, he adds, "I think it's a great loss for the country."

Arno Penzias agrees with former mentor Townes that some of the lab's special aspects have disappeared. "There is a lot in what Charlie says, especially in the physical sciences," he admits. "I have to say something has been lost. But that loss is not unique to industrial research. Nothing is what it used to be."

Bell Labs simply had to accept the fact it could no longer afford to be isolated from market realities, argues Penzias. "Isolated," he says, "is just a recipe for somebody coming along and just killing it."

The debate on where to draw the line on science will likely rage into the millennium. Meanwhile, Bell Labs has been reborn—and nobody had to kill it.

CHAPTER NINE

CHILDREN OF THE SIXTIES: XEROX AND HEWLETT-PACKARD

IN CONTRAST to General Electric and AT&T, who established labs early in the twentieth century, Xerox and Hewlett-Packard are much newer companies who forged their central research arms in the 1960s, during the heyday of science. These two enterprises followed a path diametrically opposed to that taken by the pioneers—in that they did not compete against each other for more than two decades but have recently become fierce rivals in printers and other office technology.

Both now operate research laboratories in Europe, Japan, and the United States, embodying the shift to global innovation. But as with their general business operations, the two took different routes to arrive in such similar positions. Xerox's research history has been marred by controversy, reexamination, upheaval, and renewal. In the 1980s, central research was under fire in light of the company's failure to capture the fruits of pathbreaking personal computing advances from its famous Palo Alto Research Center, or PARC, which introduced the world to graphical user interfaces, local area networking, and laser printing. Since then, under William Spencer, later chairman and chief executive officer of Sematech, and current research leader Mark Myers, the company has emerged as an innovative, research-oriented organization.

Hewlett-Packard Laboratories, meanwhile, has experienced little of Xerox's research turmoil. Until Carly Fiorina took over as chief executive in mid-1999, the company had always been run by engineers who championed the necessity of research. Unlike Bell Labs or even Xerox, HP Labs has historically attempted no basic studies, operating instead as an applied lab that has time and time again created new lines of business—in computers, printers, precision measuring equipment, and more—that have enabled its parent to maintain consistent double-digit growth. Consequently, research suffered none of the late-1980s and early-1990s turmoil that rocked other major laboratories—and took advantage of this success to triple its budget over this time period and even launch its first forays into quantum physics and other areas of basic research.

XEROX—A CYCLONE OF RESEARCH

"It's not just a technology fit. It's got to be a market fit."
—MARK B. MYERS, SENIOR VICE PRESIDENT, XEROX
CORPORATE RESEARCH AND TECHNOLOGY

Xerox Corporate Research and Technology

Research division established: 1962
Company sales 1998: $18 billion
1998 company R&D expenditures: $1.2 billion
1998 CRT budget: $250 million
Funding paradigm: 50% central funds, 50% business division contracts
Technical and scientific staff: 1,320
Principal research labs: Palo Alto, California; Webster, New York; Grenoble, France; Cambridge, England; Mississauga, Canada

---

IT'S A CLASSIC VIDEO—but not one that you'll find in any movie store. The short film opens with a spliced-in clip from the Hollywood feature *9 to 5*. Jane Fonda's character arrives for her first day of work. Cold-hearted boss Lily Tomlin guides her to the copier room, fires off the incomprehensible instructions for operating a monstrous machine, then leaves an overwhelmed Jane to her own devices.

Cut to contemporary life, where the time-lapse video shows two men in jeans trying to make double-sided copies with a state-of-the-art Xerox copier. In growing frustration, the pair huddle repeatedly to scrutinize the instructions while a mountain of single-sided copies rises nearby. After an hour, they're defeated. One sighs: "We're S-O-L."

When this video debuted before an audience of top Xerox managers, one executive scoffed at the technologically incompetent, real-life subjects. "You must have got these guys off the loading dock." That was a perfect setup for the bombshell: both men were computer scientists filmed at Xerox's famed Palo Alto Research Center. One was well-known computational linguist Ron Kaplan. The other was Allen Newell, a founding father of artificial intelligence.

The film hardly ranks as a blockbuster. But it had a big impact on Xerox. Dubbed *When User Hits Machine*, it was presented to various high-powered management groups in 1982 by a mid-level PARC lab manager named John Seely Brown. It showed clearly—as did a second, far-less-lighthearted tape of researchers trying to use Xerox's new 8200 copier—the daunting problem of making any technology truly user friendly.

It also marked one of the first efforts of a unique PARC group dedicated to overcoming that barrier: a cadre of academically trained anthropologists who spend their time studying how people interact with machines, and each other, as information flows through the workplace. Since then, John Seely Brown, or JSB as he's widely known, has risen to become the director of PARC, as well as Xerox's Chief Scientist. Under his aegis, the anthropology group eventually swelled to include a half-dozen people deployed into the wilds of government and business offices as if they were deep in the African bush observing the customs of a strange tribe. Joined by colleagues in computer science and other disciplines, their far-ranging workplace studies have zeroed in on Xerox service reps, airline operations personnel, attorneys, and civil engineers. Their growing understanding of the nature of these jobs has allowed the group to write scientific papers on the often-overlooked but important ways knowledge is informally created and shared in the office, while also providing fodder for the design of novel technologies to make work easier.

This group—this approach to innovation—is one reason Xerox has established itself as a research leader. Brown refers to such studies as pioneering research, because they start with a real business problem then reframe it to devise never-before-considered solutions—and it is through just such practices that Xerox expects its biggest payoffs in the future. "Our goal is not first and foremost to create fundamental knowledge," the director explains. "Our goal is to crack real problems that really make sense, but crack them by going to the root of those problems. In the process I believe very profound fundamental knowledge gets produced." The company calls its combination of technology and social science a "cyclone of research."

This, argues Brown, is what research should really be about—and he's had occasion to think about such matters. PARC, after all, is the place infamous for "fumbling the future." Way back in the 1970s, it was home to a remarkable suite of inventions: the graphical user interface, the laser printer, and the Ethernet, among others—just about the entire infrastructure of the modern office. Yet many of these creations wound up being commercialized by firms other than Xerox.

The whole sorry matter has been the subject of case studies and books. But the better story lies in how Xerox has used the experience to shape one of the world's most innovative research organizations. Between Brown and Corporate Research and Technology head Mark B. Myers, the company has strived to avoid repeating past mistakes by shifting the organization's orientation from mere invention to innovation, which Brown has defined as "invention implemented." This far-more-difficult job entails shepherding creations through corporate and marketplace barriers, listening to advice from inside and outside the company, and doing anything else to get products to the customer. At Xerox, innovation also involves probing the very nature of work, recognizing that technologies should change work practices—and that those shifting practices, in turn, must be used to reshape technology. While it is only a small branch of Xerox research, the anthropology group is part of a growing social sciences effort that gets at this feedback cycle in a unique way—giving the company a leg up in meeting the innovation challenge.

The Document Company grew out of the electrophotography patents—the first dating to 1937—of attorney and inventor Chester F. Carlson. To turn his ideas into a product, Carlson sought help from the Battelle Institute, the nonprofit consulting lab in Columbus, Ohio. Battelle and its development subsidiary worked out the basic principles behind Carlson's invention, made key process advances aimed at creating an automatic copier, and in 1947 sold the rights to its use to the Haloid Company. It still took the Rochester-based maker of photographic and photocopy papers and machines until 1960 before the first plain paper copier was shipped—but president Joe Wilson persevered. Along the way, Haloid coined the term xerography—a fusion of the Greek *xeros* for dry, and *graphos*, for writing. The company changed its name to Haloid Xerox in 1958. A few years later it became the Xerox Corporation.

The first product, christened the Automatic 914 Xerox Copying Machine because it could reproduce on paper up to 9 by 14 inches, proved an overnight sensation. Critical to Xerox's success was its decision to forsake sales in favor of leasing the 914. For a modest monthly fee, customers got the first two thousand copies free, then paid only four cents for each additional copy. That meant customers did not have to fork over thousands of dollars to try out an unproven technology. And it was this double innovation—in technology and marketing—that allowed business to explode. By 1968, annual revenues soared past the billion-dollar mark, making Xerox the first U.S. company to reach that plateau in less than ten years; the next would be Apple Computer.

Amidst this explosive growth—following the path of GE and other research pioneers, who started central labs only after achieving market domination—Xerox moved to form a central research organization. John Bardeen, co-inventor of the transistor and a longtime Xerox board member, pushed the idea—and in 1964 the company dedicated its Webster Research Center, near its Rochester origins in Webster, New York.

Albeit with plenty of fits and starts, research would continue to grow throughout the century. Heading into the year 2000, The Document Company annually spends a tad over a billion dollars on research and development. Corporate Research and Technology—the centralized, research-oriented aspect of R&D—gets about a fifth of that budget. The remainder goes to business groups that concentrate almost exclusively on development issues. CRT comprises four main laboratories in four countries. The original Webster facility—renamed the Wilson Center for Research and Technology in honor of the company founder—focuses on printing, collating, and other aspects of document production. PARC came next. Formed in 1970, it concentrates on a wide array of computer-related areas that run from nanotechnology to laser printing. Four years after PARC, to specialize in materials research such as the development of new inks and toners, the company opened the Xerox Research Centre of Canada outside Toronto in Mississauga. Finally, the late 1980s and early 1990s saw the creation of Xerox Research Center, Europe, with labs in the computer science strongholds of Cambridge, England, and Grenoble, France. Closely associated with these research arms are three technology and engineering centers established at the Wilson lab, PARC, and in Grenoble.

It was in the early stages of its heady growth—and intimately tied to PARC's creation—that Xerox stumbled down the path to research infamy. In the late 1960s, under attack from competitors and antitrust forces, Wilson launched a diversification campaign aimed at reducing the company's near-total reliance on copiers. To that end, to help Xerox enter the emerging field of digital computers, the Palo Alto Research Center was established. The mission: create the Office of the Future.

PARC hit the ground running. George Pake, then at Washington University in St. Louis, signed on to head the venture. One block of early hires came from the lab of interactivity pioneer Douglas C. Englebart, who had conceived of the computer mouse, overlapping windows, and many other faces of the digital future. Joining those wunderkinds came a host of Silicon Valley legends-in-the-making. Among them: Alan Kay, whose Smalltalk programming language enabled the graphical user interface. "It was like a dream," Kay stated. "There was nothing we couldn't do."

Stimulated by the freewheeling atmosphere, it took PARC's computer jockeys only about eighteen months to fashion many of the paradigms destined to lead computing into the twenty-first century. Things came together in the 1973 lab debut of Alto, in many senses the world's first truly personal computer. Within just a couple of years, PARC's systems employed an improved mouse, overlapping screen windows, an object-oriented programming language, graphics-based monitor, the first word processing program for nonexpert users, the Ethernet local area network, and the laser printer.

Then the bottom started falling out. By the mid-1970s, antitrust pressures had forced Xerox to make 1,700 patents available to competitors and reveal specifications on a variety of existing and future machines. Not only that, the company's rapid growth—from nine hundred employees in 1959 to more than thirty thousand—had spawned huge inefficiencies that made things even easier for the competition. Between 1972 and 1980, this double-whammy sliced Xerox's share of the worldwide plain paper copier market from 85 percent to 46 percent, as the company discovered Japanese firms such as Ricoh and Canon could sell machines for Xerox's cost of manufacturing. Recalls Robert Spinrad, a soon-to-be PARC director: "Xerox was totally rocked to its foundations."

So was PARC. The volumes written about Xerox's failure to capitalize on the PC revolution illuminate many daunting barriers to innovation: the balkanization of computers in a copier-dominated organization, a company besieged by competitive and antitrust pressures that focused attention on the main line just when it would have been marketing computers, a bloated middle management layer struggling to maintain the status quo and not about to take the risks associated with a novel technology. George Pake noted the tragic irony: "Research in Xerox became very risk oriented while development became just the opposite."

By the late 1970s, this split between "R" and "D" had become a gulf. In the spring of 1978, then chief executive David Kearns reportedly considered trying to shrink the divide by placing research in the hands of the various operating divisions. Research boss Jack Goldman, who had arrived from Ford in 1968 to run the company's research organization, recalled despairing: "This is unimaginable! You cannot wipe out research in Xerox! I won't let you do it!"

In the end, a compromise may have avoided Goldman's worst fears: research remained a central corporate function, but Goldman stepped down to become chief scientist, allowing Pake to take his post. But even with PARC's original director in the company's Stamford, Connecticut, headquarters, the tide could not be stemmed. At one point in 1982, the Palo Alto lab split into two facilities: the Systems Center under engineer Spinrad, and the Science Center run by Harold Hall. The next year Apple unveiled its Lisa computer—a bedazzled Steve Jobs had toured PARC in 1979—replicating the lab's overlapping windows and graphical user interface that soon became the soul of the Macintosh line. Meanwhile, the PARC exodus of topflight people and technology got so bad that *Fortune* ran a 1983 article entitled "The Lab That Ran Away from Xerox."

Although the trauma would linger for years, by most accounts the healing process began that spring, when William J. Spencer took over the reins of a reunified PARC. A physicist, Spencer had been recruited from Bell Labs to run PARC's integrated circuit laboratory. He stated flatly that at least half the blame for the computer debacle belonged to those in Palo Alto. He told lab managers half their performance reviews would be based on how well they got to know "customers," counterparts around the corporation who would actually develop, produce, utilize, and sell the creations of research. "What I was looking for," he once recalled, "was to see if you could couple PARC closer to the organization and still keep it as creative as it had been in the past."

Legend has it Xerox squandered all PARC's creations. Not true. Laser printing provides almost a third of the company's $18 billion in annual revenues. Meanwhile, three different Ethernets labor away inside the company's high-end Docutech printers, improving communications between processors and slashing the wiring connecting various elements. The highly profitable Docutech—a more modern creation of the Palo Alto lab—brought in nearly $2 billion in revenues in 1997, a remarkable figure that chairman Paul A. Allaire asserted by itself "probably justifies the entire cumulative investment in PARC."

Still, the lost opportunities from PARC's infancy continue to impact research profoundly heading into the twenty-first century. Spencer's emphasis on making his people more aware of Xerox's business side—a strategy he maintained after moving to Stamford as head of Xerox research in 1990—paid dividends by dramatically increasing research's credibility with the business units. Since then, the current head of Corporate Research and Technology, Mark B. Myers, has picked up Spencer's ball and run with it.

Outwardly, the bow-tied Myers, an engineer who joined Xerox upon receiving his doctorate from Pennsylvania State University in 1964, looks

like the stereotypic cautious business manager. But that conservative facade masks a devout student of innovation bent on vanquishing the old
linear model of research that says ideas move from the lab to development to manufacturing to market. Instead, research under Myers starts
and ends with markets. Technologies are designed with markets in mind,
and the innovation process demands constant feedback from customers
to help identify ways to improve the technology. In Myers's view, early
product iterations are almost always "wrong." Only by testing things in
the real world can researchers identify the enhancements needed to turn
even a strong technology into a powerful market force.

The failure to embody this truth was evident when PARC blazed the
personal computer trail, the CRT director says. The lab's gurus, notes
Myers, did not invent the PC, but a networked version they called "personal distributed computing." When in 1981 Xerox finally did market a
computer based on PARC's creations, it unveiled the ill-fated Star. This
was the first system to use a mouse, bit-mapped screen, graphical user interface, easy-to-use software that combined text and graphics, laser
printer, and the Ethernet. However, with businesses just starting to grasp
stand-alone computers, the power of a networked system was unsalable.
Myers calls it a case of research being "profoundly roughly right." PARC
had the right technology, but the wrong market. Steve Jobs and Apple
put it all together.

At the modern Xerox, a far more research-friendly upper management has helped fill both needs—technology and market. Approximately
every five years, to set the strategic framework for all company operations,
Xerox conducts broad reviews that lay out in writing the forces likely to
shape the next decade. The first such major undertaking, inaugurated in
1990, was Xerox 2000. In 1996, the company put out Xerox 2005, with
Xerox 2010 set to be completed by the millennium. Each is the result of
a year's worth of monthly meetings and follow-up interactions between
corporate strategists, consultants, and top level operating group managers
that examines the world in four broad domains: Economy and Geopoli-
tics; Technology and Organization; Products and Services; and Competitors. Inside each domain, planners develop a dozen or so concise
"assumptions" about the next ten years. For instance, one Technology
and Organization domain assumption for Xerox 2005 stated: "The
network becomes the computer. The Worldwide Web will support a
document-based marketplace fundamentally changing the ways that organizations work with one another, and the ways they work internally."

These well-conceived visions of the future form the context for actions by the Corporate Strategy Committee, a seven-person body co-
chaired by Allaire and president and chief operating officer Rick

Thoman that allocates funds and sets the corporation's overall strategic direction. Myers has a seat on this body. The CRT head also chairs Xerox's Technology Decision Making Board, which comprises all group presidents, the four research lab and three technology center directors, and various other representatives from research, business strategy, and manufacturing operations. The board meets monthly, with no one allowed to send substitutes, making funding recommendations to the Corporate Strategy Committee for all research, development, and engineering operations. Myers worries that bringing research into the central management fold risks rendering it less independent and objective, muting voices of radical change. Still, he insists, the setup works to avoid the mistakes that characterized Xerox's darkest hours by ensuring that top people across the innovation pipeline are on the same general page when it comes to technology and strategy.

Against this managerial backdrop, CRT's efforts span four broad research categories. In decreasing order of their percentage of the budget, but increasing order of risk and potential reward:

Core technology covers the development of specific products based on technology ready for use in existing business lines, usually in areas such as printers, copiers, and scanners, where Xerox has special expertise. Risk is minimal with these well-known technologies, so specific projects are paid for by contracts with individual business groups.

Strategic capability studies aim to create broad technology platforms or skills applicable to several businesses. Research costs are spread across business groups, each paying in proportion to what it uses. "This type of investment," Myers writes, "is clear on the possibilities of a specific emerging technology but may be uncertain about its technical feasibility and timing." Examples include working on a unique approach to printing, or developing systems software that spans the entire product base.

Emerging markets and technologies activities explore markets presently outside the scope of existing business divisions but related to Xerox's future vision. Technology from these investigations often provides the foundation for Xerox-funded protobusinesses. Early in 1996, Xerox New Enterprises formed to spin out technologies that don't fit company business lines. Through September 1999, XNE had spawned thirteen spinoffs—including flat panel display concern dpiX and knowledge management specialist InXight Software—that generated some $400 million in annual revenues.

Pioneering research, funded entirely by the corporate office, represents extremely uncertain fundamental explorations into potentially huge areas such as nanotechnology or smart materials that may not pay off for several decades.

─────── ☐ ───────

Xerox's BMWs

In May 1998, about forty-five top Xerox R&D managers descended on a resort and conference center near Santa Cruz, California, to hold the last of five Business Models Workshops aimed at giving them—and their company—a jumpstart on developing a more entrepreneurial mindset. The three-year odyssey took the group through general business strategy to finance and options analysis—all the way to designing product launch plans.

Even before the sessions concluded, the workshops helped shape several projects, including one meant to establish Xerox as a player in the low-end inkjet printer business, a feat that had long eluded the company. More important, the retreats were credited with teaching research managers to build business value into products from the start by considering manufacturability, cost issues, and likely competitor responses. "I just see the impact every time I turn around," proclaims John Seely Brown. "It has helped change the discourse, and I personally think you change culture by changing the discourse."

The catalyst for the "BMWs" was Mark Myers, with the workshops chiefly conceived by Harvard Business School strategist and Xerox consultant Richard S. Rosenbloom. The first session began in March 1995. Rosenbloom and HBS colleagues trotted out case studies—covering Nintendo, Lotus, and General Motors' Saturn cars—to illustrate business strategies. In subsequent meetings, held at or near the business school, they introduced such concepts as cost of capital and discounted cash flow analysis, as well as sophisticated tools like game theory and real options theory often used to evaluate research projects.

During the second session attendees broke into six teams. Pairs of teams were then handed real-life technological "opportunities"—in flat panel displays, network services software, or inkjet printing—emerging from Xerox labs. Their challenge: determine the best business strategy and potential payoff. The teams worked on the proposals during the third meeting and after, then presented reports early in 1997 at the fourth workshop. These "solutions" were intended as educational tools. Still, the best proposal—involving inkjet printing—won a $10,000 prize, and elements of the plan were incorporated into products.

Since large companies like Xerox are adept at developing and marketing products for existing business lines, Myers expects the biggest payoffs to come in emerging businesses and markets. "If you apply your old models to that," he notes, "then you will very likely miss the opportunity."

Xerox missed enough opportunities in the twentieth century.

─────── ☐ ───────

The hoped-for payoff to this approach is a modern day "meter click"—something akin to Xerox's original 914 copier that racked up several hundred million dollars in revenue four cents at a time. As of the fall of 1999, that breakthrough had not arrived. But Myers says one promising arena is the emerging field of Web-based electronic commerce, where scanning, translating, encoding, authorizing, and authenticating documents loom as high-volume services. "So we try to think creatively," he asserts. "How do you create new forms of meter clicks? Every time a document changes state, is there an opportunity to enable that—and can we charge?"

A lot of this comes full circle back to PARC. Brown calls his roughly $50 million budget "almost a round-off error" when compared to total company revenues of nearly $20 billion. Still, his operation, more than any other, is geared to the far-off future—five, ten, even twenty years down the road. The director portrays his researchers as laying a strong scientific base in areas such as lasers while also generating out-of-the-box ideas for novel technologies that can impact the bottom line far out of proportion to their funding. "What PARC really does is create the genetic variance for enhancing the adaptability of the species," he relates. "The species in this case is mother Xerox."

Cascading down the sun-soaked California hills, PARC is a place that seems to crackle with big ideas. Offices and labs are interspersed with pleasant lounges to allow informal get-togethers and brainstorming sessions. Descending through the complex's steep concrete staircases to the various sections, or pods, a visitor finds experts in smart machines and electronic commerce, display screens made of lightweight and inexpensive organic materials, and the lab where researchers take part in the worldwide race to perfect a blue diode laser beam that one day might enable computer printers to match the performance of the best traditional printing technologies.

One big area of focus in the late 1990s is ubiquitous computing, the idea that before too long every workplace device and appliance will contain miniature computers all networked together over the Web. VCRs can be programmed from the office, clocks dial up a server and reset themselves after a power outage, a spouse on a business trip transmits updated flight info via his laptop to an electronic calendar mounted on the refrigerator. And just as the wiring carrying electricity throughout an office or home is hidden, this computing power should remain invisible.

A variety of projects address this vision. One is Workscapes of the Future, kicked off in February 1999 to test the concept of bringing ubiquitous computing to the officeplace. To that end, Beverly Harrison and colleagues from PARC's Extreme User Interface group planted "cow tags" in books, papers, watches, copiers, and other everyday objects. Descended from electronic livestock trackers, these tiny devices transmit a radio frequency signal that extends a foot in all directions, enabling infor-

---

□

John Seely Brown

Computer scientist John Seely Brown came to PARC in 1978 to form its cognitive research group. Named director in 1990 and Xerox chief scientist in 1995, Brown has emerged as a leading thinker on innovation. Following are some of his views:

On innovation: Innovation means finding new markets for existing technologies, new technologies for existing markets, and new technologies for new markets. It means finding new ways of doing business: reaching customers, listening to the market, distributing products, managing people, managing uncertainty, and connecting to the ecologies surrounding our business. Doing any of these things requires that we see *something* differently. . . . The aim is to teach our research managers leading-edge business ideas in order to open up broader conversations between the technologists and the strategists, the marketers, and so on. By changing the discourse and enabling research technologists to traffic in these ideas, we are facilitating the movement of inventions into the marketplace. And this is what we mean by "innovation": invention implemented.

On collaboration: It was very much of an "Aha" phenomenon to realize that the sparks of an insight, a flashing insight, often come from wherever the rubber hits the road. These flashes of inspiration can start well outside of the research labs—they start from anyplace where people interact with the world.

Yet a flashing insight is not really a robust discovery from which you can make a sustainable innovation. It must often be imbued with deeper scientific engineering knowledge. There's often at least as much creativity involved in taking the invention into the world—i.e., implementing that invention—as went into the invention itself.

On attacking problems: We're looking for flips. Everybody else is looking for knowledge from which great things happen. We say, "No. Great things happen from which knowledge flows."

On the people for the job: The researchers we go after have to have a passion for impact—and they have to have deep intuitions. College transcripts are never part of the conversation.

It is rare to find people who are capable of making major scientific breakthroughs *and* of thinking in out-of-the-box ways about using technologies. . . . What we have done at PARC is to create a milieu that is ambidextrous, in that it honors the two worlds.

□

---

mation to be passed to and from "tagged" artifacts without worrying about protocols and compatibility issues.

In one experiment, the PARC team placed tags in flyers announcing details of the lab's weekly lecture series. That enabled anyone equipped with a specially programmed hand-held computer simply to move the device close to the announcement—and instantly, the event was placed in their electronic calendar (an e-mail note would be sent if the seminar conflicted with another appointment). To print copies of the flyer, researchers could then swipe the computer over any hallway copier. Because every copier and computer in the building was linked over the company's internal web, duplicates of any document could be made in the same general manner. The machine would automatically access the latest version of whatever document was open and print it out. If there was any ambiguity, users were queried about their intentions. For instance, a copier might flash a message asking about the number of copies desired, and whether they should be in color.

And that's just a taste of the fun. Books were rigged to automatically summon Amazon.com order forms when placed near a computer. Watches told time; but they also scheduled appointments. Some of these actions were preprogrammed, but mainly they depended on contextual cues such as which computer document was open. In one dramatic demonstration, "tagged" French dictionaries could be placed beside a computer—sending a signal to translate the open document from English into French. Researchers think such technology might be the future of the workplace. The group's mission statement reads: "We believe that this represents the next major revolution in user interfaces potentially as significant as the graphical user interface invented at PARC years ago."

Wandering around PARC, it's easy to find several additional faces of ubiquitous computing. One lies in Mark Yim's modular robots (see box, p. 296). Another is re:reading, a research experiment set to open in March 2000 as an exhibit at the Tech Museum of Innovation in San Jose, California. Designed by PARC's Research in Experimental Documents group and involving the lab's novel artists-in-residence program, the aim is to investigate the relationship between reading and various emerging technological forms. Exhibit-experiments include a tilty book, a table-top display whose text shifts around when it is tilted; robots that read aloud; multimedia books; and a translation station that can quickly convert documents into a number of languages.

A different take on ubiquitous computing can be found in PARC's smart matter effort. Thanks to a collaboration with University of Utah offshoot Sarcos Research Corporation, Xerox computer scientist and electrical engineer Andy Berlin totes a two-foot-long piece of steel embedded with fibers barely thicker than a human hair. An array of MEMS (micro-electro-mechanical systems) sensors connected to the fibers and a computer can detect small stress changes in the steel and order electrodes to send out impulses that grow or shrink the fibers to counteract that stress. Berlin envisions shape-shifting bridges and buildings laced with the technology instantly adapting to tornadoes or earthquakes. More along Document Company lines, though, he can show off a MEMS-based prototype printer where electrostatic charges control jets of air that can move paper along without rollers, hinting at the day when sheets can be loaded faster without buckling, and both sides can be printed simultaneously.

Another far-out form of smart matter is smart dust. Imagine tiny dandelion-grain-sized sensors that carry a microphone, video camera, and transmitter. Thousands could be disbursed to gather intelligence over a suspected terrorist site or drug lab. But each has limited range and battery power, so these motes must work ensemble to gather and transmit their data. It all poses a massive distributed processing problem—in power management, communications, transmission, image processing, data processing, and so forth—beyond anything solvable through popular programming languages. PARC has worked at cracking this distributed processing problem for close to fifteen years. One of its approaches, Aspect Oriented Programming, was generating a buzz in software circles in 1999. That's because beyond the talk of spy toys, the era of massively distributed processing looms on the horizon as an essential element of ubiquitous computing—whether it's for linking computers over the Internet, MEMS devices, robots, or just hooking up the several hundred sensors and actuators inside future Xerox copiers.

Gregor Kiczales, leader of the effort, explains that a big problem with linking lots of objects is that a change in one program impinges on many others—kicking off a nightmare of a digital domino effect. PARC's dynamic AOP language seeks to enable programmers to automatically update all affected applications simultaneously. Kiczales can't yet point to any million-lines-of-code systems where AOP is up and running. However, preliminary empirical data with smaller programs shows that changes normally requiring code writing in ten or twenty places can now be done once. If these results hold up, long before things like MEMS earthquake sensors or modular robots become reality, AOP's payoff could come in cheaply and rapidly developed software—particularly in the emerging Internet world.

So far, anyway, ubiquitous computing remains largely a laboratory vision. Other PARC researchers, however, are out in the field working with customers to deliver future generations of office products tailored to meet specific needs. The most extraordinary of those efforts revolves around the lab anthropologists. Though the core group was set to disband at the end of 1999, its pioneering "workplace studies" were part of a larger and rapidly growing social sciences effort PARC director Brown predicts will become increasingly important in the twenty-first century.

To understand the role of anthropology in this legendary den of

---------------------------------- □ ----------------------------------

Mark Yim's Modular Robots

It's an all-too-familiar image: rescue workers combing through rubble in the wake of an earthquake, bombing, or natural gas explosion, searching for people trapped in tons of twisted metal. The hunt can last days—and once in a while it pays off. But for every person found alive, think of how many others might have been saved if only help could have come more quickly.

Mark Yim does. Moreover, the thirty-three-year-old PARC mechanical engineer is laboring to provide that assistance whenever and wherever it's needed—in the form of modular, reconfigurable robots. The idea is that these motor-driven mites would scurry into the field to aid disaster crews. Yim envisions them assembling into a wheel to roll over flat ground, shape-shifting into a spider to pick their way over uneven surfaces, and maybe morphing again into a snake to slither into narrow spaces like pipes. Each robot would carry sensors that can tell whether someone is breathing. "And then, once it's found a person," he fantasizes, "it can reconfigure into something like a protective dome around the person."

It's a wild dream. But Yim has been trying to make it reality for nearly a decade, ever since his Stanford graduate student days, when he launched an exhaustive study into the ways he might get robots to move: snake-ish slithering, crawling like an earthworm, tumbling, cartwheeling, rolling, and more. The result was an elaborate taxonomy of gaits, which he then began incorporating into machines—a vision he's been refining ever since joining PARC in 1996.

Yim crafts his robots into bendable stainless steel modules, or cubes—each about two inches in every dimension. Ultimately their ability to reconfigure will hinge on accurate guidance—probably through light emitter-detectors that enable modules to sense their positions relative to each other. Once two modules steer close to one other, a docking mechanism will lock the male and

---

computing, think of a company as a nation whose ranks are composed of successively smaller groups like tribes and clans—entities called business units, districts, and customer service teams. Employees are members of all these bodies at once, but identify themselves as belonging to a particular one depending on what they're doing—and with whom. "For an anthropologist," former PARC researcher Julian E. Orr once wrote, "this is oddly reminiscent of the oppositions within segmenting lineages of the Nuer or the Afghans, or of the *nisba*, an infinitely branching Moroccan system of personal identification." Through their strong tradition of cross-cultural comparison, anthropologists strive to recognize these shifting allegiances and identify the subtle ways people learn and share their knowledge as they transform themselves from employee to boss to co-

female ends into place. Yim's modules don't have wheels. But small motors inside each unit provide the power for locomotion. Although Yim plans to program the creatures for autonomous exploration, initally they'll be radio-controlled, with a camera on the lead "bot."

Yim sees a threefold beauty in such a system. First is versatility—the ability to navigate past different obstacles in varying terrain. A second strength lies in redundancy: since modules are identical, several could malfunction without ruining the system. Finally, mass production should bring down module costs.

Of course, there's a catch. "These are the three promises," Yim notes. "In actual fact, none of these may be true." Their all-terrain abilities may not pan out as hoped. Then, too, while having more parts gives backup protection, it also raises the odds *something* will go wrong—and brings programming headaches. Finally, having lots of mass-produced parts might bring down individual module prices, but it could drive up total system costs.

Still, Yim "plugs" on. In the fall of 1998, his twelve-module robot morphed from a rolling loop to a slithering snake. While this easy shape-shift merely involved unlocking one link, Yim believes the event marked "the first instance of a robot that can reconfigure to do two different types of locomotion." A year later, he was close to getting his contraption to transform into a five-legged spider as well.

Yim isn't the only one on the "mobot" trail. Researchers at NEC, Dartmouth, and Johns Hopkins University are among those exploring similar ideas. However, the PARC scientist enjoys financing from the Defense Advanced Research Projects Agency. And he's plowing ahead to find even more uses for his robots. Planetary and deep sea exploration rank high on his list. On the disaster front, newer ideas include robots that reconfigure into splints or supporting pillars. "Since you don't know what you'll find, having a system that can change its shape to suit the task is very useful," Yim argues.

□

worker, for example. That, in turn, can pave the way for technologies that better fit the job.

Anthropologists have been part of PARC's staff since 1979, when Lucy Suchman arrived from the University of California to study everyday life in a big company. A down-to-earth researcher who feels people are too often painted with a broad brush, Suchman questioned the computer scientists' assumption that office work was so straightforward and procedural that it should be tailor-made for computerization. Therefore, she began studying the most seemingly procedural group she could find: accounting. Suchman's investigation showed that rather than simply working linearly—receiving an order, processing the paperwork, shipping the goods—clerks actually did many tasks in parallel. For instance, a customer might phone in an order, assuring the clerk the paperwork was on its way. The goods would then be shipped before all the forms had been completed. Relates Suchman, "In the end the record will have all the necessary paperwork. But if you just in a unilateral way insisted on doing things according to the rules you would actually make your customers very unhappy, and it would be an inefficient way of doing business."

Suchman's findings fit into a small movement already underway at PARC to shift research from an office automation viewpoint to the wider perspective of the "knowledge worker." The basic thrust was to tap the power of computing—for such tasks as generating, recalling, printing, and transmitting forms—to support the way work really got done. At its core were technologies like the graphical user interface. Building programs around a highly intuitive desktop metaphor provided the benefits of computerization without stripping people of control and flexibility.

Adding fuel to this movement boosted the status of Suchman's anthropology work. But the real milestone came about a year later when she and computer scientist Austin Henderson made the films showing researchers grappling with the 8200 copier. The machine boasted powerful new capabilities such as automatic feeding and double-sided copying. But what Xerox had billed as a self-evident copier was proving a disaster. As customer complaints poured in that it was too complicated, the problem was brought to PARC. Suchman and Henderson had an 8200 installed, set up video cameras, and asked colleagues to try it out.

Around the time John Seely Brown aired their results to various management groups, the researchers did the same for Xerox engineers in Rochester. The more serious film—sans Jane Fonda—was called *The Machine Interface from the User's Point of View*. Since it also featured doctorate-wielding computer scientists, dismissing it on the grounds people were technically incompetent became even more difficult. But rather than chastising engineers over the 8200's shortcomings, Suchman tried a more positive tack. "The point I really tried to make was that they should not take it as evidence of their failure, but as evidence of the difficulty of the problem they as designers had to solve."

Rochester rose to the challenge. Instead of the 8200's flashing error codes that had to be looked up in flip cards attached to the machine, a display panel on Xerox's Series 10 and 50 copier lines shows a picture of where the trouble lies. The much friendlier user interface helped slash the average time needed to clear a paper jam from 28 minutes to under a minute. Of more fundamental importance, the films opened Xerox's eyes to the potential of workplace studies. "That was what really got us going," asserts Brown, "recognizing that it's technology in use that creates value, not technology per se."

Behind that realization, PARC's anthropologists moved out of the lab to encounter machines in their natural habitat: the real world. From a lone practitioner initially focused simply on observing work practices, such "ethnographic" efforts swelled to about a dozen anthropologists, artificial intelligence experts, and computer scientists. Their twin goals—of probing fundamental aspects of work while also designing technologies to make specific jobs easier—have established anthropology as at once one of PARC's farthest-out pursuits and one of its most targeted. "The idea of a corporate research center investing in anthropology may seem exotic," admits Suchman. "But in many ways we think of ourselves more as champions of the mundane. Others dream of far-out widgets. We're saying we really have to give people *more useful* widgets."

PARC's anthropologists beat two main paths: workplace studies aimed at designing new office technologies and parallel attempts to strengthen internal Xerox operations such as service groups. Suchman heads the first effort, which really got underway in 1989 with the study of two airline control operations—handling gate assignments, meals, luggage, and the like—at San Jose International Airport. In the early 1990s, a second project was initiated at a Silicon Valley law firm.

While the airport project was limited to detailed observations of the workplace, its successor at the law firm marked an early attempt to design technology based on what the ethnographers observed: "case-based prototyping." The two-year effort focused on "M," an attorney whose file cabinet held records that served as templates for drafting other documents.

After pondering videotapes of cabinet-scouring lawyers in action, Suchman and colleagues Jeanette Blomberg and Randy Trigg copied, scanned, and digitized 862 documents—a quarter of M's cabinet. Computer scientist Trigg then teamed with other PARC researchers to design and build a prototype search aid able to retrieve not just text but document images, which could be presented as thumbnail reproductions spread across the electronic "desktop." Although the shrunken images were illegible, searchers could find the right ones by recognizing a letterhead, or even the pattern of words on a page—much the way attorneys might leaf through a pile of papers without reading each one.

Although the search technology was improved several times in consultation with M, it was never intended as anything more than a prototype. In the next field study, however, which started in 1996, the researchers were attuned to commercial possibilities from the start. The "tribe" in question this time was Caltrans, the California Department of Transportation, which was designing a replacement bridge to span the Carquinez Strait at the northeast end of San Francisco Bay.

The PARC team on this project consisted of anthropologists Suchman and Blomberg, joined by Trigg and fellow computer scientist David Levy. Their efforts centered on about a half-dozen engineers based at the Caltrans district headquarters in Oakland, across San Francisco Bay about 45 minutes northeast of Palo Alto. To complete its work, however, the Caltrans group needed to interact with consultants, contractors, and public and government organizations. So the Xerox contingent dutifully followed them through the urban jungle to a variety of meetings, making extensive visits to towns bordering the planned construction site, another half-hour north of Oakland.

The Caltrans engineers assembled information from all these venues into project files kept in three-ringed binders. The challenge for Xerox came in moving this diverse body of graphical, printed, and handwritten documents—engineering drawings, maps, surveys, letters, memos, and more—from the paper world into the digital domain. To that end, PARC staffers helped scan and digitize these documents, then provided technologies for indexing, accessing, and viewing the data on a Web-based interface.

With every modern office equipped with a scanner, this digitizing may seem straightforward. It's not. Scanned documents are typically converted from bit-mapped images to text—allowing users to cut, paste, and perform other word processing tasks just as if they had been created on a computer. However, the Optical Character Recognition process that makes this possible cannot handle such things as drawings, photographs, and handwriting—and the Caltrans records were loaded with all three.

To get around this problem and give Caltrans access to all its records in digital form, the PARC team created a hybrid application called the Integrator that allowed the engineers to search, retrieve, and peruse both text and image-based documents together. While the Integrator incorporated the novel thumbnail technology prototyped to view M's records, what really made it unique was the ability to search for images based on features such as signatures or letterheads. Suppose, for example, that an engineer needed to know the names of everyone with whom she had corresponded in a given month. The Integrator could retrieve the letters bearing her signature—and provide a summary listing all the addressees. With a click of the mouse, the engineer could call up and print any of those letters, or send them over the Web to a colleague. In late 1997, the PARC team filed patents on the image-based search and summarization features.

For Xerox, beyond helping a major customer (the state of California) maintain records, the arrangement provided a real-life test bed for a technology aimed at a variety of products. The Caltrans study was done in close collaboration with Xerox's Office Document Products Group, which eventually used the Integrator in its Document Centre Systems line of networked multifunction machines that copy, scan, store, print, and fax all from the same box. Such a strategy marks almost a reversal of normal product development. Typically, notes Suchman, developers might tailor a general technology for individual customers. In the Caltrans case, researchers began by studying a particular job, developed specific tools for helping engineers—then worked backwards to create a powerful general technology applicable to many domains.

A second thrust of workplace pursuits—focused on internal Xerox operations—can be traced to Julian Orr, a bearded, motorcycle-riding PARC technician-turned-anthropologist. Much like with Suchman's study of office clerks, Orr observed that service people relied on knowledge obtained outside their training sessions or manuals. For instance, individual copiers have various idiosyncrasies, giving rise to problems that must be divined through on-the-spot diagnosis and relayed through informal storytelling sessions over lunch or at the parts drop.

In 1992, Orr initiated a Denver-based field test that gave Xerox technicians two-way radios so they could share tips and insights without having to share lunch: all U.S. service reps have since gotten radios or cell phones. His work also helped inspire Eureka, an effort between PARC's Smart Service team and a Xerox unit in France that allowed technicians to distribute choice tidbits not by radio, but via a digital "watering hole" accessible through the country's ubiquitous Minitel electronic telephone directory system.

The brainchild of PARC computer scientist Olivier Raiman, Eureka took off in 1995 with a large-scale field effort. The Minitel-linked central data base was created to hold servicing tips arranged by category and machine. Any technician wishing to contribute a new item first sent his suggestion to a team of validators who made sure the idea was valuable. Approved submissions were placed in a New Tips category, along with the originator's name.

While Eureka included no advances worthy of reporting in a computer journal, the innovative *arrangement* of technology produced a 5 to 10 percent savings in parts and labor, catapulting the French service force into one of Xerox's most efficient. Between 1995 and the end of 1997, about 20 percent of French technicians had submitted a validated tip. Eureka was consulted seven thousand times a month, an average of almost six times for each French service rep. The electronic hotsheet proved so successful that early in 1997 it was extended to 1,500 technical Xerox reps in Canada. By fall 1999, a similar rollout had reached the 12,000-strong U.S. service force, as well as Xerox technicians in the

United Kingdom, the Netherlands, Brazil, Argentina, and parts of the Asia-Pacific region—with an improved version incorporating Web technology planned for introduction later in the year.

Even as Eureka went global, PARC prepared an offshoot called Alliance aimed at bringing Xerox service reps together with salespeople. Technicians often know when customers need new copiers. Yet because of differences in the way sales and service territories are assigned, even if one wanted to alert a salesperson to a hot prospect, it would be difficult to find the representative handling a particular account. Much like Eureka, Alliance uses e-mail and central data bases to make the connection. Tipsters also garner a modest finder's fee such as a dinner for two. Not long after its fall 1997 launching, the program was claiming over a million dollars in deals generated each month directly from service people in France alone. Says PARC research fellow Daniel G. Bobrow, who heads the Scientific & Engineering Reasoning Area that developed Eureka and Alliance: "It's easy enough to use that a service person can send a message and the sales person can be calling the customer while the service person is still there."

Because the very introduction of technology often spawns new problems, all these efforts are fraught with perils that provide grist for the social sciences mill. Bobrow reports that while it's relatively easy to get top people to sign on to such programs, effecting changes gets harder as responsibility for implementation is delegated down and budgets get disrupted. Even if managers spring for laptops, there remain the questions of how workers will accept and utilize the introduction of computer-based tools—and whether the answers can be fed back into the design of additional tools.

In Eureka's case, such obstacles—combined with the absence of a Minitel-like system in other parts of the world, which dictated a complete technical revision—delayed the technology's introduction longer than expected. Still, the enthusiasm by which it was embraced by technicians eventually enabled it to happen. Alliance also seemed to be experiencing delays, although by October 1999 the responsibility for its rollout had shifted from research to the service organization.

■

Some researchers thrive in seeing their ideas make a quantifiable difference in the real world, and willingly bear such frustrations. Others are far more content theorizing about the nature of work in a scientific paper. "One of the good things about working at a place like PARC is that you can do both those things," maintains Lucy Suchman. "I do think that PARC is quite unique in its continued commitment to having near-term problem solving and long-term research both be part of what the place is about." Brown agrees, making the distinction, though, between long-term research per se and "radical research that has a long-term, or lasting, impact."

Making that distinction more apparent, one era of Xerox's social sciences studies was coming to an end in late 1999—with another getting under way. Suchman's anthropological group was set to disband around year's end (Julian Orr had already left PARC early in the year), with its leader and other members set to take jobs in academia or consulting. But that hardly meant the lab was abandoning its pioneering workplace studies. In fact, says Brown, it's expanding them. Early in 1999, he hired Marilyn Whalen, recently of Palo Alto's Institute for Research on Learning, to head what was being called KIPA, for the Knowledge Interaction and Practice Area. By that fall, Whalen's group was up to five persons—all social scientists—who were charged with conducting workplace studies not unlike Suchman's but with the intent of turning around new technologies for Xerox's customers much faster. "It's almost like this has become so successful that we now need to move into almost a production case," says Brown. "These now all become central as opposed to peripheral to where the corporation is really going."

HEWLETT-PACKARD—WBIRL (WORLD'S BEST INDUSTRIAL RESEARCH LAB)

"Most companies die not of starvation but of indigestion."
—SAYING FONDLY REPEATED BY HP CO-FOUNDER
DAVID PACKARD

"I actually think corporate research is not very difficult. It all boils down to three things. You have to be relevant. You have to do something that's novel which adds value. And you've got to transfer the technology."
—RICH FRIEDRICH, HP LABS ENGINEER

HP Laboratories

Central research established: 1966
Company sales 1998: $47.1 billion
1998 R&D expenditures: $3.36 billion
1998 HP Labs budget: $278 million
Funding paradigm: 100% central funds
Technical and scientific staff: 800
Principal research labs: Palo Alto, California; Bristol, England

NOT LONG AFTER Joel Birnbaum took the reins of Hewlett-Packard Laboratories in 1991, he began to suspect his old darling was giving the company a stomach ache. That is, researchers were so intent on helping HP's existing businesses—its traditional sources of nourishment—that they risked missing critical opportunities to spark growth in entirely new fields. Birnbaum could not help but remember HP co-founder David Packard's maxim: "More businesses die from indigestion than starvation."

Birnbaum's fears—and subsequent actions—highlight the essence of HP Labs and its quiet ascendancy into one of the world's best industrial research organizations: an ability to reposition itself to meet future corporate needs before crisis strikes. Although Birnbaum had spent years at IBM, where he'd witnessed the myriad dangers of researchers working on problems too removed from company businesses, he felt convinced that unless the Labs expanded its horizons, HP would degenerate into a much less innovative firm.

He therefore conceived a bold strategy called MC^2 that aimed to fuse HP's core strengths in Measurement, Computation, and Communications to build new billion-dollar businesses. Then, to motivate researchers to think in such big terms, Birnbaum launched WBIRL—an initiative to make his operation the World's Best Industrial Research Lab.

The result was that throughout the turbulent early 1990s, when places like Bell Labs and Birnbaum's alma mater IBM dramatically scaled back research, the HP Labs annual budget soared from $100 million to $270 million. Simultaneously, the company launched exploratory studies into quantum mechanics and other areas of basic science designed to place it at the forefront of computing.

The smooth ride didn't last. Birnbaum retired in early 1999 under some duress—without his MC^2 campaign yet delivering on his dreams. Meanwhile, in March 1999 a temporarily slowed-down HP announced plans to split its nearly $8 billion measurement business into a separate entity—Agilent Technologies Inc.—leaving the Labs to carry on without the full synergy its maverick former director envisioned.

Still, if ever a research organization knew how to weather storms, it's HP Labs. Formed in 1966, it has always been a blue-collar—not blue-sky—place that excels at being an engine of growth for its parent. Indeed, by one count upwards of 80 percent of current company products originated in research. Past contributions include important advances in calculators, light emitting diodes, thermal inkjet printing, and laser printing—and heading into the millennium, a new wave of innovations in everything from digital imaging to information appliances and quantum computing stood poised to justify Birnbaum's expectations of future bonanzas. "We're not here to win Nobels or Turing awards," he once asserted. "Our people's rewards come in terms of the satisfaction of seeing their work reach the market—and in their profit-sharing checks."

The main HP Labs facility in Palo Alto spans four interconnected buildings midway up busy Page Mill Road—lower on the same hill as arch-competitor Xerox PARC. Featuring a cavernous central room—Building 3—with brightly colored ceiling pipes and ducts, and collegial cubicles instead of isolated offices, the place fairly embodies the HP Way, the guiding principles established by founders William Hewlett and David Packard that include teamwork, trust, and "management by walking around." Until research overran it in the mid-1980s, the edifice served as Hewlett-Packard headquarters. Even after new corporate digs went up, the founders kept their offices inside the Labs, refusing to change their original metallic green carpeting.

Hewlett-Packard—a coin flip determined top billing—was incorporated on January 1, 1939, out of a one-car garage in Palo Alto that a half-century later became a California Historical Landmark designated "the birthplace of Silicon Valley." The company's initial product was an audio oscillator important in a variety of electronic testing applications: an early customer was Walt Disney Studios, which purchased eight of the $71 units for the sound system in *Fantasia*. First year's revenues totaled $5,369, with more than $1,500 in profits. HP has run in the black ever since.

Sales exploded during World War II. But HP really didn't hit its stride until the early 1950s, when it expanded into a wide array of testing and instrumentation equipment. Its remarkably steady growth continued for the next forty years. In 1999, Hewlett-Packard ranked as the world leader in laser, inkjet, and color printing; stood first in Unix-based computers and second in workstation sales; and was the number three or four personal computer maker in the United States. In laser printing, top competitors include Xerox and Canon. When it comes to workstations, Sun Microsystems is the arch rival. However, across the broader front—from PCs to scalable servers, and e-business services—HP stacks up more closely with IBM. The dominance of its computer and printer businesses, representing some 83 percent of total sales, spurred the decision to shed the original measurement and equipment side: until its July 1999 christening as Agilent, the initially unnamed spinoff was dubbed "NewCo" around HP.

The first hard step toward creating a central research organization was the hiring of Bernard M. Oliver as director of research and development. "Barney" had been an early 1930s Stanford classmate of Bill Hewlett, Dave Packard, and future production manager Ed Porter—four friends who would become the core HP management team—in a graduate course taught by legendary electrical engineering professor Frederick Terman, a Silicon Valley pioneer. Terman let Oliver, a junior, take the class on the stipulation that if he failed the first midterm examination, he'd drop out. Oliver got the highest grade on that test—and every one that year.

After graduating from Stanford at age nineteen, Oliver completed his doctorate at Caltech. He then spent more than a decade at Bell Labs before his old friends Hewlett and Packard lured him back to California in 1952. At the time, the company maintained a small centralized R&D effort. Five years later, however, fearing the now 1,500-strong firm was growing so large it might lose the close personal ties between engineering, production, and marketing that had proven so instrumental to its growth, the founders separated into four operating groups: Audio-Video, Microwave, Time and Frequency, and Oscilloscopes. Research and development also decentralized, with Oliver in charge of coordinating this diverse body of work. But while the arrangement did generally help the operating groups innovate faster, slow growth in its mainstay areas prompted HP to diversify—and here the revamped R&D structure worked against it. As Hewlett once noted, "it was very difficult for the R&D sections of the operating divisions to carry on much innovative work outside of their own fields of specialization. The engineering staffs felt that they needed all the R&D dollars available simply to maintain or enhance a competitive position in their chosen fields."

Ultimately, the quest for verdant fields of growth led to the 1966 creation of HP Labs. The company had previously formed an Advanced Research and Development (ARD) laboratory to tackle longer-range problems than possible in the operating groups, along with a subsidiary called HP Associates that pursued transistors, light emitting diodes, and other solid-state investigations. These two enterprises formed the nucleus of HP Labs, which took root in ARD's home in Building 1, adjacent to the corporate headquarters. Initial staff included some 150 workers divided into four main areas: solid-state physics, physical electronics, electronics research, and medical and chemical instrumentation research.

Barney Oliver would direct the Labs for some fifteen years. Harkening back to the practice of GM's Boss Kettering, he kept a fully instrumented laboratory in his office and ultimately garnered sixty patents. Oliver was also an avid amateur astronomer who played an instrumental role in launching SETI, the search for extraterrestrial intelligence. Marvels Len Cutler, a charter Labs employee who became its most eminent scientist: "Barney was certainly genius quality. He was a mathematician, engineer, and physicist, and in fact almost anything you wanted to name. And he had the ability to see optimum ways to do things."

Under Oliver's tutelage, HP Labs emerged as one of the world's foremost applied labs. The first big gains—continuing earlier company work—came in quartz oscillators and Cutler's atomic clocks, which enabled precise time keeping and reliable frequency standards in a variety of products. In the late 1960s, the Labs produced the popular HP9100 desktop programmable calculator: when employees found out the machine was coveted by science fiction guru Arthur C. Clarke, they took up a collection to buy him one. The grateful writer named the machine HAL Junior, after the mainframe computer in his movie 2001: A *Space Odyssey.* The 9100 engendered the first hand-held electronic calculator—the HP 35—which redefined the calculator market. A cornerstone of its success was the bright display stemming from HP's advances in light emitting diodes. On the verge of the twenty-first century, Hewlett-Packard remains a leader in LED technology.

Thermal inkjet printing also emerged from the Labs. In 1978, John Vaught, a technician working on thin-film technology for integrated circuits, noticed that when he applied electricity to superheat his medium, droplets of fluid lying under the film splayed out. This led to the first thermal inkjet printer—the ThinkJet—then in the mid-1980s spawned the wildly popular DeskJet. Meanwhile, by combining licensed laser technology from Canon with its own innovations, the company also became the frontrunner in laser printing.

Even while helping launch the printer businesses, the Labs was busy rescuing Hewlett-Packard's struggling minicomputer business through its pursuit of Reduced Instruction Set Computing, a streamlined architecture that enabled the same operating system and software codes to run across all of HP's previously incompatible product families. RISC was invented at IBM Research, but HP Labs' early 1980s advances led to its PA (Precision Architecture) systems, spurring it to the top spot in Unix-based RISC computers.

In short, the Labs' success over its first thirty-odd years can hardly be overstated. In 1965, the year before it formed, company sales totaled $165 million. By 1970, the figure doubled to $365 million. Annual revenues then soared to $3 billion in 1980, quadrupled to some $13 billion a decade later—and crossed the $50 billion barrier by century's end. Cofounder Hewlett attributed the lion's share of HP's growth to its Labs. "A number of major new product lines have had their genesis from this source, not the least of which is the whole field of computation," he once wrote, apparently referring to its computer, calculator, and printer businesses. Indeed, after the year 2000 split, printing and computation *are* the new HP.

∎

Barney Oliver retired in 1981, devoting himself largely to SETI. The new R&D boss was John Doyle, an experienced manager who would become an HP executive vice president. But Oliver's real successor—as research director, and eventually as senior vice president for research and development—was Joel Birnbaum.

A nuclear physicist-turned-computer-maven, Birnbaum had served fifteen years at IBM Research, including five as director of computer science. In 1980, in preparation for Oliver's retirement, he was recruited to head HP Labs. But his first job—the reason he left IBM and on which his taking over research was predicated—was to rescue HP's moribund computer business by developing the RISC technology his former employer had largely ignored. To that end, he launched HP's Computer Science Center as a major new research arena—and did not take over the Labs until 1984.

Birnbaum, the first director of HP Labs recruited from outside the company, would head research for a decade. During his first stint, Hewlett-Packard's ongoing inkjet and laser printer efforts were commercialized. Then, in 1986, to ready RISC for commercialization, he shifted to development, only returning to the Labs as vice president for R&D five years later. A deep thinker on research matters, he put the defining stamp on the Labs for the rest of the century (see box, p. 310).

For almost the entire time Birnbaum was away, the Labs was directed by Frank Carrubba, who left HP in 1991 to head Philips's worldwide research organization. By the time Carrubba had taken over, though, research was so far along in fulfilling its original mission that chief executive John Young had ordered the Labs not to worry so much about creating additional businesses as improving existing ones. But when Birnbaum returned, he felt things had gone too far—that researchers worried so much about the here and now that they risked missing vital emerging opportunities.

Spurred by these fears, he did a quick survey of the future. Projecting HP's historic 19 percent annual growth rate out about a decade, he found that the company should become a $100 billion concern early in the twenty-first century. Yet when he tallied up even the most optimistic business unit forecasts, expected sales fell short of that figure by some $20 billion. Birnbaum was stunned. "The growth fuels the innovation—and if you don't get the growth you become much more of a noninnovative, middle-of-the-road company than we like to think of ourselves," he explains. How, then, to get on a faster track? One option was to start gobbling up companies. But better, reasoned Birnbaum, was for HP Labs to return to its original charter and create those new areas of growth itself.

So arose the early 1993 drive to create the World's Best Industrial

Research Lab—as well as the MC^2 concept. Birnbaum's reasoning on this second initiative was that while almost every information technology concern worked on the two "Cs" of computing and communications, only HP, Hitachi, and Siemens possessed the added measurement expertise—and so the ability to fuse these three bedrock areas represented a real strategic advantage. To augment efforts to spark the desired synergism, he created the New Business Development group, a handful of researchers charged with identifying new billion-dollar business opportunities (see box, p. 314).

MC^2 got off to a strong start. One early payoff was acceSS7, a Bristol-led software program that monitors telephone networks for signs of congestion and technical problems and was designed to enhance HP's Signaling System No. 7 computer network. Brought out around 1955, this offering addressed a glaring business hole—namely, that while HP had long sold instruments to phone companies, it had been unable to sell them much in the way of computer systems and software. AcceSS7 changed all that. The project teamed Bristol Lab computing researchers with measurement experts in a business unit division in Scotland and, later, engineers at HP's Telecommunications Infrastructure Division in Grenoble. At its core were sensors that collected real-time data about phone operations. To make sense of this shifting mass of information, Bristol researchers coupled it with HP Ease (for Embedded Advanced Sampling Environment), a unique user interface that allowed operators to easily visualize the whole network—through such details as bar charts and the use of different colors to represent traffic volume—and then click on any aspect of their screen for more particulars. An unanticipated dividend of acceSS7 was its ability to spot patterns indicative of fraud, such as a sudden volume of long-distance calls from one number. Within a few years, it helped HP's computer systems business with phone companies grow to some $1.5 billion annually. "It was a tremendous success," notes Birnbaum. "Almost all the technology that fueled that came out of HP Labs."

On the heels of acceSS7 came the New Business Development effort in digital photography—and a host of other promising opportunities for MC^2 in agriculture, bioinformation, and medicine. The Labs also stepped up the pace of innovation on other fronts—including entering into a partnership with Intel to develop what it called Explicitly Parallel Instruction Computing. The first EPIC chip, based on HP Labs technology and dubbed Merced, came out in 1998—fulfilling the dream of fashioning a computer architecture that could run software created for both Unix systems and Microsoft's Windows.

Whether it would all add up to the $20 billion in added annual revenues that Birnbaum intended remained up in the air as the century

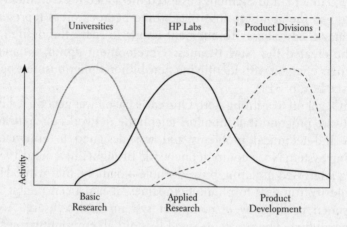

Joel Birnbaum: The Role of HP Labs

One of Joel Birnbaum's first acts after joining Hewlett-Packard in 1980 was to create a drawing that illustrates the way HP Labs sees itself—squarely between product development and basic research. Behind this long-standing philosophy, while other big labs scrambled for their very existence, his nearly tripled in size throughout the early 1990s, expanding into quantum physics and other key areas of basic research.

The crux of HP's approach to research, says Birnbaum, lies in the difference between invention and innovation. "Invention comes from a root *(inventus)* which means to discover or come upon—and it's about curiosity-sponsored research or the quest for knowledge," he explains. "Innovation has the word *nova* in its root and it really means to make new again . . . to take something that was already discovered but reduce it to practice or combine it with other things."

Basic research is mostly about invention and gaining a fundamental understanding of things without trying to create a product. This is mainly the job of universities, national labs, and private institutions. Applied research—the

wound down. In any event, his success in convincing the Executive Committee that the Labs held the key to future growth was behind its trend-busting surge throughout much of the 1990s. Once HP embraced this strategy, it made no sense to stymie Labs growth, Birnbaum once related. "It's not some capricious thing. . . . It's a natural conclusion, just like breathing."

Early in 1998—within months of being asked to stay on past his six-tieth birthday, HP's normal retirement age for upper echelon execu-

prime mission of HP Labs—is chiefly about innovation. It occupies the space between basic research and development, where innovations are readied for manufacturing. However, it's impossible to be good at the middle without strong connections to either side.

Birnbaum asserts that many labs previously practiced the "air conditioner" model of research. The analogy isn't perfect. But just as buildings of a certain size are sealed off from the outside world—that is, relying on air conditioning rather than windows for climate control—corporations that reached a critical point in their growth established central research labs that in effect isolated the company's technologists from its business divisions and customers. The typical idea was to hire elite scientists and give them a free hand—employing sharp-eyed scouts to harvest any fruits. Companies hoped these brains would bring them prestige—and maybe even generate some profits. However, says Birnbaum "what you decidedly don't have as a model is satisfaction of your customers in the product divisions . . . [or] how much money you return to the company—because that's not the purpose." Not only is this framework inherently inefficient, it falls apart during hard times. "So when you take these kind of guys and you put them in the mode that says we need results now, you've got a very dramatic cultural change."

HP Labs, says Birnbaum, was created by "pragmatic men of great vision." From the start, the model was to make money over the long haul. To that end, the Labs has always collaborated closely with development groups, while also supporting university research and conducting fundamental studies in a few pivotal areas of materials science and physics—but not things like cosmological relativity, one research area at his alma mater IBM. The Labs hires top talent. But researchers get no special titles or pay—and they're expected to produce results just like everyone else at HP.

Therein lies the key to its success. "The reason why everybody else was shrinking and we were growing was not because I have a nice smile, or they said, 'Ah, here's our chance to have the biggest air conditioner in the world,'" says Birnbaum. "It's because for thirty years a very large part of the revenue which comes out . . . started here." That is, lab windows have never been closed.

□

tives—Birnbaum was among a cadre of nonbusiness unit officials dropped from the Executive Committee. That December, chairman Lew Platt came to the Labs and noted that the research director's long-dicussed retirement would soon take place, but that a successor had not yet been identified. In the interim, the job would be split by his two top hands: Dick Lampman, who headed the Information Technology Center, the largest of HP Labs' two main divisions, and Ed Karrer, his counterpart at the Microelectronics and Measurement Solutions Center.

Even Labs insiders couldn't fathom the moves. Birnbaum would remain as HP's chief scientist, a coveted position that gave him a pulpit to espouse the vision of ubiquitous computing—a movement he had long been promoting. At the same time, the hunt for a successor had been going on for several years: why hadn't one been named? That question wasn't answered until a few months later, when the company revealed the plans to split off its measurement arm.

These events took place as HP fell under intense pressure to reverse a lackluster performance that had seen quarterly revenue fail to grow faster than 4% for close to a year. Indeed, in fall 1998, Platt had pushed through a 5 percent reduction in the Labs budget for the next fiscal year, forcing the research boss to cut about fifty employees. Moreover, although acceSS7 had proven a brilliant success, the company's digital photography efforts had gotten off to a slow start, while other faces of MC^2 had moved even more sluggishly, apparently causing management to lose faith in the vision.

In any event, Birnbaum had locked horns repeatedly with Executive Committee members who felt research needed to serve the business divisions more closely—especially its computer side. He had also vehemently opposed the split-up of HP. As he explains, "From my point of view the company's best future lay in the integration of these difficult technical capabilities. . . . I just couldn't understand any reason why that [the split] made any sense." Birnbaum even worried that breaking up the company could strip the Labs, which he felt was barely big enough as it was, of its critical mass. "I said, Whoops, that's the end of the Labs—forget it," he relates. Indeed, in the wake of his retirement, rumors swirled that the Labs' very existence had been threatened. Those rumors seem greatly exaggerated.

Still, early in 1999 HP Labs faced a challenging period. Not only was Birnbaum gone from the helm and its budget reduced, the place was dividing in two—along the lines of the company's split. It helped that both Lampman and Karrer were soon named permanent heads of the respective labs. But even by mid-year, the exact division of personnel and projects had not been formalized.

As head of the new HP Labs, which retained roughly 80 percent of its previous employees, Dick Lampman inherited the entire Bristol Lab, with its wide range of computer and networking studies. He also took over a small compiler research facility in Cambridge, Massachusetts, a group in Haifa, Israel, that specialized in mathematics and security issues, and a handful of multimedia experts from its Japan lab (the rest of the Japanese staff, working on blue lasers, went with Agilent).

One of his first acts was to reorganize this far-flung organization to reflect HP's altered structure. In tandem with forming Agilent, management had decided to operate its four remaining business areas as near-autonomous entities: InkJet Imaging Solutions; LaserJet Imaging Systems; Computer Products (mainly PCs, hand-held units, and digital cameras); and Enterprise Computing, which serves corporate computing and networking clients. Therefore, Lampman fashioned four new research programs out of the old Computer Center to address these businesses more directly. Two programs—Enterprise Computing Solutions and Computer Products—were charged with serving the enterprise and personal computing business groups, respectively. The remaining pair—Printing and Imaging Technologies and Publishing Systems and Solutions—aided the InkJet and LaserJet businesses. However, rather than having each concentrate exclusively on one business, Lampman organized them around core technologies and applications—such as ink chemistry—that applied to both fields.

Hewlett-Packard was also rethinking its research funding model. The Labs had been funded originally from a "tax" on the business units based on their sales. However, since some groups spent relatively little on R&D while their more technology-intensive counterparts invested a lot, this structure was a poor match to the differing business models within HP. In the mid-1990s, the formula was changed so that two-thirds of each group's contribution to the Labs was based on its R&D budget, not sales. The rest continued to be calculated via the old formula. Heading into the year 2000, Lampman predicts further changes in the arrangement. While he doesn't expect to start taking on contracts for specific business unit projects—the way many labs have done—he does anticipate the operating groups getting more involved in setting the research agenda.

While that could be viewed as restricting the Labs' long-cherished independence and its freedom to prospect for new veins of growth, Lampman contends that these changes—especially given the more autonomous business units—represent a fresh opportunity to make a difference. "With the new structure," he says, "we can bring more focus to the diverse needs of each of HP's major businesses. At the same time, we will commit a substantial portion of our resources to over-the-horizon work—work that goes beyond HP's current business strategies."

Against that backdrop, the Labs has big plans for the millennium. The new HP is all about providing the hardware and software infrastructure for the networked economy and the networked world, and research is active across this entire front. An account of two wildly different areas—its vision of information appliances and basic research in nanotechnology—illustrates the range of these investigations.

☐

New Business Development: Digital Cameras to Biomed Beach

Call Rick Corben and his colleagues weekday warriors. On weekends, they kick back like everybody else. But Monday to Friday, the seven New Business Development group members act like paratroopers dropping onto foreign beachheads. Since the operation arose in late 1992, landings have been staged in fields as diverse as digitizing handwriting to making microdisplays for eyeglasses. None, though, rivals its daring assaults on digital photography and biomedicine.

Corben's group is charged with what he calls "an incredibly arrogant mission": find new billion-dollar businesses. Since conservative HP likes to avoid risks, the team has two overriding principles: stand on HP bedrock and attack undefended hills. That means they don't wander into technologically unfamiliar turf—and the markets they prefer aren't infested with competitors.

First out the door was digital photography. On the surface, HP had no business taking on the likes of Kodak. But a closer examination revealed digital imaging as a great opportunity for MC^2. The key lay in seeing the camera as a measuring instrument. To complement it, HP had great high-resolution inkjet printers. It also sold the ink and paper needed to print pictures. Moreover, digital photography was not about chemicals, but image capture and processing, compression, color science, and scanning—all things in which HP excelled. In short, notes Corben, "For a nonplayer, we seemed to be a good fit."

HP's first offering was PhotoSmart—a fleet of home imaging products that included its own branded camera (made by Konica), a film and print scanner, and dedicated photo printer. These early 1997 offerings, initially running about $500 each, beat every major competitor to market. But the camera used the wrong storage technology, the printer didn't double as a standard printer, and the scanner could not handle documents. Those misfires, combined with marketing that targeted home users rather than photo hobbyists while costs were still high, cost the company its time edge.

Still, within two years HP had addressed all its major shortcomings and was poised to be a market force, notes Alexis Gerard, publisher of the *Future Image Report*, which tracks digital photography. Even if it fails in cameras, Gerard notes, HP is helping drive down costs—opening the market for greater numbers of enthusiasts and driving up sales for its paper, ink, and printer businesses: "That," he proclaims, "is the right strategy."

The landing on Biomed Beach was conceived in concert with the digital

INFORMATION APPLIANCES

IT'S BEEN A WAR of buzzwords. Both Joel Birnbaum and the late Mark Weiser of Xerox PARC aggressively espoused the idea of ubiquitous computing—battling it out with Sun's Scott McNealy and others for visionary status in describing the digital future.

photography assault, when Scott Conradson of Corben's group joined forces with Analytical Medical Laboratory director Barry Willis. The pair focused on biomedical products, a potentially huge market where precise DNA analysis might eventually enable drugs to be created based on a person's genetic makeup. Although the battlefield promised to be hotly contested, the men figured they had a key assault weapon: inkjet printers.

In their relentless hunt for new drugs, pharmaceutical firms test millions of compounds. Only the most promising go into human trials, but even at this late—and costly—stage many prove unreliable or unsafe. A drug's effectiveness, though, depends a lot on whether certain genes are on or off, or harbor mutations. So if drugmakers could easily and affordably test for a multiplicity of mutations, they might predict a drug's efficacy for different population groups with far greater speed and accuracy.

To get into this field, HP teamed with Santa Clara–based Affymetrix—and in 1997 produced a gene array scanner that performed such sophisticated DNA analyses. But these are expensive machines that require relatively large development endeavors for each unique test. That's where inkjets enter the picture.

An inkjet-based scanner promises to simultaneously test thousands of DNA segments for mutations—for a fraction of the normal development effort and time. Thermal inkjet printers employ a temperature pulse to boil bits of ink, enabling them to eject precisely controlled droplets onto paper. Similarly, tiny DNA segments—housed inside droplets of liquid—could be ejected through microchannels onto an array of chips that rapidly perform the desired analyses.

If everything pans out, DNA analysis will likely be only a staging ground for a broader inkjet-led assault on biomedicine. Prescriptions could be dispensed in dosages specific to a person's body type. Willis also envisions doctors treating a patient's infection testing on the spot for a range of bacteria—and dispensing the right antibiotics the first time. Similarly, inkjet aerosols could spray cells onto burn patients to help regrow skin far more efficiently than grafting.

HP is coy about which applications it will pursue. And others are hot on the biomed-inkjet trail. But not only is the company an inkjet pioneer, it's adept at making microchannels, measurement, computation, and laser scanning—technologies needed to build a complete solution. As Corben says: "This is the heartland of our skills."

The overriding vision is of networked computers seamlessly and securely bridging the analog and digital worlds. Cars have computerized self-diagnostic systems, data bases, cell phones, and global positioning systems—enabling technicians to plug in a laptop for a complete status readout, or stranded motorists to transmit their exact location and problem. Every home has terabit storage and bandwidth to burn. Systems respond to voice commands—or someone's mere presence. Books, pictures, medical information, newspapers, music, radio, games are all changed in fundamental ways. "What I actually see emerging is a digital society—one where most people, most things, and most activities are linked digitally, one where any thing or activity that can go digital, will go digital," says Lampman.

A lot of the Labs' effort to enable the digital society lies behind the scenes. For example, WEBQoS (for Web Quality of Service), is a software program released in mid-1999 that monitors and manages components and processes inside Web servers to try to avoid system crashes and slowdowns. Among its novel features are algorithms that bring the capitalistic view to the previously egalitarian Web that some customers are more important than others—and therefore give priority access to repeat buyers and those in the midst of completing a transaction.

A more visible face of ubiquitous computing, however, is what HP calls information appliances. These are typically small and portable devices tailored to one or two specific tasks—just like washing machines, microwave ovens, and other appliances. It's essential that they do their jobs so easily, dependably, securely, and affordably that their complex workings are taken for granted, in much the same way the motors inside a vacuum or refrigerator aren't usually given a second thought. That doesn't mean PCs are going away. However, notes Lampman, "we're going to be in a world where that's certainly not going to be the only device people use to connect up with the digital world." Alternatives include personal digital assistants like 3Com's PalmPilot or HP's Jornada, cell phones, e-books, and various kinds of wearable computers.

As with electric devices, information appliances must also be accessible through a standard interface—the digital equivalent of a wall plug. One potentially important Labs creation is Heehaw: for HEy, Everything Has A Web page. The Heehaw box looks like a modem but is actually a miniature—and cheap—Web server. Project team member Marcos Frid says it can be made small enough to fit on two chips that could be installed inside any piece of electronics. In this way, says Frid, spouses on a business trip could use the Internet to download their latest travel itinerary onto the family calendar—maybe an electronic bulletin board on their refrigerator. Similarly, people who must constantly monitor their health might place a few drops of blood into an analyzer that their doctor could monitor over the Web, obviating the need to send collectors.

Creations like Heehaw provide the backbone of ubiquitous computing. But HP is also working on the flesh and blood—the appliances themselves. Here, especially, researchers like Mark Smith see a great opportunity for MC^2. It might seem as though the measurement side of HP is disappearing into Agilent, notes Smith, who runs the Labs' Streaming Media and Platforms project. However, while that's true for oscilloscopes and other test equipment, image processing and various forms of computer-based measurement remain integral to HP Labs. "MC^2 is a true vision which hasn't failed at all—it's simply now coming into use," he asserts. "I think that to a very great degree it's going to drive the information appliances."

The early stages of that vision were evident in HP's first information appliance, a hand-held scanner called CapShare which Smith was instrumental in developing. The device, which debuted in mid-1999 at $699, enables users to scan in documents simply by swiping it freehand over text: sophisticated algorithms assemble the complete page by correlating the images of successive swipes. Captured documents can then be transmitted to a printer or PC via an infrared signal simply by pushing a button—all on two AA batteries. "There's absolutely nothing to program," exults Smith. "It just works."

That's the essence of the information appliance—but just the start of what's possible. All too often, Smith maintains, technology comes with high "negative value." That is, it takes too much effort to get things to do what they're supposed to do. For example, when making a simple credit card call—dialing an access number, entering codes, and so forth—people often have to punch 30 digits or more.

Information appliances seek to redress negative value—and that's where MC^2 can really pay dividends. Muses Smith, "What if devices knew from the minute you picked them up or got near them three things automatically, without you having to click or do anything: who you are, where you are, and now a really new one, what is going on around you?" In this scenario, people picking up their cell phones would be automatically verified through their fingerprint, voice, or other biometric analysis. They could then command the phone to "check messages," and be piped through without having to push a single button. "The phone is now an MC^2 device," Smith says. "You don't have to log on to it, you won't have to verify things with passwords and all that."

These insights are guiding HP to a number of intriguing possibilities—including incorporating the microdisplays advanced by its New Business Development group into a wide range of information appliances. Many of these ideas got an early look in a mass of electronics about the size of a calculator that Smith co-developed with Gerald Maguire, Jr., of Sweden's Royal Technical University, and Peter Beadle, formerly of the University of Wollongong in Australia. This is BadgePAD, a prototype platform for evaluating technologies that might enable information appliances to fulfill Smith's "who," "where," and "what" variables.

The first version of the BadgePAD identifies people through voice recognition. But Smith was planning future versions based on face scanning and fingerprint analysis. The device also has an array of light, temperature, and humidity sensors—as well as a triple-access accelerometer that can measure motion in three dimensions. Combined with various algorithms still under development in mid-1999, these should enable it to tell when it's been pulled out of a pocket—a signal to turn itself on, ready for use. Conversely, Smith sees BadgePAD signing off whenever it's stuck in a drawer, where it will notice darkness, a cessation of motion, and equal temperatures front and back (as opposed to the big gradient when one side rests near a human body).

Getting a bit wilder, Smith envisions every room in a building with its own identity beacon, which not only tells its purpose and location, but who is present—thanks to information from their badges. Going back to his telephone example, he says a smart communicator could then know when its owner was in a conference room and then choose not to ring—automatically relaying calls to voice mail. Not only that, if the phone's owner left it on a table and somebody else picked it up, the device would only work after verifying its new handler's identity—charging any calls to that person. This is the concept of "dynamic persona." As Smith explains, "It means that devices are mine when I have them, they're yours when you have them."

HP's core expertise—covering technologies as diverse as chips, security, optics, wireless, biometrics, and materials research—makes it extremely well suited to pursue such information appliances. Much like the practice at IBM, it plans to start with devices like CapShare that represent a small but significant improvement in the current state of the art. But Smith believes that one day low-powered sensors will be embedded every ten feet or so throughout buildings, homes, cars, and public places—enabling electronic devices to communicate and access information via the very infrastructure around them. Sound ridiculous, even scary? Security and privacy issues will have to be addressed, admits Smith. Still, he is convinced the transformation isn't that much more daunting than wiring the world for electric lights, telephones, or even cable TV. "I have stopped believing arguments that any degree of infrastructure is impossible," he says. "If people want it, it's going to happen."

THE SELF-ASSEMBLING NANOCOMPUTER

PONY-TAIL SWISHING down his back, Stan Williams strolls into the Feynman Lab, named after the late Nobel Laureate physicist Richard Feynman, one of his particular heroes. The space teems with scientific equipment. But it's the scanning tunneling microscope, octopus-like metal arms jutting awkwardly into the air, that's the star attraction this day. Somewhere in the machine's bowels lies a substrate laced with the element erbium. The microscope magnifies this surface so intensely on a nearby workstation that the entire screen represents an area roughly 250 nanometers across. That's 2 percent of the width of a single human hair—and the size of the smallest feature on most advanced integrated circuits. On this scale, it's barely possible to see individual atoms, which show up as tiny, light-colored balls.

Many researchers like to think big. But Williams ponders the small. He points to one corner of the display, an area roughly 5 nanometers square. "That's the real estate we're going after." Inside that nanospace, Williams and his small team want to build a complete electronic device. And while they've got plenty of competition, through astute observation, experimentation, and a bit of serendipity, by mid-1999 they'd already identified what Williams felt were key steps to fashioning the molecule-sized wires and switches central to success. If he's right, the work fore-shadows novel self-assembling computers several orders of magnitude more efficient than today's best—and therefore could loom as important to future computing as transistors and the integrated circuit were to the current digital age. "What we're trying to do right now is make the same type of transition . . . that occurred during the 1940s when we went from vacuum tubes to the transistor," says Williams. "Conservatively, I'll say we're shooting for somewhere between one thousand and a million times what's possible now."

Unlike AT&T or IBM, HP has never been big on basic research. But things began to change during Joel Birnbaum's 1990s quest to spur growth. In particular, the former research chief felt that HP could not succeed in blazing new trails without stronger involvement at the fron-tiers of science. Therefore, while also expanding support for university re-search, he upped the ante on in-house science. On one front came the 1995 creation of the Bristol-based BRIMS, the Basic Research Institute in the Mathematical Sciences, where a small theoretical team began probing the possibilities of building quantum computers. There is no known reason atomic-scale computers can't eventually be built. But the science involved changes from the well-understood solid-state physics governing conventional semiconductor manufacturing to the more vir-gin realm of quantum physics, which the Bristol theorists strove to under-stand and advance.

Then there was the experimentalist Williams, a chemist who had spent fifteen years on the UCLA faculty before joining HP in the sum-mer of 1995. Williams and his applied crew were trying to actually build these nanomachines—but by pushing to the limits of science at the nanometer scale rather than crossing into the realm of quantum behavior. Their efforts addressed some long-anticipated problems in the manufacture of integrated circuits. One was that the costs of building fabrication plants were becoming increasingly prohibitive—on track to hit $50 billion by around 2010. Then, too, given the rate at which more and more transistors were being squeezed onto a single chip, things would eventually become too small to obey the classical Newtonian laws of physics. Uncontrollable statistical mechanical fluctuations associated with small systems would kick in—raising the specter of switches opening and closing by themselves and corrupting data and processes.

Williams was hired to find a way around these double brick walls. Physical laws and theories advanced by Feynman and others suggested it should be possible to improve integrated circuit efficiency—the amount of computation per watt of power—by at least a billion times. However, to get anywhere near that theoretical limit, whole new computer architectures would likely need to be created that operated according to quantum mechanical principles. Even if it proved possible to construct integrated circuits on this scale through improvements in conventional lithographic techniques, the costs would be immense, notes Williams. Therefore, since these dimensions are much more common to molecules than semiconductors, the approach he favored was to fabricate the nanostructures chemically—his field of expertise. The plan was to use conventional lithography to create an overall pattern for his nanocircuits. Then, inside that framework, the various switches, memory elements, and wires would be synthesized by laying down selected nanocrystals that he hoped would bind into the desired components.

It took six months or so to assemble his team and the needed equipment in Building 26, up the hill from the main facility on Deer Creek Road, where HP Labs performed most of its materials and hardware research. Then in the summer of 1996, to help with the chemistry, he brought in former UCLA colleague James Heath as a visiting scientist. Even as they started, however, the men knew they faced a big problem. Because of the statistical yields of chemical synthesis, some fraction of their creations would not work. Moreover, unlike in chip fabrication, it would not be feasible to test them all and weed out the bad ones. This represented a serious stumbling block for any nanocomputer.

Williams and Heath were puzzling over this dilemma when they learned of Philip Keukes, an HP Labs computer scientist. Keukes and his colleagues had recently constructed a unique computer called Teramac. The essence of the complicated machine was its defect-tolerant architecture. In order to hold down costs, the HP team had fashioned it out of imperfect (and therefore cheap) parts. Yet they had used so many parts that there were enough good ones—and ways around the bad ones—for it to function. After a testbed workstation identified the defect locations, Keukes and engineer Gregory Snider programmed their computer to avoid those malformations and simulate virtually any computer architecture they wanted. Despite the presence of some 220,000 defective components, Teramac performed most calculations at a terabit per second—roughly 100 times faster than conventional workstations.

Williams, Heath, and Keukes met periodically for more than a year before they really understood what each other's groups were doing. A key problem was that chemists and computer scientists sometimes used the same word for entirely different concepts. As Heath once recalled, "We'd be happy if we came out of a four-hour discussion and all of us agreed on the definition of a noun." Finally, though, the trio realized that a defect-tolerant architecture might hold the key to a successful nanocomputer. It wouldn't matter if chemically synthesized components contained lots of defects, so long as enough good ones were created to get the job done.

The three scientists, joined by Snider, described the concept in a June 1998 *Science* article. The paper, says Williams, "set off a firestorm." Soon thereafter, the U.S. government's Defense Advanced Research Projects Agency moved to accelerate the creation of defect-tolerant computers by chemical assembly. DARPA's challenge, for which it would provide seed funding to qualified contenders, was to build either a 16-bit memory or two coupled logic gates—both essential computing components. All computational elements had to fit in an area no larger than 100 nanometers on a side—and be completed within two years. Those who succeeded qualified for four years of additional funding.

Until then, the HP team had run on basic science time. Recalls Williams, "We figured we could have something in about five years, but we didn't know what." DARPA's gauntlet, though, spurred the group to actively consider how to build the switches, memory and logic circuits, and other nanocomputer components. Even before the agency's start-up funding began in mid-1999, progress was coming fast and furious. One serendipitous HP discovery was that a combination of erbium and silicon would cause the wires needed to carry current through a nanocomputer to form by themselves—sans lithography. A July 1999 follow-up paper in *Science* also reported on the success of Heath and a UCLA team in synthesizing and testing the switches needed to perform logic operations. The devices were much larger than required by DARPA—about a micron wide—but the behavior of the molecules in them represented a critical proof of concept.

The work led to the filing of six patent applications that Williams feels could form the basis of a new technology. To the extent he's right, his investigations are likely to be vital to both HP and Agilent. Consequently, while it appeared possible in mid-1999 that Williams's group would become part of the measurement company, both it and HP were negotiating about how best to share any fruits of his research.

As it strives to fashion wires with the right form and spacing, shrink switch sizes, and integrate these molecular-scale creations into a working system, the HP team faces a world of fundamental chemistry, physics, and engineering challenges—not to mention competitors. Still, notes Williams, "we actually have all the components in our hands that we need."

Even if HP's approach doesn't win out, however, the news is almost universally good. Williams tells how Richard Feynman on his deathbed asked a friend to turn some of his lectures and notes on computation into a book. In that treatise, *Feynman Lectures on Computation*, he presented the theory that shows how mankind was still at least eight orders of magnitude away from reaching the theoretical limits of computers. Therefore, notes Williams, no matter who wins the race to build a nanocomputer, one thing is certain. "We haven't even begun the age of computation yet."

■

HP Labs scored a real coup in mid-1997 when it hired Donald Norman, former director of Apple Computer's Research Laboratories and an evangelist for making technology invisible—through information appliances designed from the start to fit people's everyday lives and needs. Indeed, his signing on at the executive-level position of senior technical adviser seemed to signal HP had made a major commitment to achieving this goal. A year or so later, however, Norman was gone. As he puts it, "I quit after a year of complete frustration."

Norman has a lot of good things to say about HP, its people, and his treatment there. But as a "friendly critic," he notes that he should have recognized that Joel Birnbaum had already been pushing the ubiquitous computing-information appliance concept for five years with little success—and that therefore the company itself might be incapable of bringing such a radical notion to reality. "HP can't react," he later concluded. "It's too many small divisions independent of one another feuding with one another." This spawns a culture of excessive attention to short-term results, with executives unable to sustain innovative, further-out ideas that take several years to turn a profit. "On a quarter-to-quarter basis it's always better to spend money on existing products than on the future," says Norman. "You get promoted and rewarded based on what you did last quarter, not what you did five years from now."

This, then, was the Achilles' heel in Birnbaum's MC^2 vision. Even presuming the Labs could pull off the envisioned synergism, there still lay the far greater issue of getting the corporation to implement the fruits of research. Almost by definition, backing an MC^2 business or product—and that included most information appliances, let alone a new fault-tolerant computer architecture—meant that at some point the computer group was going to have to suborn workers to the measurement group, or vice versa. Birnbaum wasn't fooling himself this would be easy. "So whether this turns out to be successful or not," he said in late 1996, "is going to be much more a matter of whether we can reinvent ourselves as a corporate organization." Still, he reasoned, HP had done it before. It could do it again.

As of mid-1999, Hewlett-Packard had not been able to pull it off. The upcoming year 2000 split marked a determined attempt, but one that entailed stripping out of HP—and the Labs—a lot that had been unique about it. As Norman notes, "The future is in small, flexible, spe-cial-purpose devices. This is exactly what the test and measurement peo-ple are excellent at. So HP took this set of skills and separated them from the company: they threw the future away. Crazy." Then, too, Dave Packard died in early 1996 and Bill Hewlett, already director emeritus, stopped going to board meetings around the same time. Without their long-range vision to keep the Labs insulated from the rest of the com-pany, Norman predicts that more and more of the research agenda might be determined by the business groups, further putting the brakes on radi-cal innovation.

But by that fall, on the heels of two consecutive quarters of more than 30 percent growth in net earnings, rising revenues, and the selection of hard-charging former Lucent Technologies executive Carly Fiorina as president and CEO, even naysayers were gaining renewed faith in HP. Norman notes that the measurement and instrument engineers were among the most conservative groups at HP. "So maybe shedding them is actually an important move in the right direction. HP needs a much more customer-centric focus, a human-centered design focus. It has to shed its engineering roots. Maybe they are beginning to make that transi-tion. . . ."

Birnbaum was also wavering in his convictions. The failure of MC^2 to really take off had been "really disappointing." But finally he came to believe that the barriers between the various HP groups were insur-mountable. "I had basically given up, to tell you the truth," he says.

Then came the split announcement. A long list of businessmen, among them Lucent chairman Rich McGinn, had told Birnbaum that the very act of splitting would cause people to think and act differently—but he hadn't really bought it. To his astonishment, though, the increased clarity of focus and appreciation of the situation's seriousness sparked a sea change in people's attitudes. Even seemingly doomed ideas—like the MC^2 concept of information appliances, and the related e-services strategy of providing the software and hardware infrastructure that linked them—found fresh momentum as previously uncooperative groups in measurement and computation began working together. Birnbaum had worked hard to bring about this cooperation, with limited success—and he had felt certain that what progress had been made would evaporate under the split. "I always had a feeling that without a very strong hand on the tiller . . . these guys would already have gone their own way," he relates. "Now I observe that with no structure and no reason to talk to each other, people from one organization are making deals across the other organization."

So could he have been wrong to oppose the split? "What I've been thinking about and saying to myself is I guess the dope was really me in some sense. . . . Humans are pretty complicated creatures. It was probably extremely arrogant of me, it was definitely arrogant of me, to imagine I could be a group psychologist and figure out what people were going to do."

As for the future of MC^2, HP Labs, and the new Hewlett-Packard, the jury is still out. But Birnbaum, for one, is no longer pessimistic. "There's a new spirit in the land. . . ," he relates. "I think there's a very, very good chance that the printer guys and the PC guys are going to home in on the utility services strategy. If they do, HP will have a very different look than most computer stories. So it is quite possible that this story is going to have a very happy ending."

CHAPTER TEN

THE NEW PIONEERS: INTEL AND MICROSOFT

RELIVING THE experiences of pioneers like General Electric, AT&T, and DuPont, modern-day powers Intel and Microsoft waited until they had achieved dominance in their industries before creating centralized research arms in the 1990s. In both cases, this act didn't take place until the 1990s—making them the new pioneers.

The combination of these two West Coast giants has been dubbed Wintel, a reference to the seemingly invincible pairing of Microsoft's Windows operating environment and Intel's microprocessors, which together drive more than 90 percent of all personal computers. Despite this shared legacy, however, when it comes to research Intel and Microsoft have already followed remarkably divergent paths, reflecting their vast differences in philosophy and corporate style.

Intel has taken the cautious, more engineering approach. It established a small central lab very quietly in 1995, promoted managers from within, and built it slowly—never dreaming of creating a university-like lab where researchers pondered theoretical problems and basic science.

Microsoft, on the other hand, has been far more aggressive and higher profile, fashioning what Bill Gates envisions will be the Bell Labs of computer science. Research is run by Nathan Myhrvold, a mathematician and protégé of famed scientist Stephen Hawking who dabbles in the culinary arts and dinosaur studies. The lab itself, while built on Microsoft's main campus and hardly in the tradition of an academic haven architecturally, follows the old tradition: hire the best and let them strut their stuff.

INSIDE INTEL

> *"I gather that Microsoft is starting to get into fairly*
> *fundamental research—you know, hire the best people, let*
> *them do what they think is interesting. We really haven't*
> *done that at all. We've hired the best people and suggested*
> *what direction ought to be interesting."*
>
> —GORDON MOORE

Intel Microprocessor Research Laboratory

Company sales 1998: $25 billion
1998 company R&D expenditures: $2.5 billion
1998 MRL budget: $20 million to $30 million
Funding paradigm: central
Technical and scientific staff: 160
Principal sites: Santa Clara, California; Hillsboro, Oregon; Beijing, China

---

SERGE RUTMAN is on the fly again. Born in Moscow, moved to Israel, Great Britain, Canada, and finally the United States, it's all anyone can do just to stay up with him walking down the hall. His friendly voice, a smile on sound waves, seems to scurry ahead of his fast-moving feet: words come out rapid-fire, laden with *bon mots* delivered in a slightly clipped accent of mixed origins. The young émigré put himself through college as a glassblower. He lives on a ranch in the Santa Cruz mountains, in a log cabin lighted by the electrical plant he fashioned with his own sweat.

Right now, though, the corridor Rutman steams down is at his workplace. Through a doorway into a large room, and over to a vast table. Don a special visor, and wham!—flat images projected from under the tabletop, reflected off a standard bathroom mirror, and passed through a plastic screen jump to three-dimensional life. A rose hovers in space, but wielding a pen-like device users can drag it around the way a mouse manipulates two-dimensional objects: magnetic sensors track the position of both pen and viewer in space so that the model can be rendered realistically. Click the pen and a 3-D volcano rises up. It's even possible to peer behind the stereoscopic image, watching lava flow down the backside into the sea.

Rutman has plans for the table—maybe as an interactive educational tool that helps kids learn at their own pace. "I went to school in several countries, so I can't blame anyone in particular for the misery," he relates.

"But one thing I know about education is there is no throttle—you take information at the rate they want to give it to you. I want a throttle."

Back down the hall to another room and a bank of five television sets. One plays a CD-ROM game. Its neighbor exhibits photographs of the Grand Canyon. Japanese dancers glide in incredible resolution on a third screen, while another runs a Power Point program, and an advertisement airs on a big Sony screen. All five "shows" are broadcast over a single standard channel. No HDTV here. But it's still a specially broadcast digital television signal—DTV—from Rutman's rooftop station that is fed into a jazzed-up personal computer and piped to each box simultaneously. Consider the marvelous possibilities: children play an on-line game while Dad takes educational courses and Mom channel surfs: with the right memory tweaks, it's possible to fiddle with the live broadcast. "My kid asked me how to rewind TV," explains Rutman. "I said, 'Not yet. Wait two years.' "

So goes the future—Intel style. Until mid-1999, when he took his small group and vision lock, stock, and barrel into development mode, Rutman was part of a growing core of talented engineers and computer scientists inside the chipmaking Goliath's Microprocessor Research Laboratory. Formed in 1995, some twenty-seven years after Intel itself got started, the lab probes the boundaries of microprocessors and personal computing five or ten years down the road, and is likely the first long-range research enterprise of its kind in the semiconductor industry. But while its creation places Intel in the ranks of other industrial powerhouses such as General Electric, General Motors, and IBM, which established far-ranging research operations only after coming to dominate their fields, the carefully managed company created by Robert Noyce and Gordon Moore takes a unique approach to its venture that may herald a new model for conducting corporate research in the twenty-first century.

The picture that emerges defies easy categorization. MRL is not a central lab, but it is. It does blue-sky research, but it doesn't. It's small—just 160 researchers, of whom three-quarters (or 120) are divided roughly equally between the Santa Clara headquarters and a large Intel R&D complex in Hillsboro, Oregon. Yet it already boasts a global presence, with offshoots in Israel and China, plus scores of collaborations with other companies and universities. In a hot-wired world where text, video, audio, movies, records, and still images can stream into any living room from Timbuktu, MRL pursues applications in things like interactive multimedia games, speech recognition, and user interfaces—anything to burn mips and keep chips rolling off assembly lines. But here's another twist. If its creations succeed—great. MRL's deeper and more novel aim, though, is to forge the framework that ensures no matter where tomorrow's big applications come from, Intel's chips can handle the job. Asserts Richard Wirt, the mathematician who headed MRL until May 1999, when he turned it over to fellow Intel Fellow Fred Pollack to concentrate on running a much larger software support effort, "We're laying the track in which the train will go. We don't have to design the Killer Ap, but we'll get close enough in that direction that we'll be ready when the Killer Ap comes along."

Pitfalls abound on the frontier. MRL is young, its researchers still largely inexperienced, and it belongs to a company not exactly used to gazing beyond the next generation of chips, or maybe the one after that. On the other hand, its people teem with talent and possess a clear and focused mandate. Elsewhere inside Intel, folks hurry toward the day when a billion transistors are crammed onto a single chip, calculating how to make things smaller and smaller. The Microprocessor Research Laboratory is one place where people get to think B-I-G.

Call MRL the anti-lab. A bit to the south, IBM's serene Almaden research center lords over parklands populated by eagles and an occasional mountain lion. Pollack's group inhabits a section of the third floor of a new six-story structure erected next to Intel headquarters, surrounded by hotels, office parks, and traffic. Whereas Big Blue spends more than half a billion dollars a year on its 2,900-strong research division—eight labs worldwide—Intel says it forks over no more than $30 million a year for its "blue-sky" operation. No university-like campus or chalkboards covered with cosmological equations for Intel. Outside of a few computer-packed lab rooms, it occupies the same vast open area—gray carpets, gray walls, and gray cubicles—found throughout the building, with nary a sign spelling out the research area. Asserts senior vice president Albert Yu, general manager of Intel's Microprocessor Products Group and the man to whom MRL reports: "The chance of us having a Maui lab—where you lay on the beach and contemplate the future of computing—is not in the future."

Such a bottom-line approach to research may seem surprising for a $25-billion money-printing machine that pioneered the microprocessor and claimed a hefty $2.5 billion in R&D spending in 1998. What's more, perhaps alone in the computer and electronics industries, Intel has always been steered by Ph.D. scientists—co-founders Robert Noyce and Gordon Moore, chairman Andy Grove, and chief executive Craig Barrett—with strong research backgrounds. Yet over its thirty-year history the company has hardly spent a dime on long-range scientific or exploratory studies, and has never formed a research division. Sums up Moore, sitting in a room just across from where his National Medal of Technology hangs: "In my view Intel has a very small 'R' and a very big 'D.' The research we do is very strongly directed and very close in to the areas in which we're operating."

It's not that Intel lacks innovation. The Pentium Pro, for which MRL head Pollack served as the chief architect for its first two years of development, stands as a masterpiece of design and manufacturing power. Moreover, the company conducts important and sometimes fundamental research in chip manufacturing at its Semiconductor Components Research Lab in Santa Clara, while the Intel Architecture Laboratories outside Portland explores new uses for personal computers. But while both these research arms serve vital roles, their work rarely, if ever, focuses on problems beyond a two-to-five-year time horizon. And they don't do what MRL does—pursue microprocessor designs and circuits that can handle tomorrow's workloads.

The key to Intel's views on research lies in its Fairchild Semiconductor roots. In the late 1950s and 1960s, co-founders Moore and Noyce, along with Andy Grove and senior vice presidents Yu and Leslie Vadasz, chairman of Intel's Research Council, all worked in the big Fairchild R&D laboratory in Palo Alto, the scene of Noyce's 1959 invention of the first commercially practical integrated circuit.

Moore, who holds a doctorate in chemistry and physics from the California Institute of Technology, served as director of research and development for much of that time. When the lab really got going in 1956, he recalls, "The processing was more witchcraft than it was science-based really. So we really had to do quite a bit of research." Beyond manufacturing issues, the lab's chemists and physicists opened basic investigations into such areas as physical phenomena and even theory. So fundamental was the thinking, Moore remembers: "After we did the first integrated circuit, a group of the senior people out in the laboratory sat down and said, 'Okay we've done integrated circuits—what'll we do next?' Like we'd finished that."

Other strides came fast and furious, especially in the area of metal oxide semiconductors—MOS chips. But as Fairchild Semiconductor grew, organizational barriers formed that made it increasingly difficult to transfer the fruits of research to the production arm in Mountain View, a mere ten miles away. Says Moore, "Essentially, as they got more technical capability in the receiving organizations they got less interested in having us tell them what to do—and the end result was it got to the point where they had to kill everything and start over again. It was a very inefficient transfer."

The semiconductor industry's extremely short product cycles of

□

Intel's Other Forays into Research

In addition to its Microcomputer Research Laboratory, Intel maintains two research operations generally focused on shorter-term needs in semiconductor architecture and manufacturing. Meanwhile, the Research Council chaired by senior vice president Leslie L. Vadasz—Intel employee number three—oversees extensive ties to academic research.

Intel Architecture Labs. This operation in Hillsboro, Oregon, formed in 1991, hosts some eight hundred computer scientists and engineers dedicated to finding new uses for the PC and advancing the Intel architectural platform.

One focus area is home automation—integrating the PC throughout the household through such strategies as speech synthesis or controlling appliances and lights over the Internet. IAL is also a major player—along with Microsoft, Compaq, and major phone companies—in working to standardize high-bandwidth communications for consumer Internet access.

Components Research. Based in Santa Clara, this effort, launched in 1985, includes about fifty technical staff members who pursue "generation-after-the-next" semiconductor manufacturing technologies. These include lithography, improving transistor functions and interconnections, developing low-voltage circuit elements, and modeling failure modes. The basic job: anticipate fundamental technology limits that might hinder the fulfillment of (Gordon) Moore's Law, which states that the complexity and density of microprocessors will double approximately every eighteen months.

Intel Research Council. Supervising some two hundred university projects, the council meets quarterly and is divided into specialized committees assigned to a particular technical area. Almost all its twenty-odd members hail from research. "I want the same people who are involved in the internal research to be the people who are linked to the external environment," explains Vadasz. That arrangement aims to maximize cross-fertilization of ideas and minimize friction in assimilating findings into Intel operations. Project examples include studies of graphics rendering at the University of North Carolina, wearable computers at Carnegie Mellon, parallel rendering at Stanford and Princeton, and compiler technology at the universities of Illinois, Minnesota, and Michigan.

□

around eighteen months—with each new wave of chips obliterating the previous generation—dictates a steady transfer of technology from research and development to production. The frustrating memory of the failure to make that transition never left the Intel founders. "So when we set up Intel," Moore continues, "we decided we would get rid of that inefficiency, that we would do the R&D that we had to do right in the production facility. The research we did was very task-oriented and very short term. We certainly weren't out exploring for curiosity's sake. We were doing what was necessary to solve particular problems—and Intel has continued to operate more or less that way."

Moore acknowledges the decision might have come at the cost of sacrificing efficiency in development and manufacturing individually. However, the end result was to avoid the far greater problem of spending time on ideas and technology unable to find a home in the organization. The overall strategy generally worked like a charm. Outside of lean times in the mid-1980s, when Japanese dumping of cheap chips helped force Intel out of the memory market and into microprocessors, the company has never looked back. Nor did Moore, Noyce, Grove, and the rest have reason to tinker much with their original research blueprint.

But by the mid-1990s, important changes were sweeping the industry. Basic semiconductor research had become largely the domain of universities, and with federal funding for such pursuits leveling off, programs were feeling the pinch. Meanwhile, traditional industrial leaders of further-out studies—IBM and Bell Labs—had their own financial woes. With Intel clearly the industry kingpin, senior managers determined to fill the void by creating their own long-range research program to complement the two shorter-term efforts in manufacturing and microprocessor architecture.

The decision likely went down in headquarters Room 528, the spacious fifth-floor conference room near Andy Grove's impressively cramped cubicle where senior management routinely confer. In this case, senior management meant then-chief executive Grove, chairman (now chairman emeritus) Moore, and current CEO Craig Barrett, then chief operating officer. It wasn't exactly Gordon Moore's style, he confesses. "Both Andy and Craig have a tendency toward action. My philosophy tends to be let's leave the organizations lie." Still, he understood the rationale. "We were starting to run out of ideas from the academic computer scientists. We wanted to continue to see how we could improve our processors—so we got to the point where we had to do more of it ourselves. When you run out of plowed ground, you got to plow some yourself."

So MRL was born.

Corporate research had come a long way from its House of Magic days. Unlike the creation of centralized research at General Electric, IBM, Xerox PARC, or even Microsoft only a few years earlier, Intel's fledgling lab never signified a radically new research paradigm for Intel. Instead, MRL was conceived and organized as a natural extension of the company's long-standing R&D philosophy. Tightly coupled to development and extremely small when compared to its counterparts, the lab is also highly leveraged—helping fund close to one hundred university studies (about the same number funded by the rest of Intel). "Frankly, I really believe this is the future of industrial research. This is the way it will have to be in order to get the best benefit of the capabilities," states Vadasz, who as the Research Council chair oversees Intel's ties to academe. "Universities will have to play an increasingly major role in the longer-term research topics."

True to Intel's determination to keep its neophyte research effort tightly aligned with business operations, the lab was originally given space in the headquarters facility, close to the Microprocessor Products Group: both operations later moved almost ensemble to the adjoining building, SC (for Santa Clara) 12. To ensure that the research was relevant and could find its way into the products groups, half the original recruits came from other parts of Intel. Serge Rutman was one of these. When it came to the high-speed circuitry increasingly needed in personal computers, the lab also got a big shot in the arm by snagging the core of Intel's aborted supercomputer effort. A few additional researchers—among them IBM graphics expert Bob Liang and Jesse Fang, a compiler specialist from Hewlett-Packard Labs—signed on from other major research labs. The rest, whom Wirt called "aggressive new Ph.D.s," came straight from college. "We haven't gone for the big names as much as people on the rise," the original director explains.

From the start, MRL has been about computer brainpower—setting the pace for speed and the nature of the tasks a machine can tackle. Moving from bottom to top, managers identify four main areas of pursuit: circuits, microprocessor and compiler, platforms, and compute-intensive applications such as 3-D graphics, video, speech recognition, and other human interfaces (see box, p. 333).

Though only a small part of what MRL does, applications drive the lab. That's because the rise of the Internet has heightened the need to handle information in a variety of forms—text, still images, video, 3-D displays—while just over the horizon hover wearable computers, speech recognition, and other novel user interfaces. By 2005, Intel's gurus foresee a billion installed clients and maybe 100 million servers, all interacting in new and often intermixed ways. Accommodating them in microprocessor design is a formidable task—all the more so since changes in one area can affect the whole box.

In order to put the pieces of the puzzle together and determine what far-off microprocessor generations should look like to meet the challenge, researchers need prototypes of future applications. Traditionally, Intel receives early copies of Microsoft programs being readied for mar-

---

The Basics of MRL's Microprocessor Investigations

Circuits control connections between a microprocessor's millions of transistors. As the world moves rapidly toward the day around 2005 when a billion transistors cohabit a chip, as opposed to the Pentium II's "mere" 7.5 million, designers chase two fundamental quests: low power and high performance. Because it involves work on the level of the transistor itself, designing low-power circuits resides the closest to basic science of anything done in MRL. Increasing performance is also extremely tricky, but more straightforward. "The concept here," says Wirt, "is how fast can you push electrical signals over a wire."

A step up the stack come the *microprocessor* and *compiler*. The microprocessor—the aggregate collection of all those transistors—forms the brains of a computer. Compilers are software instructions that translate high-level programming languages such as Java or C to binary instructions the microprocessor can understand. The main challenge for MRL is that memory chips have not kept pace with microprocessor speed—creating bottlenecks when it comes to executing stored instructions. Chip wizards have to dream up tricks to make memory look fast when it really isn't—a devilish problem.

Beyond the microprocessor, personal computers include chips for memory and graphics. Together these form the *platform*. As PCs learn to tackle images, video, 3-D, speech recognition, and a coming tide of other marvels, designers must rethink the division of labor between the various chips—as well as the highways (they're called buses) that connect them. One strategy might be to speed things up by building some of this functionality into the microprocessor. Intel took a major step in this direction in 1997 with its MMX graphics and speech instructions extension for the Pentium II. But the world has likely only seen the tip of the iceberg.

Compute-intensive applications are best thought of as the neat stuff computer users actually get their paws on: such wonders as virtual reality, data mining, and video games.

---

ket—using those as guidelines in designing next-generation chips. However, those applications are typically conceived several years before arriving at Intel and are aimed at the installed base of PCs—meaning they are meant to run best on microprocessors at least three years old. As Pollack notes, "We can't design microprocessor and platform architectures that are five to seven years out looking at what will then be antiquated software."

Even though Intel is not a software house, that realization led MRL to dive into creating its own prototype applications. In hot pursuit of the main goal, it then employs special simulation software to analyze what these programs require from the microprocessor and tries to accommodate those demands in future chip and platform designs. Recent offerings like the Pentium II can also be refrigerated to facilitate their running at much faster speeds than normal—close to where the future products will come in. From these two tests researchers get a good picture of needed modifications—as well as their estimated costs—and those changes can also be simulated. In the past, future architectures were tried out on expensive mainframe computers. More recently, it has become much cheaper—and better—to run the simulations on networks of desktop computers. Some four or five hundred engineering workstations will be co-opted in this manner, dedicating full power to the test at night or when the normal users are away, while putting the simulation in the background during working hours.

The lab seems not to have missed a beat under Pollack, a gifted engineer with diverse experience in operating systems, computer architecture, and product planning who joined the company in 1978 and was named an Intel Fellow in 1993. Pollack's last job before taking over MRL was as head of measurement, architecture, and planning for the Microprocessor Products Group, where his major focus was to plan for Intel's future 32-bit and 64-bit microprocessors. Already attuned to the future, he hit the ground running at MRL. One of his first acts was to bring into the fold a seven-person research operation from Intel's microprocessor development group in Haifa, Israel. He also continued Wirt's plan to dramatically expand MRL's small lab in Beijing. The China arm, directed by former Sun Microsystems employee Robert Yung, was started in late 1998. By the following fall, it had reached some thirty employees, with plans to double that figure in 2000. The lab specializes in speech recognition, text-to-speech processing, and natural language understanding—initially for Chinese and English—and aims to avoid the restrictions of inputting complicated Chinese characters through a keyboard and bring the power of computing to China's vast population.

Pollack also has big plans for the Hillsboro and Santa Clara labs. Among other things, he wants to move things more aggressively out of research and into development. Following is a brief look at some of the ideas being examined, at least one of which is already out the door:

THE FUTURE FACE OF TV

WHEN SERGE RUTMAN gazes into his crystal ball, he's watching TV. After all, when it comes to the microprocessors used in personal computers, he explains, "The difference between having our market share and all of the market is not particularly interesting." So one way for Intel to keep growing in leaps and bounds is to redefine personal computing. "Beyond evolving the microprocessor you're evolving the PC," says Rutman, standing in his television-packed lab. "And now you're looking for a place for the PC to go—and that's why you see a TV here."

Rutman knows no revolution succeeds unless it's what the people want. Looking over his five TV sets, each airing a different type of data, he thinks people might be very interested. Take the CD-ROM game. It could be broadcast directly to a buyer. It's good for the environment: no plastic, no cardboard. And vendors should love the shelf space. Or take his startlingly clear pictures of Japanese dancers, courtesy of compression technology from New York-based Duck Corporation. They look like HDTV, feel like HDTV, but they're not. "That's regular definition done right," boasts the thirty-nine-year-old engineer. What's more, it's all possible with a standard 6-megahertz television channel, run-of-the-mill TV, and hardware extensions anyone can buy for about a hundred bucks. No $30,000 sets needed here. Quips Rutman, "You can start now."

Of course, first TV has to go digital. But that's already underway—and Rutman has taken a hand in laying the groundwork. He jumped into the fray in the mid-1990s, as the Federal Communications Commission debated standards for digital TV. Already by then, it had been determined that digitized television signals will travel in 188-byte data packets. Of that, 4 bytes contains coding about the type of broadcast individual packets contain. The rest is programming. The problem for Rutman: the FCC was addressing only standard video and audio signals. Meanwhile, the world had gone Internet, and powerful programming languages like Java could adapt data intended only for computers to broadcasting. All it takes are fairly simple algorithms and a multiplexor that can sort through packets and send them to the right place in the right order.

"To broadcasters this is shocking news," relates Rutman. "Their first reaction is, 'Let's run four *I Love Lucy* shows.'" Out to change that, he convinced Intel to become the first computer company to join the FCC's Advanced Television Systems Committee, where digital TV standards are being hashed out. After a lot of meetings, the move paid off. Early in 1998 the committee began ratifying specifications for how to deliver all sorts of data—a process that was expected to be finalized by year's end.

So will we really be able to rewind TV? In the summer of 1999, Rutman's entire seven-person operation moved out of MRL and into Intel's Home Products Group. It marked the largest such technology transfer in MRL's brief history. Rutman's charges are striving initially to deploy some of their basic TV technology, while also working with the broadcast industry to identify viable business models for "datacasting" via digital television. As for the rest of it, stay tuned.

MANAGING THE MEDIA

TAKE MONICA LEWINSKY. Working for ABC News as the scandal breaks, you vaguely recall seeing the President and his former intern in public—hugging. But where? You spend hours hunting through news tapes before finding that now-famous scene where a crowd of well-wishers greet the President—and one of them is Monica! The Prez embraces her warmly and whispers a private word.

If that's how newshounds dug up that sequence, it was too much for Boon-Lock Yeo. Even as Serge Rutman hunted for new ways to bring data inside the home, Yeo and colleagues were figuring what to do when all those bits arrive. Simply receiving information is not enough, says the manager of MRL's video technology branch. "We ought to have better ways to handle the material." In particular, Yeo is concerned with capturing video and still pictures concisely, storing them in a data base, browsing quickly through thousands of images, and passing any items of interest on to others—all while guarding against unlawful use.

Ultimately, the goal is to allow computer users to avoid the sequential playback—with occasional fast-forward and rewind—that characterizes video viewing today. To that end, Yeo's team is developing ways to sift through hours of video the way people thumb through an index or browse the table of contents to find things rapidly in a book.

The initial stage involves digitizing, capturing, and compressing the video—all pretty standard today. But scanning that material is another matter, since the sheer volume can quickly overwhelm anyone. Yeo's answer is first to provide still image summaries of the video footage. Such capsules are possible because most film centers around relatively few basic scenes. In the case of a news segment, you might see repeated shots of the interviewer, interviewee, and maybe some scenery. That enables several sequences of images to be summarized in one image icon—say of the reporter.

This main image is called the video poster. Grouped around its borders are much smaller, postage stamp-sized images that sum up the various sequences involving the poster child. Say the TV reporter had been detailed to cover a traffic accident. One subimage might show her talking to a witness. In another little box, she stands in front of the wreckage. In a third subimage, she could be speaking with a police officer dispatched to the scene.

In this manner, the MRL group reduced five minutes of *Frasier* to one head shot of star Kelsey Grammer and its subimages (Newsflash: sitcoms aren't very interesting)—condensing 60 megabytes to 100 kilobytes of data and greatly facilitating sorting and browsing. If at any point more details are needed about what's happening in any of the subimages, the viewer simply clicks on one and a brief snippet of video magically starts to roll.

The job is hardly done there. In this increasingly all-digital era, perfect replicas of TV shows, movies, or photos can be easily moved around the planet without anyone knowing whether users have the legal right to employ the images. Here Yeo is aided by his wife and fellow Intel employee Minerva Yeung, an authority in digital copyright protection, or watermarking. The central idea is to add data—either visibly or invisibly—to an image in a way that is readily scanned for copyright information but makes it hard to delete without impairing the picture. Once a watermark is in place, images can be safely distributed over the Internet.

Yeo envisions the technology helping to archive and sort through videotaped lectures, classes, speeches, and training films—as well as banks of still images. When it comes to TV, intelligent agents could scan broadcasts all day and night, presenting capsule summaries of shows they think you might like. Freed from their VCR, people would then watch television whenever it was convenient. "Maybe there's no more concept of prime time," predicts Yeo. "It might change the way advertising works."

FACE TRACKER

EXCEPT FOR FIRING periodic salvos at guards and attack dogs, the guy playing the legendary shoot 'em up game Quake keeps his arms still as his torso contorts in stiff robot-like maneuvers: forward, down, left, up. You worry about the guy before realizing it's his body language—not a mouse or keyboard commands—that enables him to navigate the labyrinthine Strogg crater-fortress, almost as if he's been turned loose inside the game itself.

Meet Face Tracker, one of Intel's forays into better user interfaces. "Computers right now are blind, deaf, and dumb," explains MRL pattern and vision recognition specialist Gary R. Bradski. "So we adapt ourselves to them, not they to us." Face Tracker attempts to get computers to do the adapting. A camera resting on the monitor tracks players' head movements with four degrees of freedom—meaning users can move their heads up, down, left, right, forward, and back—and even roll them a little. Relatively simple software enables those movements to replace the mouse or keyboard input, allowing users to roam Quake's scenario far more naturally than the game normally allows.

But while games are fun, Face Tracker's real thrust belongs to a larger effort to get computers to track humans by video, taking cues from gestures and facial movements. Consider teleconferencing, long cited as a killer application but always falling short of the mark. The way the technology currently works, participants usually go to a room outfitted with expensive equipment—a real pain. But suppose the same inexpensive cameras used to play Quake sans hands rode atop everybody's monitor—and all it took to join a videoconference or chat with an offsite worker was a gesture. Suppose, too, that facial movements and gestures could be monitored by the system, so that if you looked to the right of your computer screen, to where the person on the other end had drawn some figures on a blackboard, the camera would pan over to the chalkboard.

You might even be in the middle of a group meeting and open a private channel with one of the participants—just as if you were actually sitting next to that person and leaned over to share a few whispered comments. The screen might then show the shared videoconferencing scene, along with a smaller box depicting the person with whom you're engaged in the one-on-one. Says Bradski, "There is a huge loss of productivity because people don't tend to interact if they are not on the same site, or even on the same floor. If videoconferencing from your desk can be made as natural as being there live, productivity would soar from the resulting collaborations and flow of ideas. Giving computers visual awareness is one of the keys to this."

Already Bradski can describe a world of enticing scenarios. People arrive home from work, stand in front of a big screen depicting an orchestra, and unwind by conducting a virtual symphony—wielding a baton that would cue the violins to jump in, or the horn section to tone it down. In the same way, you could walk "inside" a game that involved running and jumping and get exercise while you played—a lot more fun than jumping on a stationary bike or treadmill. All this is far out. But, Bradski relates, "The pieces are starting to come together."

SMART COMPUTING ANYTIME, ANYWHERE

IT'S BEEN A BATTLE of visionaries. On one front come gurus like Sun Microsystems chief executive Scott McNealy, proclaiming that the PC is on its deathbed—that the network and so-called information appliances represent the future of computing. Defending more traditional turf: the folks at Microsoft and Intel, who maintain the personal computer will remain the standard-bearer of the digital age. Lately, though, Intel has indicated that it isn't about to ignore the information appliances and networking ends of the market, among other things snatching up or investing in a host of companies that produce or service data and voice communications networks. Therefore, while sticking to the idea the PC will always be around, not long after taking over MRL Pollack launched a research effort to explore the alternative vision. After all, he explains, "Our objective . . . is to look for new uses and new users for our microprocessors. So in this case we want to look beyond PC computing."

Just underway in the late summer of 1999, the fledgling effort was tentatively dubbed Smart Computing Anytime, Anywhere. But while it reflects the general view of pervasive computing—in which people interact with machines on human terms such as speech, and the computers do the right thing—as with everything else in MRL, the intent is not so much to pioneer new trends as to make sure Intel's chips can handle them. Therefore, Pollack dispatched his forces to first probe just one representative face of the vision—the home of the future. The basic premise is that just as many households possess a central heating unit, an increasing number will soon include a central server. Set up in a garage or closet, this machine will tout broadband Internet connections—usually through a cable television system or the digital subscriber lines that boost the capacity of conventional copper telephone wires. All around the home, a host of devices—hidden microphones, speakers, cameras, displays, appliances, TVs, and more—will be linked wirelessly to the server, moving many inputs and outputs away from the PC and thereby creating a fundamentally new human-computer interface.

Pollack says initial investigations will concentrate on building the software infrastructure for this vision. The primary concept is to create a "knowledge bus" for connecting various interface functions such as speech, vision recognition, video, and 3D graphics—which are already MRL projects—to both the distributed computer network and an information data base that can mine the Web for needed data.

Although the goal is to make this "bus" flexible enough that it will be easy to plug in new functions and services as they become available, MRL will test this software backbone initially by prototyping three aspects of intelligent services: financial, communications, and entertainment.

To demonstrate the role of each, Pollack imagines himself returning home from work. A front-door camera sends his image to the server, which performs rapid-fire pattern recognition to identify which occupant has arrived. The system then greets him by name and . . .

Financial: . . . posts the closing price of Intel's stock—and other stocks in his portfolio—on the foyer wall display. Important related news is also shown, including a cheery item about Advanced Micro Devices having problems with its newest chip. Pollack can ask the system for more details, even as he's hanging up his coat. The new items will be either read aloud, displayed—or both. Satisfied he's still solvent, he then enters the living room, greets his wife, and begins to unwind. Only . . .

Communications: . . . the computer informs Pollack that his boss, Albert Yu, tried to reach him at the office just after he left. All voice and e-mail communiqués are integrated, so Pollack doesn't care which mode Yu used. He just says. "Play that for me," and out comes a message asking him to get in touch. Pollack then tells the system to get Yu on the phone, prompting the computer to access his electronic Rolodex and place the call. After briefly going over the MRL head's idea to raise the recruiting bonus for a key strategic hire, Yu gives his approval, which the system then forwards to human resources. Pollack and his wife settle down to dinner. They get talking and . . .

Entertainment: . . . decide to see a movie. They announce their intention to the system, which knows not to select *The Matrix* (which Pollack might see alone) but a romantic comedy given at least three stars by a trusted reviewer. What's more, since it's already getting late, the show must be at a nearby cinema and start at a convenient time. Such information is readily accessible via the Internet, notes Pollack, but it takes many clicks to get it—a real pain. Far better to let the computer do it automatically, while he slips into something more comfortable.

The key technological elements for making this scenario real are already close at hand: the Internet, wireless and broadband communications, speech recognition, and natural language processing. "The real challenge here, and it's a major challenge," Pollack notes, "is the integration of these technologies." If it can be achieved, what's called the TAM (total available market) for the company's chips should expand nicely. Helping that along is what MRL is about.

"Smart Computing Anytime, Anywhere is going to happen . . . ," Pollack relates. "In business, it will increase productivity even more than computing has done to date. In the home, it will enrich our lives. . . . By prototyping this vision of computing we have two goals in mind—to learn how to optimize our future microprocessors and platforms for it and to act as a catalyst to both hardware and software companies so that this vision is realized sooner than it would be otherwise. The first will result in better microprocessor products, and the latter will greatly increase the market size for them."

■

The future is a tricky place—it rarely looks like people thought it would when they get there. That truth almost doesn't matter to Intel's newest and farthest-looking research arm. As somebody once said about life, it's the journey that's important.

Take the case of Serge Rutman's stereoscopic 3-D interactive table—a potential windfall for Intel because it chews up processing power. To test his hunch that the technology might find a role in interactive education, Rutman invited two high school physics teachers to the lab to see what can be dreamed up. Maybe they'll hit on something incredibly useful that can be spun off to operating groups or Intel's aggressive venture capital arm—but that's not the main point. "I'm not making a strong statement that this is relevant," explains Rutman. However, he is making a strong statement that if the technology goes critical, Intel will need to make changes in microprocessor architecture. Only by testing the technology far ahead of time will Intel be prepared for that eventuality.

Experiments like Rutman's also give researchers a chance to bounce ideas off each other. For instance, the magnetic sensors that track the movements of both pen-mouse and users might be replaced with cameras. That could be important, because Face Tracker and other lab technologies also employ cameras. "When you integrate those together," notes former MRL director Wirt, "you have a more powerful product."

Whether Intel can realize such lofty ambitions is another question. From MRL's earliest days, it became evident that the mold-breaking thinking demanded of the lab ran counter to the bent of many staff members—particularly engineers recruited from other Intel operations. Explains Albert Yu, "In products we try to be 50 percent better, or 2X better. But in the research area, I'm looking for ten times better. If you come in with two times, I'm not interested."

That tendency to think in incremental terms led Yu to begin meeting each month with a different research team in informal sessions designed to encourage "free-floating" ideas. Close ties to scores of university projects also help. In addition, about a dozen professors spend summers at the lab or come to Santa Clara on sabbatical—while the company periodically holds symposiums on various far-out research topics such as visual computing. All these efforts are at least partly designed to extend the vision of MRL's researchers.

Yu and other senior managers still vividly remember the lesson of their Fairchild days and are not about to tinker much with the original concept. The way things stand, MRL's visions have the benefit of being shaped by the product and development people down the hall or around the corner—and vice versa. That extremely close coupling of research and development, Yu asserts, "is the magic we have I think that nobody else has."

That doesn't mean big advances will come easy, notes Les Vadasz.

"We learned how not to do it, and we continue to struggle on how to do it. You have to feel what works and doesn't work and continuously increment and modify—and it's so much people dependent." He continues, "This is really what defines our goal. The guideline is ensure relevance, make sure that the research is broad enough to handle the end-to-end needs—not just technology, but from technology to application—and establish an efficient technology transfer between research and development groups. That is really the key. And then once you say that this is my key, you always struggle with the organizational needs of how to achieve this."

Given this hard-learned view, it's unlikely the lab will ever turn into a research haven along the lines of the great old labs of yore. Nor will Intel follow the footsteps of fellow computer industry giant Microsoft, which is growing its eight-year-old centralized research operation to upwards of six hundred members by early in the millennium and opening a basic research center in England, in the backyard of Cambridge University. Says Gordon Moore, "What we don't do is really exploratory things in other areas. I gather that Microsoft is starting to get into fairly fundamental research—you know, hire the best people, let them do what they think is interesting. We really haven't done that at all. We've hired the best people and suggested what direction ought to be interesting."

Wide-ranging basic research, Moore contends, rarely helps the corporation that supports it. Take IBM's Nobel Prize–winning invention of the scanning tunneling microscope. "They've done some great stuff. The scanning tunneling microscope is really a great tool. But IBM is not going to get anything out of it," states Moore. "I strongly believe that society benefits tremendously from that kind of research. Therefore it is the ideal thing for government to support. And I think government support of peer-reviewed research, particularly the stuff instigated by individual investigators, is a phenomenal investment in the future." However, he stresses, it's rarely for companies.

So while MRL should grow steadily as first-rate researchers are found, don't look for it to hire theoretical physicists who ponder fundamental, even potentially revolutionary, issues such as quantum computing.

"If quantum computing turns out to be the way to do things, we're going to be on the outside looking in," sums up Moore. "We're doing a kind of research, but we're not broadly looking for something to replace the digital computer. We kind of like digital computers around here."

MICROSOFT—NO SOFTWARE IVORY TOWER

*"Even in the late 1970s we were watching the work being
done at Xerox PARC and talking about when we would be
able to have a research group like that."*
—BILL GATES

Microsoft Research

Research group established: 1991
Company sales 1998: $18.5 billion
Fiscal year 1999 company R&D expenditures: $3 billion
Fiscal year 1999 Research budget: $100 million (estimate)
Funding paradigm: centrally funded
Technical and scientific staff: 500
Principal research labs: Redmond, Washington; San Francisco, California;
   Cambridge, England; Beijing, China

THE BOXES had hardly been unpacked when the move was set to start all over again. It was the fall of 1999. Only a year earlier, the inhabitants of Building 31 had vacated their longtime home in the center of the Microsoft's big quad for this roomier three-story facility on the northeast end of the action, an area known as Pebble Beach. Now, new headquarters again bulging at the seams, they were gearing up for yet another shift—this time to the west side of campus and a pair of four-story buildings in a shaded ten-acre enclave known as Cedar Court. Over the course of one long and busy weekend, offices would again be cleared out, desks and file cabinets emptied and restocked, boxes of equipment packed, hauled away, and uncrated. If it was anything like the last move, the relocation portended several days of an even worse triple horror: no computers, no e-mail, no Internet.

Another moving day for Microsoft Research (MSR). Upheavals are common at the titan of software—but the ride Research has taken counts as wild even by Microsoft standards. Since its 1991 inception, this foray into the future has gone full-out to woo some of the world's foremost computer wizards, snowballing to some five hundred staffers, twice overrunning its headquarters, and sprouting offshoots in San Francisco; Cambridge, England; and Beijing—on target to reach six hundred researchers by mid-2000. Among corporations, perhaps only IBM and Lucent Technologies boast larger computer science efforts, and the lab rivals programs at top universities like Carnegie Mellon and M.I.T. While their names might not ring bells in Peoria, a swarm of digital legends now pad MSR's carpeted halls. Among them: laser printer inventor Gary Starkweather, personal computer pioneer Butler Lampson, multimedia visionary Linda Stone, and graphics gurus James Kajiya, Jim Blinn, and Alvy Ray Smith.

Nearly a decade into the lab's existence, though, it's getting to be put-up time. The Cedar Court–bound digerati are busy shaping a future of startling 3-D images, machines that talk and respond to a person's expressions, and lifelike Web-roaming agents. But while researchers have placed software code in a wide range of Microsoft products, the lab has so far fallen short of its stated aim of turning out a true breakthrough. That gives rise to some hard questions: Can Microsoft's wizards—more than a few already multimillionaires—find the edge to make computer science history? Will a giant assailed for its lack of innovation—where cornerstone products like DOS, Windows, and even Internet Explorer spring largely from purchased technology—find a way to innovate over the long haul from within?

Not surprisingly, research managers answer "Yes." Critics paint the lab as an intellectual showpiece, where researchers pursue their whims out of touch with the real world. But while keenly aware they have yet to earn the full respect of their peers, Microsoft's brain trust claim to be creating a unique blend between the old-style, free-thinking corporate research haven and the modern day mantra of shipping products. If they're right—and while the issue remains very much in doubt, there are indications that MSR's gurus are beginning to back up these assertions—the company's expanding galaxy of computer science stars stands to become one of the world's most influential forces in shaping the future of personal computing.

BUCKING THE TIDE

A CENTURY AGO, the brilliant mathematician Charles Steinmetz, a cigar-smoking, hunchbacked inventor and theoretician who held court as a national scientific and technical oracle, memoed his General Electric Company superiors suggesting that with Thomas Edison's key light bulb patents expiring and competitors assailing GE's electric lighting monopoly, a scientific lab should be established to defend the company's turf and create new avenues of growth. Galvanized by Steinmetz's arguments, the company recruited M.I.T. chemistry instructor Willis Whitney—and, in December 1900, the nation's first great corporate research laboratory was born in a barn alongside the Erie Canal.

Knowing the genesis of GE's House of Magic, it's hard not to draw comparisons to MSR. Wags call the enterprise "Bill Labs," a reference to Bill Gates's presumed attempt to re-create Bell Laboratories. But despite its roomy new digs, Microsoft's fledgling research arm so far smacks more of the early GE facility than AT&T's famous research arm. As with its illustrious predecessor, the lab traces its origins to a brainstorming mathematician—Nathan Myhrvold—out to battle complacency and blaze new business trails. Tapped to lead the operation was Whitneyesque university scientist Rick Rashid, who sought to build the facility around the best and the brightest. And like GE's initial foray into research, while the lab's promise remains bright, its first few years have proven less than spectacular.

It all started with wunderkind Myhrvold, Microsoft's chief technology officer and a mathematician every bit as colorful as Steinmetz. He graduated from UCLA at nineteen with a bachelor's degree in mathematics and a master's in geophysics and space physics. Four years later, he earned a doctorate in mathematical and theoretical physics from Princeton, then won a fellowship under Stephen Hawking at Cambridge University, where his studies included quantum field theory in curved space time. More recently, the Hawking protégé—he turned forty in 1999—has dabbled in mountain climbing and formula car racing, and launched his own investigation into the bird-dinosaur link. He's also a gourmet cook who once won first prize in the world barbecue championship and who still serves sporadically as an assistant chef at Rovers, a top Seattle French restaurant.

Myhrvold hooked up with Gates in 1986, when Microsoft purchased Dynamical Systems, a Berkeley software company he founded. Myhrvold's active and fertile mind—he uses a Dvorak keyboard to keep up with his thoughts—resonated with his new boss. Within five years he was running Microsoft's advanced development unit, charged among other things with overseeing longer-range graphics work and Microsoft's ill-fated interactive television venture. In mid-1991, acting in that capacity, he proposed allocating $10 million a year to support a research laboratory designed to help Microsoft take charge of its future.

For three decades, Myhrvold's "vision statement" argued, the technical agenda for computing had been set by big hardware companies and university computer science departments who chiefly developed applications and technologies for mainframes. Minicomputers were fast bringing the power of mainframes to the desktop. But the old players continued to set the technical agenda—concentrating mainly on creating ways to help mainframe applications work on smaller computers. They just weren't interested in the things Microsoft's customers wanted, like ways to make computers more friendly and easier to use. "The only way to get access to strategic technologies," wrote Myhrvold, "is to do it yourself." Put another way, he notes that Microsoft took just twenty-five years to reach a market valuation of $200 billion. That growth showcases "the power of technology. Technology in the future could make us redouble that or it could drive us to zero. And so it would be crazy for us not to be involved."

The idea of a research lab warmed Gates's heart. He had long admired the pioneering personal computing work done at Xerox's Palo Alto Research Center in the 1970s and had dreamed about creating a similar lab. Besides, the Microsoft chairman harbored a healthy paranoia. Even as his company grew dominant—ultimately provoking antitrust investigations that led to the fall 1999 finding that Microsoft was indeed a monopoly—he worried that two guys in a garage could put it out of business with an unforeseen innovation. Apparently since longer-term research represented a way to guard against that possibility, the green light came quickly—and Microsoft announced the creation of a research organization in July 1991. The move would hardly catapult it into the ranks of great industrial research houses like Bell Labs or IBM. Still, it marked a first for the software industry, which had always concentrated on developing its next product generations, and signaled that Microsoft meant to be a player in fields such as speech recognition, futuristic user interfaces, and 3-D graphics—cutting edge technologies that might not bear fruit for years.

The first step was finding a day-to-day director for the lab. For help, Myhrvold sought out C. Gordon Bell, the guiding light of Digital Equipment Corporation's VAX minicomputer and a Microsoft consultant who later took a full-time position in the research organization's small San Francisco arm, where he joined data base expert and Turing Award winner Jim Gray. The men agreed they needed someone skilled in computer science and administration—who could write code *and* run a spread sheet. The lab's leader would also have to possess the cachet to draw in first-tier researchers, key people who would set the tone for the entire enterprise. The two weren't thinking of Willis Whitney, but they could have been.

The pair brainstormed up a list, with Myhrvold flying around the country interviewing potential hires. By that September, they had settled on Carnegie Mellon University computer science professor Rick Rashid. A respected networking and artificial intelligence specialist, Rashid had directed CMU's acclaimed advanced Mach operating system project, whose core technology was adopted as the basis for DEC's 64-bit Unix offering and Apple's Rhapsody and MacOS X systems, among others. Rashid joined Microsoft in September 1991—the date typically viewed as the formal creation of Microsoft Research.

"Basically what excited me about starting a basic research organization at Microsoft was the idea I could really create a world-class research organization within the context of a software company," says Rashid. The original charge was to grow to about one hundred people in five years—with the idea that accomplished senior researchers would make up about a third of the staff. Another third would be mid-level computer science veterans, with the remainder raw recruits right out of school. Such a mix, Rashid explains, combines experience with fresh views and energy. The early emphasis was on finding the senior people, because they act as beacons for attracting younger talent. At the same time, Rashid decided to follow the CMU model and restrain the number of groups created. "We wanted to have critical mass groups, not onesies and twosies," he recalls.

Those goals in mind, Rashid joined Myhrvold in building the lab with a vengeance. In a way, the erstwhile professor had a head start: he actually arrived after the lab's first wave of recruits, a trio of IBM's linguistic experts: Karen Jensen, Stephen Richardson, and George Heidorn. Then, early in 1992, he brought in his former undergraduate roommate at Stanford University, Dan Ling, a rising star research manager at Big Blue who had co-invented the video RAM and led departments in everything from workstation architecture to data visualization. Ling would eventually take over as Microsoft's research director, when his old roomie moved up the ladder to research vice president.

The lab quickly gained momentum. Prodded by the fact that several rival West Coast research organizations were losing their luster, a number of computing legends moved to Washington state. Signing on from Apple Computer were multimedia wizard Linda Stone and laser printer inventor Gary Starkweather. From Digital Equipment Corp.'s Silicon Valley lab came Butler Lampson and Chuck Thacker, co-inventor of the Ethernet: both men, along with Starkweather and fellow MSR recruit Charles Simonyi, had been colleagues at Xerox PARC in the 1970s. Graphics got a big shot in the arm in 1994, when James Kajiya arrived from Caltech. On his heels came Andrew Glassner and Alvy Ray Smith, a founder of Pixar, the computer graphics house made famous as the force behind the movie Toy Story. Rashid also raided his old institution Carnegie Mellon, as well as U.C. Berkeley, M.I.T., and other top computer science programs so vigorously that when graphics expert Michael Cohen left Princeton for the Microsoft lab, a colleague e-mailed him: "Last one in academe, turn out the lights."

Thanks to the blitz, lab managers met their goal of reaching around one hundred staff members in the first five years. Even Myhrvold was a little surprised. He'd wondered initially whether anyone of note would agree to come to remote and rainy Redmond. Many digerati consider Microsoft a pariah that hawked an inferior version of Apple's Macintosh

□

Daniel Ling

Research director Dan Ling joined Microsoft in 1992 after five years at IBM Research. His views on the future of computing—and his organization:

On tomorrow's PCs: One of the big issues we're interested in is how can we make it more natural for humans to interact with computers? Being able to interact via speech is clearly one element of that—normal, natural language that humans are comfortable with. And also having the computer be more aware of you and its surroundings. So being able to see you, being able to identify you, maybe being able to see whether you're happy or sad—and being able to respond to those things.

A whole other set of things is more related to companies and the government and society in general. People are trying to deploy very complicated, very critical information infrastructures in a very dynamic and changing environment that's quite unstable in the sense that the technology is moving very quickly. So how do we, Microsoft, build our own software to let people do that?

On the information tsunami: People almost expect you to be able to respond to a message, wherever you are, in a relatively short period of time. One consequence of all of this is that people are being flooded with information, whether it be e-mail messages or news, and people need techniques to help them manage that flow of information—whether that be visualization techniques, or agents of some kind that can pre-filter information.

On humans and technology: There is a cycle in technology, where you enable a certain set of things, but because it's enabled, there are some consequences that you hopefully can control by using some more and different pieces of technology. The interesting thing about this is the variable that doesn't change is the human. And that's why it is so important to come back and think about the person in the middle, because the technology is changing and everything else is changing around us; but the capabilities of an individual human remain roughly constant from year to year. And so managing human attention is becoming one of the critical resources that we have to worry about.

On areas where Microsoft Research needs work: The amount of time and effort that goes into testing software, and getting bugs out of software, is enormous. And yet, as we all know, there's still bugs. One thing that we're trying to think about in Research now is: are there some interesting breakthroughs that we can make in that area of design? It's something Bill has repeated a number of times that he wants to see more progress on.

On the perception Microsoft lacks innovation: That's very distressing, because we've done a lot of things that are really very innovative. And our commitment to doing research is a corporate commitment. The fact that we are the only software company making that investment, I think, is very significant.

□

operating system called Windows. Then, too, when stripped of its bells and whistles, the Redmond giant made software for PCs, hardly the cutting edge of computer science.

It helped to be able to offer intellectual freedom, six-figure salaries, and stock incentives that promised to make recruits millionaires. But freedom and money were hardly driving factors: the stars, anyway, already had both. Instead, people tell of being blown away by the intellectual energy of the place—and the chance to create software for millions of people. Take Linda Stone, who joined the company in late 1993 after making it almost a "religious statement" to have nothing to do with Bill Gates's products. "I never imagined I would go to Microsoft—I was an Apple person," she relates. But her resistance wore down in the face of Myhrvold and his dynamic team. "I thought, 'Wow. I could come in to work every day and work with this group of people.' " (See box, p. 350.)

To hear staff members tell it, that same basic scenario of courtship, resistance, and then captivation—with both the intellectual climate and the chance to make a difference in the way the world did computing—was repeated many times. Each successful recruit added to the lab's momentum. "What's happened is that we've grown geometrically," notes Rick Rashid. "Your capacity to grow increases dramatically with your size." In the three years following 1996, research swelled to include some four hundred computer scientists (of that number, two-thirds had worked at the lab less than two years), spurring a new mandate to reach six hundred staffers by mid-2000. In 1997, to augment the central facility and the tiny San Francisco wing formed two years earlier, Microsoft announced it had hired renowned computer science pioneer Roger Needham to help open a lab in Cambridge, England. In taking the step, the company pledged to commit $80 million to its new branch over the next five years: the lab had some forty staff members by the spring of 1999. It followed up on that with the November 1998 announcement that Research was expanding to a third continent—by wooing former Apple and Silicon Graphics speech and multimedia expert Kai-Fu Lee and opening a five-person lab in Beijing that is expected to reach one hundred staffers within a few years.

In the grand scheme of things, even a six hundred-person operation represents a small gamble. Microsoft's fiscal 1999 R&D budget hit $3 billion, the vast majority of which was dedicated to developing its next generation of products and services. Therefore, spending a fraction—a good guess is $100 million, though officials won't say—on longer-range and in some cases blue-sky research seems reasonable.

And the potential payoff is immense. While never wavering from its

□

HutchWorld

The center is always open. The friendly concierge doesn't object as visitors wander freely. So look around the school, chat, or check out the pigeonhole mail slots, where community members pass and receive notes: envelopes jut from many boxes—and some hold flowers or chocolates.

So went the vision behind HutchWorld—the virtual presence of Seattle's Fred Hutchinson Cancer Research Center. It's still a viable idea. But the original emphasis in this virtual space was on graphical reality. It was laid out like the real "Hutch": people could choose avatars that looked like them, and the entire setup was designed as a realistic immersive experience. It is turning out, though, that graphical reality may not be what people always want most from their virtual worlds. Microsoft's response to this surprising finding may hold major implications for making on-line interactions more truly social.

HutchWorld is the brainchild of Linda Stone, head of Microsoft's Virtual Worlds Group. In 1996, suffering from severe allergies and experiencing the benefits of on-line information and support, Stone realized her group could make things even better by creating software platforms that offered greater sense of community, social presence, and interaction than current Web sites and chat groups. Graphics could be 2-D or 3-D. But the combination of synchronous and asynchronous communications in distributed objects—enabling users to share the same settings and events—was critical. Another key idea was "persistence," allowing people to leave each other notes and treats like virtual chocolates. She pondered a test involving her own ailment. "But then I thought that AllergyWorld—that doesn't cut it."

Stone approached the "Hutch" early the next year. While she ultimately hoped to adapt multimedia-multiuser virtual reality to forums such as education, entertainment, and commerce, the Hutch constituted a compelling test bed because she could bring her group's expertise to bear where it could quickly make a difference. The world's foremost bone marrow transplant facility, the center performs some four hundred procedures annually. Patients stay at least three months—first in the hospital, then in nearby apartments. And entire families sometimes journey to Seattle to help care for them. People need maps, store locations, entertainment options, you name it. The Hutch even runs a school for patients' siblings and children. It tries to prepare people for the move, explains Ann Marie Clark, director of the center's Arnold Library Internet Services department and the face of HutchWorld's concierge. "But we thought, 'Wow, we could do a lot more with this technology.'"

A scaled-back HutchWorld debuted in August 1998 using V-Chat, Microsoft's graphical chat product. In trials throughout the fall, users chose avatars from a cartoon-character stock or created their own likenesses from photos. They were able to wander the lobby and stop by the concierge desk, which provided links to information about Seattle and patient apartments. The hope was to attract a multitude of visitors by providing a truly immersive virtual experience—and Hutch volunteers and staff planned to sponsor a variety of social support conversations with patients and their families.

But the impressive idea didn't fly in the real world. Few people were ever on line simultaneously—and those who did come were less interested in the multiuser format than in individual exchanges and finding answers to specific questions. Planned forums and even prearranged individual meetings weren't popular because people could never be sure when they'd feel up to going on-line. Instead, they would check in briefly when able—staying longer only when the right person was present. The "reality" of the interface proved impractical for the kind of instant and ad hoc communications they sought.

Stone sees the experiment as a tremendous research success—one that forced her to question basic assumptions about the immersive graphical experience. "We initially believed that these Snow Crash–like virtual worlds would be quite interesting for people—that immersion and rich graphics would enhance the user experience and contribute to a sense of context, place, identity, and community," she says. But while Stone thinks such immersive experiences remain highly desirable for entertainment, it appears that people looking for information, social interaction, and community prefer something that allows them more control over their level of immersion. When folks do want to talk, she adds, it's usually one-on-one—not in a group.

This realization prompted big changes for a second take of HutchWorld scheduled to go up in 2000, when the cancer center opens a state-of-the-art hospice for its transplant recipients. One job is to retool the old interface to reflect what was learned in the 1998 trial. The screen will be divided in half vertically. The right side will be reserved for such productivity tasks as Web browsing and e-mail. The left side will hold the HutchWorld conversation space. A small center window will showcase the 3-D world originally envisioned: users can enter it with a mouse click. Stone's team plans a chat area below the window, with the top section hosting a variety of navigational tools. The intent is to allow people to easily monitor the virtual space and move rapidly between it and their normal full-focus activities—interacting with others only when they choose. Instead of forcing an immersive interaction with HutchWorld, Stone notes, users can give it their "continuous partial attention."

This revamped HutchWorld will still pursue the original concept of persistence—offering things like mailboxes, virtual candy, and flowers. Also planned are virtual school and auditorium tours, as well as a graffiti wall where users can leave inspirational messages. Stone feels this opportunity to be with other people in a safe and supportive way will be especially nice for hospice patients, whose weakened immune systems force them to minimize outside contacts.

A second, related experiment being considered involves a much-scaled-down space—HutchWorld Lite—which people would access through lightweight palmtop devices. Instead of a 3-D virtual world, it will likely offer the core features people want most—the on-line status of friends and family members, e-mail, instant messaging, bulletin boards, and maybe the ability to browse the Web without a lot of graphical bells and whistles. That way, if people spot someone they want to interact with, Stone notes, "they can just lie in bed and decide if they are strong enough."

Both experiments seek to let people use HutchWorld on their terms—not in the way technologists think would be nice. This idea resonates with Hutch staff as well. "Our intent is to create an environment that's comfortable for people and that they want to share with each other, but not to sort of overwhelm," says Clark. "We think that will be an exciting possibility for them."

belief in the personal computer, Microsoft has modified its vision and is now seeking also to extend its reach far beyond the desktop, into the age of ubiquitous computing. It also seeks to create a rich, Web-based communications medium, where PCs respond to gestures and eye movements and rapidly redraw screen images to depict where people are pointing or looking. Getting there could take decades and a slew of breakthroughs, notes Ling. "And that's what we're chartered to do: look at technologies that we believe will be really fundamental to computing over the next thirty-five years or so."

REPEAT AFTER ME: "MICROSOFT IS NOT XEROX PARC"

IN THE ANNALS of computer science studies, only Xerox PARC has gotten started with a similar rush. Formed in 1970, that lab took just a few years to fashion the graphical interface, laser printer, and other boilerplate technologies that still define personal computing—and have enabled Microsoft to make its fortune.

But there's a legendary catch. PARC was isolated from Xerox's East Coast headquarters, where copier-oriented suits didn't grok computing. The computer jockeys themselves didn't understand what could be *sold*. They envisioned a $20,000 networked machine, the Xerox Star, and it wasn't until Steve Jobs usurped their ideas for Apple's $2,000 personal computer that PARC's inventions took off.

As Microsoft shapes its hallmark lab, the question arises whether it is destined to make similar mistakes. Given the huge customer base locked into its existing products, would something truly revolutionary stand a prayer of being embraced by Microsoft's business groups? Some observers say no—at least not without the company's experiencing the competitive hardships that rocked other labs in the early 1990s and forced researchers to work hand-in-hand with developers, marketers, and customers to turn lab inventions into meaningful products. "My impression is that Microsoft Research is much more like a sandbox for playing," says one top IBM Research manager. "I wonder if you can learn without going through the trials by fire."

There *is* a sandboxy feel to the lab. All computer industry research labs are casual: sports shirts and blue jeans. But this is a shorts and tee-shirt kind of place. Rashid is a die-hard *Star Trek* fan who keeps a picture of himself with James "Scotty" Doohan on his office wall. Staffers hang with sci-fi writers like Neal "Snow Crash" Stephenson and Greg Bear, who created a character named Nathan Rashid for his book *Slant*.

And then there's Myhrvold himself, who must have found it hard to keep his edge after reportedly netting $104 million from 1997 stock options sales and being featured in an aircraft ad because he was the hundredth person to order a Gulfstream V jet. Indeed, it was to give himself even more time for his wide-ranging pursuits that Myhrvold left Microsoft in July 1999 for a one-year sabbatical and shortly thereafter forked over millions for a spread on Maui. In *Barbarians Led by Bill Gates*, former Microsoft developer Marlin Eller and co-author Jennifer Edstrom complained that the fast-talking visionary threw out ideas without understanding the real challenges involved: in particular, they cited his misplaced early 1990s evangelism for interactive TV, which they felt came at the expense of projects that pushed Internet technology. "Talking to Myhrvold was a little like smoking dope," they wrote. "It could give you "insights" but in the light of day those insights often didn't make any sense."

A palpable indicator of that criticism may lie in the World Wide Web. It wasn't until December 1995 that Gates proclaimed "a second PC revolution—the Internet." To astute observers, being that late in the game smacks of big problems for the way research fits into the organization "They already have manifested to me a problem that we had been criticized for in the early days of PARC," says John Seely Brown, the Xerox lab's current director. "And that is, take the hypothesis that they have great researchers—and they've had them for some time. How is it conceivable that Gates did not understand the Internet until he did?"

The researchers, says Brown, certainly grasped the Internet's importance. Therefore, the indication is that mold-breaking ideas must be funneled through others to Gates before they can diffuse out to the ranks. That kind of funnel, he says, can curtail the effectiveness of a research organization—and innovation in general—by slowing the spread of good ideas. "That was, of course, what happened at Xerox at one level—back in the seventies," he says. "Ideas have to get appropriated from all sources in the organization. Watch out for funnels."

Despite vigorous Microsoft denials, speech recognition also seems to have progressed more slowly than hoped. The company put a lot of mental muscle into speech research almost from the beginning. But by late 1998, while it could point to the development of some important backbone technology that it insisted was second to none, Microsoft had yet to turn out any speech recognition applications. "If their research had delivered what they might have preferred, then I would expect them to have a product in the marketplace today," industry analyst Amy Wohl said at the time. Meanwhile, IBM, as well as smaller rivals like Dragon Systems and European speech recognition leader Lernout & Hauspie (in which Microsoft has a minority stake) all offered products compatible with Word and Microsoft Office.

Yet another ominous warning sign might be seen in Talisman, a Kajiya-led system for rendering high-quality PC graphics. Microsoft touted it as a new standard. But Talisman proved a bust when it rolled out in late 1997. A big reason: the novel technology required both software companies and graphics chip-makers to adopt different new procedures. Then, too, the Talisman architecture was directed toward improved image quality that would open new horizons for 3-D graphics. However, admits Kajiya, now the lab's assistant director, "The market was focused on games, and it was just not natural for the vendors to service any other market than games."

Similar shortcomings—missing emerging trends, lagging behind in core areas, and failure to appreciate market realities—have done in labs. But even with such strikes against it, Microsoft research officials warn that competitors should not easily dismiss their effort. For one thing, while Myhrvold has admitted attaching too little importance to the Internet in the early 1990s, Rashid and Ling insist that the company was only able to catch up as rapidly as it did because technologies developed in the lab and other parts of the R&D organization were close at hand.

That may be viewing the Web through rose-colored glasses. Still, it's hard to find a major firm that wasn't behind the Internet curve. Then, too, no good lab can escape at least a few Talisman-like disappointments: otherwise it's not thinking big enough and taking risks. In any case, the lab has since rebounded from its disappointment by moving some of the Talisman technology into Microsoft's own products—a more humble approach than its first forays, but one that is ultimately more realistic.

What's far more important to Microsoft's research brain trust than a few isolated successes and failures is the long list of proactive steps taken to avoid the worst sins of places like Xerox PARC and to prevent the creation of a software Ivory Tower. Establishing the facility on the main campus—rather than near an academic center à la Xerox PARC, or in pastoral seclusion like IBM Research—was intended to keep business objectives high in researcher minds. Staffers routinely lunch with product group colleagues, and when technology is transferred, researchers often join development teams or business units until products are complete. Those close ties help ensure the lab tackles problems important to the product side—and that product personnel know what research can offer, notes Richard Draves, manager of the lab's operating systems group. Says he, "You can pick up a phone and call up a friend in just about any group in the company because you've worked with them in the past."

Microsoft's commitment to such fusion was readily apparent in the summer of 1998, when virtually the entire research speech effort— twenty people in all—was transplanted over to Building 25 to help what was later called the Intelligent Interface Technologies group evaluate certain speech recognition technologies for possible inclusion in a variety of applications. The two-day transfer took place almost like an airlift: movers boxed up the research contingent's equipment and leader Xuedong "X.D." Huang had a core team up and working even before everything was unpacked and remodeling complete in the offices and labs flanking the long corridor that houses the effort. The pace of such transfers picked up steam in 1999. By that summer they included a data mining group called Aurum that was helping incorporate its technology into an upcoming version of Microsoft's SQL server software, as well as a massive eBooks venture (see p. 359).

While a lot of this synergy comes naturally, starting in mid-1998 the company set out to cultivate it by appointing senior program manager Kevin M. Schofield as research liaison. Schofield is a veteran of Microsoft's product groups—and he and his small staff seek to use their extensive network of friends and former colleagues to foster even closer ties between MSR and Microsoft's various product groups. That involves encouraging these groups to consult researchers about general problems, such as which software architectures might enhance Windows NT's security features. It means helping set up joint projects, where researchers and developers write code hand-in-hand to solve specific problems. And it extends to large-scale personnel transfers, as in the case of the speech work.

Up close and personal interactions with product groups is just one part of the puzzle, however. Despite any talk of funnels, lab personnel cite their strong bonds with Microsoft's senior management as another key factor to keeping research on track. Myhrvold (and Rashid while Myhrvold enjoys his hiatus) reports directly to Gates, not through a chain of command. What's more, as opposed to most established firms, where top managers invariably come from marketing and business backgrounds, executives like Myhrvold, senior vice president Jim Allchin, and Developer Group vice president Paul Maritz hold doctorates in mathematics or computer science. That expertise, insists the research hierarchy, helps pave the way for the fruits of research to be integrated into products and overall corporate strategy. "At the very top ranks of the company people are technologists at heart—they love technology," says Daniel Rosen, general manager of Microsoft's New Technology group, which supplements Research by investing in or acquiring technologies from outside the company. That goes double for chairman Gates, Rosen adds: "He knows every research topic we're doing and what it's about—and name the CEO of another major company who could identify many of the researchers by name off the top of his head and could also identify all the topics."

Another key element in the Research strategy is that unlike Bell Labs and IBM, where studies range from hardware to software, biology to physics, Microsoft's sights are relatively narrow. Outside of a six-person theory contingent led by the wife and husband team of Jennifer Chayes and Christian Borgs—and including Fields medalist Michael Freedman—its thirty-odd core groups focus on speech, natural language processing, graphics, and other areas central to the future of personal computing. "We are a software company and we can never forget that," Myhrvold once asserted. That gels just fine with researchers, who as computer scientists typically want their innovations out in the real world—a stark contrast to the old-style physicists who once roamed Bell Labs' halls, content to write research papers for their entire professional lives.

The result, lab managers believe, is a centrally funded blue-sky lab with an "it's-cool-to-ship" attitude. Rashid and Ling proudly cite the fact that MSR has become a major force at prestigious computer science conferences such as SIGGRAPH (graphics) and SIGMOD (data bases). But they're just as quick to boast the lab has contributed code—and important code—to virtually every company product. Beyond the large blocks of researchers that have transferred to the business side, MSR claims some twenty to thirty contributions to Windows 2000. Other proprietary creations include compression algorithms that allow more data to be stored on a disk, speech recognition technology, an inference engine in Office 97 that makes troubleshooting easier, Microsoft's core on-line chat products, a video server, techniques for "streaming" media through computers, and a host of development tools to facilitate programming. Rashid deems the effects of these and other creations on company products "so pervasive, it's hard to extract them."

Such feats—seemingly mundane in contrast to discovering laws of nature—mark a far different orientation from the old Bell Labs–IBM–Xerox PARC style of throwing things over the wall to development groups. Research managers say their product-centric mindset stems from the dual philosophy governing the lab. On the grand level, researchers are mandated to do for their fields what DOS and Windows did for the PC: set the standards on which everything else rests. But home runs may take years to realize. So in the meantime, the aim is to turn out smaller spinoffs that help justify the lab's ongoing existence. As Myhrvold has said, "You get more small ideas by thinking big than by thinking small."

Three hot areas of investigation—natural language processing, eBooks, and a Microsoft concept called Flow—illustrate the hopes for that approach.

THE PRODUCT RAINBOW

IF YOU EVER want to know the future of computing, lab members joke, just watch *Star Trek*, where users interact with their machines in normal, everyday lanaguage. The trouble is, achieving that aim is extremely difficult. One of the biggest pushes comes in natural language processing (NLP), a keystone group led by ex-IBMer Karen Jensen that has swollen to some twenty researchers. NLP overlaps somewhat with speech recognition—another big Microsoft effort that focuses primarily on getting computers to translate the spoken word into text. But it is nevertheless a distinct group that involves enabling computers to grasp the meaning of any text—typed, e-mail, or voice converted to text—and then take action based on that understanding.

Handing computers such powers is essential to passing the *Star Trek* test. "If you can do that, you can get the feeling that the computer is behaving intelligently," says NLP theoretical linguist Lucy Vanderwende, an avid hiker and world traveler whose love for her work permeates nearly her every word. Still, it's a gnarly problem that involves knowing grammar, style, semantics, word morphology, even context. Only then can computers understand the myriad ways of saying the same thing and make sense of poorly constructed utterances, such as: "I ate the fish with bones." Such a statement wouldn't faze a human, but how does the computer decide you didn't wield bones like a fork to eat the fish? Going a step further, suppose you'd forgotten a file name but could simply tell your trusty computer: "Find that note I took a few weeks ago after Bill Gates called."

Achieving such human-machine symbiosis will likely take decades. So why should Microsoft support Ph.D. linguists when it could just wait and buy out whatever programmer solves the problem? Well, aside from the fact the solution might involve an entirely new operating system that renders Windows obsolete, the plan is to use what's learned along the way to improve a slew of applications.

Enter the NLP Product Rainbow. Few things illustrate the lab philosophy better than this graphic laying out milestones Vanderwende and colleagues hope to achieve. The inner arc of the "rainbow" depicts general scientific goals—word breaking, morphology, syntactic sketch, logical form, discourse, and meaning representation. Its outer companion, meanwhile, lists products the team hopes to spin off as it reaches the scientific goals: grammar and style checker, information retrieval applications, and programs that scan and summarize documents.

The group hit the word breaking, morphology, and syntactic sketch milestones a few years ago. The achievement involved creating a simplified dictionary that contained most words in the English language, then writing codes that enabled the computer to parse any sentence desired. This brought researchers closer to the Holy Grail of meaning representation. But Vanderwende and her NLP colleagues also saw in their creation the foundation for a grammar and style checker that would suit their needs better than the one Microsoft licensed from Inso, formerly a Houghton Mifflin division.

The team built a program they hoped would outperform Inso's product, adding such features as the ability to change passive sentences to active. After product development counterparts conducted tests that confirmed their program's viability—and pointed to further improvements—Microsoft dropped the Inso contract and installed its home-grown checker in Word 97. With that one move, notes research director Ling, Microsoft got far more than an improved product and some extra cash. "The grammar checker, per se, is obviously no home run," he admits. "But what's much more important is the fact that built into every copy of Word now is the parser, which has the ability to parse English sentences of substantial complexity and length. And it is the foundation for building lots of different natural language tools. Once we've started down that path, it becomes much easier to add new things to it, to do much more sophisticated things over time."

That's already happening. Recently, NLP researchers created Mindnet, a vast information repository automatically built from the contents of two dictionaries by analyzing their text with the same parser that lies at the heart of the grammar checker. The first-of-a-kind system can be accessed while the parser dissects grammar and logical form, enabling it to forge links between words with similar meanings. This gives computers the smarts to treat "I saw the hatchet" and "The ax was seen by me" as two ways of saying essentially the same thing. Related technology opened the door for an improved search and retrieval program incorporated into Encarta 99, an updated version of Microsoft's CD-ROM encyclopedia rolled out for Christmas 1998. In addition to scanning the table of contents or launching key word searches, Encarta users can find information by entering a natural language query, such as: "What animals eat insects?" The system doesn't provide answers—yet—but points up likely articles that Vanderwende says often wouldn't have been found with more traditional searches.

Such a capability—at least in name—has already appeared on big search engines such as Alta Vista. But these are essentially key-word technologies. That means that if you ask "How many hearts does an octopus have?" you'll be referred first to articles that mention both heart and octopus, followed by those with one of those two words—and, in some cases, thousands of cites for articles with the word "how." By parsing the query and doing further processing of the logical form, the Microsoft product avoids this trap. Explains Vanderwende, the logical form says that "octopus" is the subject of having, and that "hearts" is the object of having. It also parses the entire text of Encarta—then looks for full or partial matches between the logical form of the query and the logical form of any sentence in the encyclopedia. In most cases, she says, users will get back a listing of all the articles in Encarta related to the query. But even if some are missed, she guarantees one thing: "We don't bring back a lot of crud."

EBOOKS

WITH HIS SCOTTISH BURR, quick wit, bushy beard and pony tail, Bill Hill is hard to forget. Particularly if you were at the November 1998 Comdex show, where he took the stage dressed in a kilt alongside Bill Gates. Hill was there to trumpet ClearType, an innovation that dramatically improves screen character readability. But while the Bills discussed ClearType's eye-popping qualities, they only hinted at its key role in forging a major new avenue of growth for the twenty-first century. Indeed although some Microsoft bashers quickly disputed its originality—citing a similar discovery by Apple's Steve Wozniak in the early 1980s—the technology may hold the key to the Research arm's first real breakthrough.

It all revolves around electronic books. Microsoft views ClearType as an "enabling technology" that will propel eBooks to reality by finally making it palatable to read on-screen material. By early 2000, the software giant was planning to announce its strategy for staking a big claim to virtually every non-hardware aspect of the nascent industry—from writing the software that lets electronic books run on diverse platforms or helps vendors sell eBooks more easily to publishing digital tomes themselves. Forget MS-DOS and Windows. Think MS eBooks. "A hundred years from now," asserts project head Dick Brass, vice president of technology development, "there's a fair chance Microsoft will be remembered for creating the first readable electronic book." And producing it and selling it, too.

Along with researcher Hill, Brass is the mainstay of Microsoft's eBook strategy. A collector of plastic sushi so real a janitor once left him a note fretting it might attract ants, he was an early-1980s pioneer in electronic dictionaries who eventually became Oracle's senior vice president for corporate affairs. He retired from that position in May 1997. But that summer, over lunch with Nathan Myhrvold—a kind of farewell salute between two old rivals—Brass mentioned he felt convinced eBooks were finally poised to take off. When Myhrvold countered that Brass should come to Microsoft to launch a venture, the former Oracle exec couldn't resist. "I really believe this is one of the few companies on earth with the resources and the vision and the commitment and the interest to create this new paradigm—to change the world, really," he relates. The effort got going the following November.

From the start, Brass had two main goals. One was to join other eBook hardware and software companies in creating a uniform standard that ensured every eBook will play on every machine: the first version of the standard was announced in late September 1999. The second push aimed to find ways for Microsoft to profit from the industry. For starters, that meant creating reading software. Although people can read on Word, notes Brass, it's really a writing tool. To make eBooks truly viable, there needs to be a comparable reading application that can easily adapt fonts, justifications, margins, and such to the different-sized screens on a PC, laptop, palm device, or dedicated eBook machine. It must include copyright protection to prevent people from illegally copying and distributing eBooks. It should offer a search function, plus ways to underline or highlight text, write marginal notes, and catalog itself in an electronic library. Above all, it must make eBooks more book-like. That's where Bill Hill enters the picture.

A charismatic figure who tools around in corduroy shorts, Indian belt, and phone holster, Hill is a former newspaper reporter who's been involved in some aspect of desktop publishing since the early 1980s. He came to Microsoft in 1995 as head of its typography group—where he worked to make screen fonts more readable—and one of his true passions is books. "The book is a magical thing," he enthuses, that transforms "sooty marks on shredded trees . . . [into] a three-hundred-page-long waterslide for human attention."

Hill harbors strong convictions that conventional approaches to eBooks are doomed because they fail to offer a similar "immersive reading" experience. "My whole thrust is that readability is the core piece of electronic books. I don't care how sexy the device looks, how many titles are on it," he states. "The rubber will meet the road when somebody opens their electronic book and starts to read."

In May 1998, almost as soon as he heard about the new eBooks venture, Hill switched to Brass's research group to see if he could help find a better way. He immediately immersed himself in a study of readability to try to figure out why paper-based books beat screen reading so decisively.

The key, he decided, was easy word recognition. Briefly, the way ink spreads on paper leads to a phenomenon called dot gain, where the ink bleeds a bit to round out hard edges and make the letters appear sharper. A liquid crystal display, by contrast, puts hard lines at the corner of every pixel—so bleeding is impossible. This leads to a no-win situation. Normal type utilizes a single pixel for stems—the main upright parts of each letter—and serifs, the smaller strokes that finish off the details. But due to the lack of paper-like bleeding, one pixel is too light, leading to what Hill calls the eye-fatiguing "jaggies." Adding a second pixel sharpens things up a tad, but then the characters become too broad and bold, wearing down the eye in a different way. The end result was that typical screen resolutions of between about 82 and 110 dots per inch (dpi) were barely a third as good as low-cost laser printers—and far worse than quality printed books—making reading "more like interpretation than recognition."

Although some high-end LCD prototypes got up to around 220 dpi, Hill figured it could take five years before those displays were affordable enough for eBooks—and even then they didn't solve the core pixel problem. Besides, he didn't want to wait five years to make screens readable. Enter ClearType.

Joined by Bert Keely, who had worked on eBooks at Silicon Graphics, Hill focused on the fact that each pixel consists of three sub-elements—red, green, and blue lines—that were being treated as one unit to produce a single black screen character. In a burst of inspiration sometime that June, the pair realized that if they could find a way to utilize the subpixels as individual units rather than en masse, they could escape the traditional whole-pixel requirements of creating type. That, in turn, would allow them to fashion characters with something in between one and two pixels—sharpening things up dramatically without overloading the eye.

Achieving this goal was a difficult challenge that involved developing algorithms to exorcise the color fringing that results from tapping the power of the red, blue, and green lines. There was also the issue of optimizing existing fonts for the new ClearType rendering technology. But with the help of type expert Greg Hitchcock and display whiz Turner Whitted, the pair reached their goal in a matter of months. ClearType enhances screen characters with unheard of subtlety—like being touched up with an extremely fine electronic paintbrush. And talk about sharp. With an estimated resolution of 200 to 300 dpi on a small display, even italics emerge crystal clear—and it does still better on sharper screens. As Hill recalls, "Bill [the other Bill] was so excited. We recognized this was a breakthrough, because if you can really read on a screen, it's going to change everything."

From there, it was on to Comdex. Almost as soon as Microsoft made its announcement, however, a wellspring of challenges surfaced. The major complaint is that the basic technology goes back at least until 1976, when Apple co-founder Wozniak made a similar subpixel discovery that was incorporated into the Apple II personal computer. One Web site, posted by legendary graphics guru Steve Gibson, included a detailed account of the technology that seemed to confirm at least many of the anti-Microsoft claims. Wrote Gibson: "So while I agree with Microsoft that this is indeed 'breakthrough technology,' it is in fact very old 'breakthrough technology' that has always been owned by the public. I'm quite excited that something Steve Wozniak may have first invented so many years ago has returned in a form that will be able to benefit all users of personal computing platforms, whether they're running Windows, Mac OS, Unix, Linix, or other portable color LCD-based displays."

But Microsoft hasn't budged. "If somebody really did invent Clear-Type twenty-two years ago, why haven't they implemented it? Where has it been for two decades?" counters Brass. He says to some extent the controversy reflects the fact that Microsoft has not yet been free to reveal the full details of its creation. "The patent process, peer review papers, and the analysis of fair-minded people will bear us out—when more details can be made available," he predicts.

Meanwhile, the race to develop ClearType thundered on. For Microsoft, even with the core readability issue solved, that still left a host of obstacles to overcome before people did any real water-sliding on electronic devices. To go with sharpness, Hill explains, people want the look and feel of books. That includes page leaves, bright and ample margins, and the like—"all the techniques the typographers have been using for two hundred years that we've never bothered to put on the screen."

Microsoft won't say where exactly all this work is in its pipeline. But early in 1999, Brass's entire forty-person team transferred from Research to product development—the Business Productivity Group—an indication it was coming sooner rather than later. The transfer marked the first time such a large body of researchers had moved en masse to the product side. Brass says it was prudent, in part because developing eBooks requires a deep knowledge of such diverse areas as typography, display technology, annotation, copy protection, library management, and publishing not normally found in any single software development group.

Only months after the transfer, at a publishing industry conference in late August 1999, the company announced that its reading application—Microsoft Reader—would start shipping in the first quarter of 2000. While it was the vanguard product to incorporate ClearType, Brass's group also planned to add the technology to some existing Microsoft products: after all, merely sharpening up spread-sheet and word-processing documents would be a nice and potentially valuable feature. And that's just for starters. Microsoft's gurus also envision applications in areas such as graphics and computer-aided design—where additional resolution should be a godsend.

Still, it's in helping open the infant field of eBooks that the company sees ClearType's biggest payoffs. Beyond creating reading applications, Brass notes, Microsoft Press is already a successful publisher of computer books and CD-ROM reference material such as its Encarta encyclopedia. It would only be natural for it to publish eBooks as well—and to develop software that helps retailers sell them.

Microsoft was planning to announce more details of its eBooks strategy by January 2000. But whatever those plans entail, the company is banking on helping shape a new world of reading. Neither Brass nor Hill see eBooks totally replacing the paper variety. However, they note, the electronic form offers lots of advantages—from search and archiving to the ability to store hundreds (and eventually thousands or millions) of books in one small device—that will appeal to many readers. As Brass puts it, "Wood has taken us about as far as it can go with books, just as it did with buildings and weapons and ships. We're about to step beyond wood, and it's a really interesting step."

FLOW

AROUND THE LAB, people point with pride to a variety of smaller spinoffs like natural language processing and even some potential big ones more in the eBooks vein. But a key part of the lab's ambitious plan is to fashion core technologies that might underlie a variety of product families—from Encarta to eBooks and far beyond.

For instance, creating interfaces more appealing than today's flat screen icons occupies labs worldwide—and Microsoft is no different. So its gurus are seeking to combine natural language processing with other technologies that together might give computers human-like abilities to respond to gestures and facial expressions, as well as spoken commands. Indeed, it's by fusing advances from a variety of research areas that managers expect to realize the full potential of their interdisciplinary lab to take personal computing ever closer to that *Star Trek* ideal—or beyond. That hope, anyway, is embodied in one of the lab's farthest out notions: Flow.

The idea is that a new computing paradigm is emerging. Initially, computers were behemoth calculators: white-coated scientists fed in programs and machines crunched away. By the 1980s, personal computers emerged and evolved into office productivity tools. Today, argues Jim Kajiya, an Asian-American whose pony tail and huge muttonchop sideburns make him look like a sci-fi samurai, "The PC role is shifting yet again to be primarily a medium—a medium for information flow."

The signs are in the ether: e-mail, increased video conferencing, Web surfing—things that go beyond document preparation to communication and learning about the world. For a research lab, Kajiya notes, "The question to ask is what are the relevant technologies? What are the Killer Aps when the PC is not a productivity tool any longer but an information flow device?"

Microsoft's answer lies in novel combinations of natural language processing, speech, graphics, user interfaces, animation, vision, mathematical theory, and more. Two of these fields—vision and graphics—are converging particularly rapidly to address the flow concept. The basic idea behind the vision part of the challenge is to give computers the power to interpret and extract information from such cues as stored images or live camera feeds. Microsoft's fourteen full-time vision researchers follow two main pursuits. Roughly half the group is interested primarily in the problem of constructing fresh viewpoints of a given scene from just a few images of the same setting, as shot from different angles. The remainder focus on vision-based user interfaces. This involves such problems as watching faces through a camera mounted atop the computer and determining whether there's someone in front of the machine, where he or she is looking—even his or her facial expression.

Graphics, of course, is heavily involved in techniques for assembling 3-D pictures. But it's also big on faces. Senior graphics researcher Brian Guenter is building an elaborate data base of faces, facial muscle movements, and expressions. When combined with speech recognition, natural language processing, and vision technology, he hopes this archive will enable him to create virtual characters who communicate on-line via the spoken word—not text—and whose lips and expressions move in perfect sync with their voices.

Vision and graphics technologies might be combined like this: users select a face to be their on-line persona. They then sit at their computers, speaking into a microphone while a camera captures their expressions. Words, grimaces, smiles would be broken into bits, transmitted over phone lines, and reconstructed in the virtual environment so that they appeared to emanate from the on-screen characters. Such technology, Guenter believes, will prove a boon to on-line chat sessions and game-playing, where people still want a degree of anonymity but desire far more realistic interactions.

And that's hardly all. Kajiya speaks of similar video-voice connections—coupled with advances in graphics, computation, and processing—that will bring long distance, person-to-person communications into an entirely new dimension. The system would utilize programs that track gestures and eye movements and instantly redraw the screen image so that the person on the other end gets a different representation of the same scene, taking into account where his colleague is pointing or looking—just as they would in real life.

Achieving such goals will require new algorithms and other wizardry. What's more, while Microsoft is interested in producing software, the ultimate solution might involve new camera and chip designs, meaning hardware manufacturers have to be brought on board. Just leaving all that for others to hash out is not a palpable strategy for a company with as much at stake as Microsoft.

■

The determination to resolve—or at least take part in resolving—fundamental issues in computer science is a prime motivation behind the growth of research. When it comes to thinking big, denizens of Building 31-Cedar Court certainly aren't falling down on their jobs. But while concepts like Flow and the advances in natural language processing can impress and even charm, they do little to dispel the cloud of doubt that hovers over Microsoft Research. A more fitting question as the company seeks to define the future of computing is whether it's already too fat to be truly innovative. Even if researchers strike gold, will the Internet-funnel syndrome stifle innovation? Will researchers continue seeking Talisman-like coups that require too much of the market—or find better ways to fulfill their big dreams?

Beyond such Microsoft-specific issues lies the more general question of whether a scientific-style lab can work in the personal computing–software industry with anything close to the efficacy it does in more physics- and engineering-oriented arenas such as computers, telecommunications, and electrical studies. As former IBM research executive John Armstrong observes: "The kind of software Microsoft does is the part of computer science for which there ain't much science. There's nothing like the laws of thermodynamics which tell you what the best operating system is. Or, to take it even further, there's nothing scientific really about the design of user interfaces. . . . A lot of this stuff is just arbitrary—and then it's just good engineering."

The real question, Armstrong notes, is whether the more rigorous aspects of mathematics and computer science can be brought to bear on personal computing. And that's an area where even Microsoft's research managers have entertained a few doubts. It's true that a lot of what the lab does is based on a researcher's experience rather than some immutable law of nature, admits Rashid. That means, in part, that no two people will create software or operating systems the same way twice. "So is it research—or is it science?" queries Rashid. "That's an issue that does get raised."

His answer is that even hard sciences like physics cover a lot of fuzzy areas. For example, he notes, theorists may look down their noses at experimenters, viewing them as merely recorders of events rather than scientists. The same might hold true for biologists, who could be perceived as sophisticated notetakers. Still, Rashid argues, "When you're clearly moving the state of the art forward, when what you're doing really gives people a new insight on how to solve a problem—that's research. And it really doesn't matter if it's engineering or purely theoretical."

Rashid stands convinced his lab is advancing the state of the art. Besides, if Microsoft can't pull off a personal computing research coup, who can? Rashid notes that budgetary pressures have forced his old school Carnegie-Mellon, the Defense Advanced Research Projects Agency, and others to rein in research in areas such as speech technology whose biggest payoffs still seem far down the road. Microsoft, meanwhile, has the wherewithal to keep similar studies going for ten or fifteen years. "We can do that," he asserts.

Time, of course, will be the ultimate judge of Microsoft's new venture. But Microsoft has already done what no other software company has managed: offer everything from programming languages to operating systems to Internet services and applications that span the gap between spread sheets and games. Largely in light of that evolution, few observers seem willing to dismiss the lab's future prospects.

"Nathan is a very imaginative guy, and he and Rashid are hiring some incredibly excellent people," Xerox PARC's Brown noted not long before Myhrvold left on sabbatical. "And so there's no question in my mind that they're building up a first-class research operation." Even those who take a harder line give the Soft at least a fighting chance of pulling off its research dreams. "We're seeing more and more that we're running into them as a competitor for hiring particular candidates that we want to hire," says Michael D. Garey, director of Mathematical Sciences Research at Lucent's Bell Labs. "They've got a good story about the environment they're trying to build—we'll see if they can make that all come true. If they do it right, if they hire people and let a decent number of them do research that gets published in serious journals, then maybe one of these days the reality will match the PR story, and they will take their place among the premier industrial research labs."

Nearly a century ago, when establishing the General Electric research lab, Willis Whitney got off to a rocky start before unleashing a flood of innovations that helped bring the world everything from X-rays to radio. While experiencing their own ups and downs in the early going, Myhrvold, Rashid, and Ling—backed by Bill Gates—hope to follow a similar rise to glory in the rapidly expanding arena of personal computing. If they can pull it off, Microsoft may be building the next great lab for the twenty-first century.

THE
INNOVATION MARATHON

THE SCENE WAS A Cape Cod waterfront mansion in the summer of 1987—the National Academy of Engineering's annual governing council meeting. The afternoon session had just ended, and the twenty or so officials spilled out of the conference rooms onto a broad lawn to make their way to the main house for dinner. General Electric senior vice president and chief scientist Roland Schmitt, who about six months earlier had turned over the research reins to Walt Robb, strolled next to Ralph Gomory, senior vice president and chief scientist at IBM.

Schmitt and Gomory knew each other well. Their research organizations had periodically collaborated on projects—and the men themselves had recently co-authored a paper about Japanese technological innovation. During the short walk to dinner, they got chatting about their new jobs. As Schmitt recalls, "We began to say why are we senior vice presidents of just science?" After all, the men noted, industry was mainly concerned about technology. Shouldn't that word also be included in their titles?

As it happened, GE's Jack Welch served as the current NAE chairman—so he was among that predinner crowd. Relates Schmitt, "So I said, 'Let me catch up with Jack and we'll clear it with him right now.' " Cutting Welch out of the herd, he explained his reasoning and said, "How about it?" The chairman hardly blinked an eye. "Fine." Soon thereafter, Gomory returned to Armonk and won the same approval from his boss, John Akers. Both men thus became senior vice presidents for science and technology.

Change, as the adage goes, doesn't take place in a vacuum. Although the full-scale crunches that would rock labs to their cores were still a few years off, the Japanese threat was already at hand and a new day dawning. And while their reworked titles marked a largely symbolic step in coming to grips with this reality, the moves still spoke volumes about the changes ready to sweep corporate research. Neither Schmitt nor Gomory engineered the wholesale revamping that their old research organizations would soon endure. But that doesn't mean they didn't see the coming storm and take vital steps that helped their organizations weather it. Especially in the case of IBM, the near unanimous sentiment at the company is that without Gomory's heroic efforts—most notably the Joint Programs, but also a host of other steps that also built the division's reputation as a valuable corporate contributor—the central research organization might not have survived.

All this underscores the fact that any judgments of individuals or organizations should be done against the fuller context of the times: not just what they did, but what was possible at a given moment—and what those actions made possible later on. To that end, while this picture of corporate research is far from complete, one of its main points is to provide the context needed to better assess the past and understand the future—whatever it will be. An organization like research, after all, exists almost in two dimensions. There is the here and now, and maybe with it the recent past and foreseeable future. But there's also the larger dimension of evolution and growth. The crises and upheaval of the late 1980s and early 1990s were real. At the same time, though, it does not follow that what came before was a failure. Sometimes it was. But more reasonably, the upheaval formed part of a continuum, the natural evolution of a vibrant organism called industrial research, itself a piece of the larger corporate picture.

This evolution was readily apparent even during the three-plus years spent writing and researching this book—as all the labs examined went through significant and often profound changes. People retired or otherwise moved on. Reorganizations took place. Companies like Hewlett-Packard even laid plans to split in two. Some projects that looked good petered out. Others on the back burner blossomed. Some of the profiles written for this book were updated or swapped for other ones throughout the spring and summer of 1999 in an attempt to stay on top of this moving target. Still, there is no guarantee that all will still be viable when this account finally appears—though at least these profiles should serve to illustrate the spirit, philosophy, and mission of their progenitors.

An even bigger transformation took place on the macro level. This book began at a time of caution and doubt in corporate research, when labs were still emerging from the dark days of the early 1990s. It ended

during an apparent heyday, at least for the United States. Heading into
the fall of 1999, American firms were clearly back on top, the economy
still booming. R&D expenditures were growing nicely. Basic research—
proclaimed to be dead in corporate labs a decade earlier—was alive and
increasingly vigorous. In short, in terms of being vital to corporate competitiveness, it seemed that central research as a whole was healthier than
at any time in the last fifty years.

All of which leads back to the National Innovation Index mentioned
in Chapter 4. Harvard Business School professor Michael Porter and
M.I.T.'s Scott Stern wanted to probe whether the United States was laying the foundation necessary to sustain its productivity edge ten years
down the road or further. To that end, they examined some key determinants of innovation in the United States and twenty-four other nations—
ultimately ranking countries according to real and projected number of
international patents per million residents. International patents are considered a better measure of productive output than total patents, because
firms presumably would not incur their tremendous costs—upwards of
$50,000 to file in the United States, given translation fees and the like—
unless those inventions were unusually promising. Porter and Stern factored in eight variables—ranging from R&D spending and personnel to
higher education outlays and the strength of intellectual property protection—they say can be statistically weighted to explain 99 percent of the
change in a nation's patent rate over time. Assuming a country's commitment to these areas would stay on its same trajectory, they then projected
future performance.

When the numbers were crunched, the United States fell from the
top spot in 1995 to sixth place a decade later. Over the same period, the
study projected that Japan will rise from third place to first. The Scandi-
navian countries of Finland, Denmark, and Sweden, joined by tiny
Switzerland, all surpassed the United States, as well. In explaining the
trend, the men noted, among other things, that many foreign companies
were spending a higher percentage of sales on R&D than U.S. firms, and
that America's scientific and technical work force was declining relative
to other nations.

A host of criticisms have poured out about the study. For one thing,
notes Harvard Business School professor emeritus Richard Rosenbloom,
"While it can be argued that patents are somehow related to innovation,
nobody has ever established that." Therefore, while otherwise praising
the work, he asserts, "it is a misnomer to call it an innovation index." Former Lucent–Bell Laboratories research director Arno Penzias adds that
in today's knowledge economy, many innovations involve the integration
of existing technologies in ways that may not even be patentable—and so

might not turn up in any patent-based analysis. "We are still using metrics based upon reductionist thinking," he stresses. "Easier to measure stand-alone inventions than integrative innovations."

Yet another indication that the index is missing something fundamental—even presuming patents are a good measure of innovation—comes from the opposite end of the spectrum. As Porter noted at the National Innovation Summit, it turns out the actual number of patents granted to U.S. residents and corporations is running significantly ahead of what the index predicts. In fact, patenting by this group in the United States relative to the nation's size and spending had been flat or declining throughout much of the 1970s and early 1980s. In 1984, however, it reversed course and has risen at an accelerated pace for the last dozen years.

Several possible causes for this mysterious patent explosion have been identified. These include a recent strengthening of the patent protection system that presumably encourages more filings, as well as a rush of software and biotechnology patenting that might be skewing the overall data. However, Harvard Business School professor Josh Lerner and Boston University economist Samuel Kortum examined these hypotheses and others—and came up short. For one thing, the pair noted that if U.S. legal changes were driving the surge, there should also be an uptick in patenting by foreign inventors. Not only was the big rise confined to the United States side, American inventors were also patenting more abroad, where laws had not been changed. Similarly, Lerner and Kortum found a spike in filings that extended beyond software and biotech to roughly 70 percent of all patent classes.

Their preliminary conclusion is that the United States is winning the innovation race—at least as it relates to patenting. The men theorize that changes in research technology, such as the widespread proliferation of computers and information technology, more efficient research management, or a concentration on the applied activities likely to generate patents could be fueling the trend. In any case, sums up Lerner, "What we are arguing is that, even if the shift in patent policy had a significant impact on firms' willingness to patent . . . there seems to be some fundamental shift that has happened in the innovative activities of the U.S. economy."

Of course, it's an innovation marathon, not a sprint. Despite some strong evidence that many U.S. firms—including most profiled in this book—are on a new innovation track, all organisms experience periodic ups and downs. Just as people are often too quick to claim death and destruction when bad days come, they can be overly hasty in claiming victory during the good times. No matter how bright things look for the United States, rest assured that Japan and Germany are not out of the game. Neither are a host of developed countries and emerging economies from Scandinavia to Southeast Asia. In this at the very least, Porter and Stern must be considered correct.

Still, while no company or research organization can avoid periodic slumps, it can minimize the effects of any downturns. In assessing the experiences shared in this book, some common threads emerge about how to survive the storms—and thrive. First, as a foundation for research, it is hard to beat the central tenets laid out by Ralph Gomory for IBM.

- Excel technically
- Know IBM (the company)
- Know the technical world
- Provide technical leadership

As a supplement to this framework, to take into account a far faster-paced world rife with competition, one might add:

- Stay relevant
- Solve real-world problems in simple ways
- Remember that research is part of a team

Such prescriptions aren't ends unto themselves. Everyone agrees that staying open to new ideas is more crucial than ever. Companies and their research arms must also be prepared to cannibalize existing products, because, as Bell Labs' Arun Netravali has noted, it's going to happen anyway. The best research organizations spend some significant portion of resources on longer-term, riskier projects that may not bear fruit for five, ten, even twenty years. A small amount of blue-sky scientific research can even be supported—so long as it is part of a specific business strategy to build organizational prestige and attract customers and talent.

The list could go on—but mainly to amplify these points. If there's anything to add, it's the watchwords of Charles Kettering. Above all, he admonished some eighty years ago, maintain a friendly and welcoming attitude toward change.

ROBERT BUDERI
January 1996–September 1999

APPENDIX

TOP TWENTY-FIVE U.S. PATENT EARNERS, 1990

RANK	COMPANY	NO. OF PATENTS
1	Hitachi Ltd.	1,064
2	Toshiba Corporation	1,004
3	Mitsubishi Electric Corp.	901
4	General Electric Co.	898
5	Canon Inc.	875
6	Fuji Photo Film Co. Ltd.	773
7	Philips Electronics N.V.	738
8	North American Philips Co.	721
9	Eastman Kodak Co.	696
10	Hoechst AG	688
11	Bayer AG	675
12	Siemens AG	625
13	IBM	588
14	E. I. DuPont de Nemours & Co.	521
15	NEC Corporation	489
16	Novartis AG	458
17	General Motors Corporation	455
18	BASF Group	454
19	Nissan Motor Co. Ltd.	452
20	Lucent Technologies*	438
21	Daimler Chrysler AG	419
22	Matsushita Electric Industrial Co.	419
23	Motorola Inc.	413
24	Ciba-Geigy AG	410
25	CBS Inc.	409

SOURCE: CHI Research, Inc. Utility patents only.
*Adjusted to reflect Lucent as it would have appeared had AT&T's 1995–96 trivestiture already taken place.

TOP TWENTY-FIVE U.S. PATENT EARNERS, 1995

RANK	COMPANY	NO. OF PATENTS
1	IBM	1,389
2	Canon Inc.	1,108
3	Hitachi Ltd.	1,088
4	NEC Corporation	1,077
5	Toshiba Corporation	1,039
6	Motorola Inc.	1,032
7	Mitsubishi Electric Corp.	1,003
8	Matsushita Electric Industrial Co.	986
9	Sony Corporation	854
10	General Electric Co.	791
11	Fujitsu Limited	780
12	Eastman Kodak Co.	741
13	Lucent Technologies	676
14	Bayer AG	628
15	Hoechst AG	601
16	Siemens AG	585
17	Philips Electronics N.V.	570
18	North American Philips Co. Inc.	566
19	3M (Minnesota Mining & Mfg. Co.)	563
20	Xerox Corporation	558
21	Texas Instruments Incorporated	543
22	Fuji Photo Film Co. Ltd.	532
23	Samsung Group	521
24	Samsung Electronics Co. Ltd.	489
25	Hewlett-Packard Company	485

SOURCE: *CHI Research, Inc. Utility patents only.*

TOP TWENTY-FIVE U.S. PATENT EARNERS, 1999

RANK	COMPANY	NO. OF PATENTS
1	IBM	2,789
2	NEC Corporation	1,853
3	Canon Inc.	1,800
4	Samsung Electronics Co. Ltd.	1,544
5	Sony Corporation	1,429
6	Fujitsu Limited	1,231
7	Toshiba Corporation	1,225
8	Motorola Inc.	1,207
9	Lucent Technologies	1,156
10	Mitsubishi Electric Corp.	1,089
11	Matsushita Electric Industrial Co.	1,065
12	Hitachi Ltd.	1,020
13	Eastman Kodak Co.	993
14	Micron Technology Inc.	934
15	North American Philips Co. Inc.	866
16	Hewlett-Packard Company	855
17	Advanced Micro Devices Inc.	825
18	Intel Corp.	735
19	Siemens AG	731
20	General Electric Co.	700
21	Xerox Corporation	671
22	Texas Instruments Incorporated	604
23	Sharp	566
24	Sun Microsystems Inc.	565
25	Proctor & Gamble Co.	551

SOURCE: *IFI CLAIMS Patent Services. Utility and statutory patents only.*

PRELIMINARY NATIONAL EXPENDITURES FOR RESEARCH AND DEVELOPMENT, BY PERFORMING SECTOR AND SOURCE OF FUNDS, 1998

			SOURCES OF FUNDS			
PERFORMERS	TOTAL	INDUSTRY	FEDERAL GOVERNMENT	UNIVERSITIES AND COLLEGES	OTHER NONPROFIT INSTITUTIONS	PERCENT DISTRIBUTION, BY PERFORMER
All R&D: Basic Research, Applied Research, and Development (millions of current dollars)						
Total	220,617	143,714	66,636	6,819	3,449	100.0
Industry	163,328	140,847	22,481	—	—	74.0
Industry-administered FFRDCs	2,418	—	2,418	—	—	1.1
Federal government	16,936	—	16,936	—	—	7.7
Universities and colleges	25,672	1,829	15,247	6,819	1,778	11.6
U&C-administered FFRDCs	5,529	—	5,529	—	—	2.5
Other nonprofit institutions	5,928	1,038	3,219	—	1,671	2.7
Nonprofit-administered FFRDCs	807	—	807	—	—	0.4
Percent distribution by sources	100.0%	65.1%	30.2%	3.1%	1.6%	
Basic Research Only (millions of current dollars)						
Total	34,426	8,795	19,523	4,314	1,793	100.0
Industry	7,845	7,161	684	—	—	22.8
Industry-administered FFRDCs	745	—	745	—	—	2.2
Federal government	2,867	—	2,867	—	—	8.3
Universities and colleges	17,606	1,157	11,009	4,314	1,125	51.1
U&C-administered FFRDCs	2,688	—	2,688	—	—	7.8
Other nonprofit institutions	2,564	478	1,418	—	668	7.4
Nonprofit-administered FFRDCs	111	—	111	—	—	0.3
Percent distribution by sources	100.0%	25.5%	56.7%	12.5%	5.2%	

KEY: FFRDC = Federally funded research and development center; U&C = Universities and colleges

NOTES: Some figures rounded. State and local government support to industry is included in industry support for industry performance. State and local government support to U&Cs is included in U&C support for U&C performance.

SOURCE: National Science Foundation, Division of Science Resources Studies

U.S. R&D AS A PERCENT OF GDP

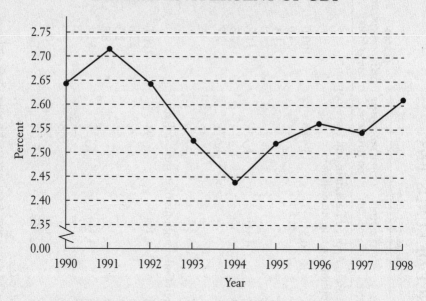

NOTES: *These data are based on reports from R&D performers. Data for 1997 and 1998 are preliminary.*
SOURCES: *National Science Foundation, Division of Science Resources Studies.*

U.S. R&D FUNDING, BY SOURCE

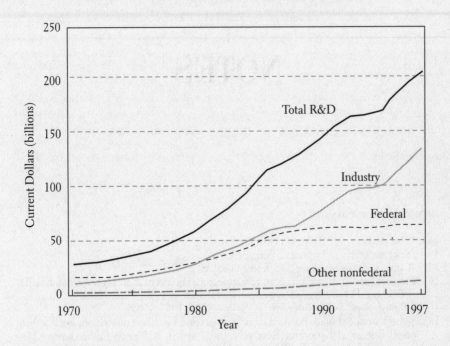

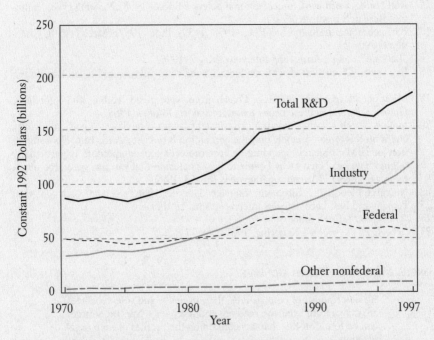

SOURCE: *National Science Board,* Science & Engineering Indicators *(Arlington, VA, National Science Foundation, 1998)*

NOTES

INTRODUCTION: CHANGE

15 "The value . . . it": Cohen (1948), p. 67.
15 "Science today . . . business": ibid., p. 3.
15 "I am . . . change-making": Kettering (1961A), p. 77.
16 Industrial research organizations first appeared: For a very short overview, see Basalla (1988), pp. 124–27. For a longer one, see Chandler (1990), throughout.
16 Solow showed: Solow interview, cf. Solow (1956, 1957, 1960).
17 Solow's work debuted: For general economics and invention discussion, see esp. National Bureau of Economic Research (1962) and in it, Nelson (1962A). Also: Minasian (1962); Arrow (1962B); Klein (1962).
17 R&D budget and work force: National Science Board (1998); Payson (1998); Industrial Research Institute (1997).
17 Early chemical industry: See Beer (1958, 1959); Ihde (1957); Mees (1950); Chandler (1990).
18 "Let's say . . . air": Armstrong interview, May 2, 1996.
18 In a sense: Discussion of semiconductors, solid-state physics, and radar: See Buderi (1996).
19 Misconceptions about research: Drawn from widespread reading and interviews. However, several of these issues are addressed in Gomory (1983).
20 more research is better research: Armstrong (various interviews) is particularly eloquent on this topic—namely how the perception is not necessarily true. His comparison of IBM's research spending against those of top competitors is particularly telling. Says he, "That's why I came to the conclusion that you can spend too much money on R&D." (Armstrong interview, July 13, 1999).
22 "research bloodbath": Birnbaum interview, Dec. 12, 1996.
25 Penzias and Bell Labs: Penzias interview, April 10, 1997.
25 "U.S. cars . . . fins": ibid.
25 "The Lord . . . trailer": Kettering (1961A), p. 77.

ONE: A MATTER OF DEATH AND LIFE

All descriptions of corporations, their projects, and research styles are drawn from extensive readings and interviews. Specific sources can be found in the chapters containing the profiles of each organization.

26 "Research is . . . change": Kettering (1961B), p. 91.
26 warm August night: The story of Penzias's awakening and experience is drawn chiefly from Penzias interviews, supported by visits to Bell Labs and interviews with various employees.
26 "My gut . . . grab": Penzias interview, June 5, 1998.
27 "I just . . . thought": ibid.
27 "reverse the . . . research": Draper (1990), p. 1.
27 "AT&T's senior . . . wisely": ibid.
27 similar story unfolding at IBM: McGroddy's experience is drawn chiefly from extensive interviews and correspondence with him, verified with other top IBM officials.
28 IBM's total R&D budget: I use figures from *BusinessWeek*'s annual R&D Scoreboards.
29 "by my count . . . Lucent": Penzias interview, June 5, 1998.
30 "People bitch . . . output": McGroddy interview, Jan. 22, 1997.
30 "Basic Research . . . Results": Uchitelle (1996).
30 "shackle the economy": ibid., p. D6.
30 "invention implemented": Brown (1997B), Introduction, p. xii.
31 "spectacular shrinkage of space": quoted in Chandler (1990), p. 607.
31 "In the . . . used": Anthony interview, Nov. 12, 1996.
31 U.S. industrial research spending climbed: NSF Science Indicators 1998.
32 "bold but grounded": J. S. Brown interview, Dec. 10, 1997.
32 "You get . . . necessary": ibid.
32 "We were . . . happen": ibid.
33 "It's no . . . do": Lee interview, Nov. 12, 1996.
33 "You have . . . concept": Netravali interview, Dec. 16, 1997.
33 "I visualize . . . on": Spinrad interview, Dec. 16, 1996.
33 Corporate labs evolved: widespread reading. Besides early chemical industry material cited above, see Wise (1985); Reich (1980).
34 "research scientists . . . goals": Basalla (1988), p. 125.
34 World War II fueled another era: see esp. Kevles (1979); Buderi (1996).
34 President Eisenhower spoke: see Associated Press (1953); Stetson (1956); Raskin (1956); Anon. (1956).
35 "The welcome . . . things": Raskin (1956), p. 15.
36 corporate research poised for outright disaster: funding figures and related details: *BusinessWeek* R&D Scoreboards; Buderi (1991A, 1991B, 1992A, 1992B, 1993A).
36 "The genie . . . town": Penzias (1995), p. 75.
36 "These assumptions . . . policymakers": Odlyzko (1995).
37 "The success . . . research": Rosenbloom and Spencer (1996A), p. 69.
37 "Specific industries . . . products": Phillips (1996).
37 "Basically, we . . . center": Grandke interview, Nov. 7, 1996.
37 "What was . . . researchy": Brinkman interview, Dec. 16, 1997.
38 "a crazy . . . opportunities": Hounshell (1996), p. 54.
38 Lee Davenport spent: Davenport's story and box are drawn from interviews with author. Cf. also Buderi (1996).
38 "Research is . . . favor": Davenport interview, Sept. 13, 1997.
39 "You must . . . accordingly": ibid.
40 "I said . . . no": Fuhrer interview, Nov. 25, 1996.
40 "The market . . . stronger": Ishiguro interview, March 6, 1997.
41 "almost instant . . . it": Kephart interview, Jan. 16, 1997.
42 Arun Netravali: Netravali interview, Dec. 16, 1997.
42 "Let us . . . anyway": ibid.
42 "The fact . . . altogether": Bromley interview, April 23, 1997.
43 "Without a . . . impaired": Rosenbloom and Spencer (1996A), p. 69.
43 Edward E. David, Jr.: David (1994).

43   These views—or fears: for more on Motorola Labs, see Hardy (1999).
43   "reinvent the corporation": J. S. Brown interview, Dec. 10, 1997.
43   "genetic variance": ibid.
43   "but the . . . strengths": Ishiguro interview, March 6, 1997.
43   Mark Myers: Myers interviews; Myers (1996). Cf. Myers and Rosenbloom (1996).
44   "binary syndrome . . . zero": Armstrong interview, April 8, 1999.
44   "I do . . . research": Weyrich interview, July 3, 1997.
46   "That's really . . . physics": Gunawardena interview, March 21, 1997.
46   "We don't . . . years": Birnbaum interview, Dec. 12. 1996.
46   Ralph Gomory: Gomory interviews.
46   "It was . . . changed": Gomory interview, April 8, 1997.
46   "There's no choice": ibid.
46   "The proper . . . error": Rosenbloom interview, Jan. 27, 1998.
47   "advancing waves . . . progress": Kettering (1961A), p. 77.
48   "I think . . . forward": Tyson interview, June 1, 1998.

TWO: THE INVENTION OF INVENTION

49   "The greatest . . . invention": Jewkes et al. (1958), p. 32.
49   "We can't . . . bee!": quoted in Kuhn (1962), p. 455.
49   Story of Perkin, organic chemistry, and early dye industry is drawn from these chief
      sources: Beer (1958, 1959); Chandler (1990); Haber (1958), esp. pp. 63–84; Meyer-
      Thurow (1982), pp. 364–365. Cf. Ihde (1957). For Perkin background, see esp. Beer
      (1959), pp. 5ff. Cf. Birr (1966), p. 57; Jewkes et al. (1958).
51   *Meisteren* and egg whites: Beer (1959), pp. 40ff, 62–74ff.
51   the struggling firm of Höchst. The company's story is drawn chiefly from: Beer
      (1959), pp. 62–64, 77, 97–99; Chandler (1990), p. 478; cf. Haber (1958), pp.
      170–80; Haber (1971), p. 131.
51   "as a . . . Lyon": quoted in Beer (1959), pp. 63–64.
52   BASF: The company was the later incarnation of what had started out in June 1861
      as Chemische Fabrik Dyckerhoff, Clemm & Co., a partnership registered in
      Mannheim. In 1863, the name was changed to Sonntag, Engelhorn & Clemm, as
      the concern bought a disused zinc smelter and began making Lebanc soda and
      fuchsin. But after repeated complaints of chemicals polluting local wells prompted
      city officials to order it out of town, the struggling business landed in the hamlet of
      Ludwigshafen, opposite Mannheim on the west bank of the Rhine. That same year
      the partnership dissolved, but armed with additional financial backing incorporated
      anew as Badische Anilin- und Soda-Fabrik.
52   *technische Hochschulen:* For the rise of German higher education and its contribu-
      tions to organic chemistry and corporate research, see Beer (1959), pp. 57ff; Haber
      (1958), esp. pp. 63–73; Bernal (1953), pp. 137–44; Mees (1950), pp. 81ff; Chandler
      (1990), p. 499.
53   "Almost overnight . . . end": Beer (1958), p. 133.
54   Demise of British and French dye industries: Beer (1959), pp. 19–48; cf. Chandler
      (1990). In the case of France, another factor was the country's loss of Alsace-Lorraine
      to Germany in the Franco-Prussian war of 1871, a defeat that deprived the nation of
      a rich reservoir of organic and dye chemists.
54   AGFA: Based in Berlin, AGFA was the only major German dye house located out-
      side the Rhine.
55   BASF and indigo synthesis: see esp. Beer (1958), p. 66.
56   Bayer initially a small concern: For the company's story, I rely on Beer (1958); Beer
      (1959), pp. 70–93; Haber (1958), esp. pp. 134–36; Meyer-Thurow (1982). Supple-
      mental facts gleaned from Bayer's Web page.

56 "invention of new colors": quoted in Meyer-Thurow (1982), p. 366.

56 embrace of research came haltingly: basic factors leading to corporate research in the dye industry, and early stages at German dye houses: Beer (1958, 1959); Haber (1958); Meyer-Thurow (1982).

57 A common practice that undoubtedly played a role in technology transfer was the recruiting of chemists from other firms, with many jumps coming just after new products were launched. Companies worked hard to keep their best people—so most defectors proved second-rate and were not widely trusted. But Bayer experienced some success in luring two alizarin specialists and one fuchsin expert from Höchst.

57 Duisberg background: Beer (1959), pp. 78ff.

57 discovered three important colors: The azo group, discovered by Peter Griess in 1856, had no dye properties itself. But compounds readily linked with coal tar products to give useful dyes. Moreover, these had the advantage of being able to dye cotton directly, obviating the need for prior treatment with a mordant. See Ihde (1957). Duisberg's other two creations included a blue azo and sulfonazurin, the first artificial dye for woolens to challenge indigo. The blue azo was developed in conjunction with Friedrich Bayer, Jr., son of the company founder: Beer (1959), p. 81.

59 "The locus . . . industry": Meyer-Thurow (1982), p. 372.

60 Leverkusen: Duisberg's memo and drawing provided by Bayer archives, with explanatory note to author from Dr. Margarete Busch, Nov. 5, 1997. See also, Chandler (1990), pp. 476–77; Haber (1958), pp. 135–36; Beer (1959), p. 146n. The site included so much room for expansion that it was not until around 1954 that Duisberg's original blueprint needed any substantial revision.

60 "that works . . . engineers": quoted in Chandler (1990), p. 477.

60 Duisberg pressed to free scientists: Chiefly from Meyer-Thurow (1982), pp. 372–73.

61 Bayer chemists' salaries and bonuses: ibid., p. 374. By contrast, General Electric at almost this same time offered researchers with doctorates a starting salary of $1,500 a year, or about 6,300 marks—significantly less than their Bayer counterparts and without any comparable royalties: Wise (1980), p. 418.

62 Kaiser Wilhelm Gesellschaft: chemical industry role in creating the institute: Beer (1959), pp. 112–14; Haber (1971), pp. 49–50.

62 "We esteem . . . work": quoted in Meyer-Thurow (1982), p. 380.

63 Duisberg and Parke, Davis: Chandler (1990), p. 161.

63 "These tires . . . automobile": quoted in Beer (1959), p. 161. The announcement was met with great applause, as it had been in Germany, due to the high costs of tires. However, early hopes for synthetic rubber proved premature, since a price war broke out that within a year forced Bayer to abandon its manufacturing plans. Production was revived during World War I, but again only temporarily.

63 In one critical way: For the state of U.S. science and industry before 1900, I rely on widespread reading. See Bartlett (1941); Bernal (1953); Birr (1966); Chandler (1990); Layton (1972); Howard S. Miller (1972); Jewkes et al. (1958); Kevles (1972, 1979); Mowery and Nelson (1996); Pursell (1972); Van Tassel (1966B); Wise (1985). Cf. Hoddeson (1977); Hounshell (1996).

64 "heroic" invention: Jewkes et al. (1958), pp. 53–54.

64 "quack" who ran: Carnegie (1948), p. 174.

64 "and great . . . knowledge": ibid., p. 175.

64 "What fools . . . competitors": ibid., pp. 175–76.

65 merger mania: Smith (1996), p. 41; Mowery and Nelson (1996), p. 193.

65 "A man . . . him": Mees (1950), p. 34.

66 "If we . . . do": quoted in Chandler (1990), p. 228.

66 By 1900 . . . thirty-nine corporate research facilities: Wise (1985), p. 78. Overviews to the growth of American industrial research can be found in Birr (1957) and Bartlett (1941A, 1941B).

66   "I have . . . College": quoted in Wise (1985), p. 79.
66   On a typical Friday evening: The account of Whitney's train ride and early days in Schenectady comes from Wise (1985), pp. 4ff, 81ff.
67   "cluttered, overdecorated . . . apartment": ibid., p. 82.
68   GE and Edison: main sources for the events leading up to the creation of GE's research lab are: Wise (1980, 1984, 1985); Hawkins (1950). Cf. Gorowitz (1981). Whitney background: Wise (1985). Steinmetz background: Wise (1985); Kline (1992). Edison background: Gorowitz (1981); some basic details in Basalla (1988) and McAuliffe (1995).
68   "a minor . . . so": quoted in Basalla (1988), p.125.
68   "Oh these . . . wrong": quoted in Kevles (1979), p. 8.
69   "Before many . . . practical": Bartlett (1941A), p. 31.
69   Edison invented carbon filament incandescent lamp: British inventor Joseph Swan independently discovered the incandescent lamp as well. The Russian A. N. Lodygin is also sometimes credited with this invention: see Basalla (1988), p. 60.
70   Westinghouse Electric: a good synopsis of George Westinghouse's strategy and the early battles with GE can be found in Wise (1985), pp. 69ff.
71   "It seems . . . avoided": quoted in ibid., pp. 76–77.
71   "It does . . . principles": ibid., p. 77.
71   "We all . . . laboratory": quoted in Wise (1980), p. 414.
71   "burying my . . . industry": quoted in Wise (1985), p. 81.
72   "Come in, rain or shine": ibid., p. 131.
72   "Are you having fun?": ibid.
72   "conditions affecting . . . him": quoted in Wise (1980), p. 418.
73   "These men . . . fail": quoted in Wise (1985), p. 88.
73   "The only . . . here": ibid.
73   "Although our . . . discovered": quoted in Wise (1980), p. 408.
73   "bear pit": quoted in Wise (1985), p. 101.
73   "I am . . . it": quoted in Hughes (1989), p. 67. For more on the German university-industrial model of research that influenced Whitney, see Hughes (1989), p. 160.
74   Weintraub's mercury rectifier: Although it turned out Weintraub had been edged out in his discovery by independent inventor Peter Cooper Hewitt, the two claims were quickly tied up in litigation. This showcases another reason corporate laboratories exist: not necessarily to be first, but to be close enough.
74   GEM lamp: Wise (1985), pp. 111ff.
74   GE research's darkest hour: for the whole saga, see ibid., pp. 122ff.
74   "shameful . . . been asleep": quoted in Kline (1992) p. 150.
75   "It affects . . . intercourse": quoted in Wise (1985), p. 140.
75   "The times . . . dollar": ibid., pp. 140–41.
75   "With active . . . made": ibid., p. 146.
75   Coolidge background and discoveries: see esp. Wise (1980); Wise (1985), pp. 119–37.
75   "hot swaging" process: Though upheld many times in court against outside challengers, Coolidge's 1913 patent would cause an internal split over the credit for the ductile tungsten process: General Electric researcher Colin G. Fink, later president of Columbia University, claimed he was the true inventor and apparently left the lab over the matter. See Reich (1983), p. 209.
75   Langmuir background and discoveries: see esp. Reich (1983); Wise (1980); Wise (1985), pp. 149–61.
76   "free to . . . wish": quoted in Reich (1983), p. 205.
76   "looking around . . . university": Letter dated July 16, 1909. Quoted in Wise (1980), pp. 421–22.
76   Three years later a practical result: For a brief but excellent description of this work, see Kevles (1979), pp. 99–100.
77   "He has . . . eye": quoted in Hawkins (1950), p. 45.

THREE: HOUSES OF MAGIC

78 "Large corporations . . . small": quoted in Vagtborg (1975), p. 106.

78 "I often . . . oftener": quoted in Wise (1980), p. 413.

78 The Chicago crowds poured past: Description of the 1933 World's Fair and GE expo are drawn from House of Magic videotape and supporting documents and photos provided by the Schenectady Museum. Later descriptions of performances and attendance numbers come from Anon. (Undated), *Out of the Research Laboratory.*

79 "So now . . . today": Schenectady Museum videotape.

79 A lot had gone down: Whitney retirement and lab accomplishments are drawn from wide reading. See especially: Wise (1985).

79 RCA patent tangle: Over the period 1912 to 1926, at least twenty important patent interferences involving radio equipment erupted between the companies involved in RCA's formation. To resolve the issue, RCA was created in 1919 to pool some two thousand patents held between the various parties after Marconi's American operation was bought out by General Electric. RCA then served as the marketing outlet for consumer radios made by GE and Westinghouse, with 60 percent of RCA's products coming from General Electric, the rest from its competitor.

    The agreement was expanded the next year to include AT&T, whose own indispensable triode work paralleled that of GE's. The result was an elaborate cross-licensing agreement. The gist was that Ma Bell bought $2.5 million worth of RCA stock and received exclusive licenses to all patents dealing with wire telephones and telegraphs, as well as some radio rights that pertained to its phone network. RCA and GE, meanwhile, possessed the rights to use the various patents in wireless telegraphy and international two-way radio communication—including the ability to provide wireless telephone apparatuses for the amateur market. RCA chief executive David Sarnoff later persuaded GE and Westinghouse to let his operation take over radio manufacturing. In spring 1930, three weeks after the agreement was signed, the Justice Department filed an antitrust suit that forced GE, Westinghouse, and AT&T to make RCA an independent corporation. Sarnoff expanded quickly, pumping funds into research and development and establishing RCA as a leading-edge electronics force. Ironically, in 1980, RCA was bought out by GE. Many operations have since been sold or spun off, including its famed Sarnoff Lab in Princeton, NJ. Ownership was transferred to SRI International in 1987, and the lab was turned into a for-profit subsidiary and private research consulting house: Birr (1957), pp. 46ff, 98ff; Chandler (1990), pp. 219ff; Reich (1977), pp. 215ff; some modern day details from SRI background publications.

79 Floyd Gibbons: For some background and House of Magic story, see Anon. (Undated), *Out of the Research Laboratory.* Cf. Hawkins (1950); Birr (1966). For Gibbons background, see also *Who Was Who 1925–47.*

79 "This is . . . Magic": Anon. (Undated). Wise (1985), p. 214, puts the House of Magic broadcast in 1925.

80 325 campus newspapers: Wise (1985), p. 214.

80 GE's magic set the standard: General facts and trends in pre-WWI industrial research are pieced together from: Chandler (1990); Hounshell (1996); Basalla (1988), pp. 125–26; Birr (1966), pp. 59–68.

80 Westinghouse and Kodak: Mees (1950), pp. 137–49; Bartlett (1941B), pp. 54–55. Although Kodak's lab opened in 1913 with a staff of twenty, the lab dates its founding to 1912.

80 Duisberg and Kodak: Hounshell (1996), p. 25.

81 "Almost everything . . . investigations": Mees (1950), p. 147.

81 initial job of research: Though limited in scope, by far the best analysis of the rise of research and its role in overall corporate evolution and structure is Chandler (1990).

81 a firm's bedrock patents: The best assessment of the role of patents can be found in Reich (1977, 1980).

81 government's toughened antitrust stance: Mowery and Nelson (1996), p. 193, note that although it discouraged big firms from horizontal mergers, pre–World War II U.S. antitrust policy did not discourage research arms from acquiring outside technologies.

81 early AT&T research: Hoddeson (1980); Hounshell (1996), p. 23; Hughes (1989), pp. 150–59; Kevles (1979), pp. 98–99; Reich (1977, 1980). Cf. Bernstein (1984).

81 De Forest's triode: De Forest had invented the triode, or audion as he called it, back in 1906. He had already sold various rights to GE before AT&T began its licensing negotiations in 1912, when it was realized the triode could amplify signals as well as detect them. Work began in 1913. See esp. Reich (1977), pp. 212–27.

82 "This condition . . . arts": quoted in ibid., p. 213.

82 first transcontinental telephone line: Anon. (1915G); Bernstein (1984); Hoddeson (1977).

82 "Mr. Watson . . . now": quoted in Anon. (1915G), p. 9. This re-creation came near the ceremony's end. Earlier, Bell had formally opened the transcontinental line with the words: "Ahoy! Ahoy! Mr. Watson!" (ibid., p. 5). What's more, Watson's direct quote was lost to posterity. His words were relayed to the Dey Street audience by Carty, who noted to great laughter: "He says it would take him a week to get there now." (ibid.).

82 "Doing Without Europe": The first article was called "Government Co-operation With Our Industries." The rest ran under the banner: "Doing Without Europe." See Anon. (1915A–F).

82 "If the . . . benefaction": Anon. (1915A), p. 99.

83 U.S. corporations doled out $50 million: Vagtborg (1975), p. 51.

83 World War I and industrial research: War's impact on corporations in general and research in particular: Scientific American series anon. (1915A–F); Chandler (1990), esp. pp. 84, 139ff, 289ff, 341–69, 502ff; Hawkins (1950), pp. 59ff; Hounshell (1996), pp. 29–33; Kevles (1979), pp. 99–116; Mees (1950), for British see pp. 92ff.

83 "The plain . . . research": quoted in Bartlett (1941A), p. 36.

83 Trading with the Enemy Act: Bartlett (1941A), pp. 35–37; Hounshell (1996), pp. 29–30. The act enabled the Alien Property Custodian to appropriate the German patents. Licensing was coordinated via the Chemical Foundation, Inc.

83 Kodak and DuPont served as templates: Hounshell (1996), pp. 24–25, 30; Hughes (1989), p. 176; Mees (1950), pp. 123–25.

83 National Research Council: the best overview of the NRC's creation is in Kevles (1979), pp. 111ff.

83 "War should . . . research": quoted in ibid., p. 116.

83 Corporate research continued to explode: See WWI sources above. Also, Bartlett (1941B). Kevles (1979), pp. 102ff, explains the various steps taken in the U.S. to spur research, both during and after the war. Chandler (1990) has the best overview of how the war affected research at European and American firms. He notes (pp. 295ff) that compared to other European companies, British concerns moved less aggressively to expand research during and after WWI. United Alkali Company, Nobel's Explosives, Ltd., armament maker Vickers, Ltd., and drug house Burroughs Wellcome were among the few with substantial prewar research investments; the war hardly changed things. Despite the removal of Siemens and AEG from the country, GEC was the only electric company to push research. Instead, many companies, often family-run, opted for current income over investments in production, distribution, and R&D. From 1925 through 1927, Courtaulds, Ltd., the country's rayon pioneer, paid an average 30 percent dividend as it disdained research spending. The family stunned the stock market in February 1928 by declaring a 100 percent share bonus. See Chandler (1990), pp. 307–11.

83 Treaty of Versailles: Under the treaty, Germany suffered heavy war reparations. France took charge of the Lorraine's coal and ore fields, as well as Alsatian textile industry and potash beds. Besides agreeing to sell off its foreign properties expropriated

by the Allies, Germany also had to pay stiff reparations that ultimately included $35 billion in pensions to Allied military personnel. It turned over millions of tons of coal, and large amounts of coal tar, benzol, and other chemicals and dyes to France, Belgium, and Italy. Finally, in March 1921 the heartland of industrial Germany—the Rhineland and the Ruhr—was occupied by French and Belgian troops, helping spark several years of hyperinflation. The net result was that the war had caused the country to lose roughly 13 percent of its population and European territory, all colonies, and about 15 percent of total productive capacity. It was no wonder that German firms lost their edge in research and everything else, temporarily.

84 "Science is . . . business": quoted in Kevles (1979), p. 173.

84 NRC's inaugural survey: Figures for research labs and total staff come from Bartlett (1941A), p. 37. Details on their concentration are in Cooper (1941). Cf. Chandler (1990), p. 84.

85 "The research . . . industries": Chandler (1977), p. 375.

85 AT&T was one prime mover: Bell Labs details from Hoddeson (1980); Mees (1950), pp. 121–23, 155.

85 Westinghouse recovered from bankruptcy: Bartlett (1941B), pp. 54–55; Chandler (1990), pp. 214–16, 222; cf. www.westinghouse.com, corporate history and milestones listings.

85 Westinghouse Research Fellowships: Graham (1985A), p. 52.

85 Chemical concerns: I rely on Bartlett (1941B), pp. 42–45; Chandler (1990), pp. 69–70, 170ff.

86 Rubber and tire manufacturers: Chandler (1977), pp. 435–38; Chandler (1990), pp. 105ff.

87 The oil industry: I rely generally on Bartlett (1941B), pp. 45–49; Chandler (1990), pp. 101ff; Mees (1950), pp. 42–43; cf. Enos (1962); Hounshell (1996).

87 "The General . . . thing": quoted in Chandler (1990), p. 102.

88 heavy crude is cracked: An alternative cracking method came from Jesse A. Dubbs, owner of a small refinery in Obispo, California. It led to the formation of the Universal Oil Products Company in Independence, Kansas. Backed by some $6 million from the meatpacking fortune of J. Ogden Armour, Universal's research staff—including Dubbs's son Carbon Petroleum Dubbs—grew quickly. A lab was built southwest of Chicago in Riverside, Illinois. From this base, Universal pioneered ways to produce higher octane gasoline. Some eighty years later, its name changed to UOP and the headquarters moved northwest of the windy city to Des Plaines, the company remained a leading developer of process technology for the refining and petrochemical industries. The pathbreaking UOP work, done in the late 1940s, was led by Vladimir Haensel. In 1997, Haensel received the prestigious Charles Stark Draper Prize—a gold medal and $450,000 honorarium—for his revolutionary chemical engineering process, called Platforming: see oil industry sources, op. cit. Some details from UOP and Draper Laboratory materials.

88 Gulf Research & Development: Mees (1950), pp. 127ff; cf. Anon. (1937); Mellon Institute (1971); Hounshell interview, Aug. 27, 1997. Chevron Corp. took over Gulf Oil in 1984. Research was essentially seen as redundant and disbanded, with surviving parts sent to Richmond, California. The Gulf research site became an industrial office park.

88 steelmaking: Adams (1982); Bartlett (1941B), pp. 58–60; Chandler (1990), pp. 138–39; Hounshell (1996), pp. 37–38.

89 pharmaceutical makers: I rely on Bartlett (1941B), pp. 60–61; Chandler (1990), pp. 152–55, 161–64.

89 early carmakers: general details from Bartlett (1941B), pp. 56–57; Chandler (1990), pp. 205–11.

89 General Motors: a few additional details taken from Sloan (1964), p. 249. Background on Kettering from *Who's Who in America, 1946–1947*; National Inventors Hall of Fame (1995), p. 35.

89   Delco: Delco was sold in 1916 to United Motors Service Company, which shortly
      thereafter was acquired by GM. The new Dayton Research Laboratories Company
      was merged in 1917 with the Dayton Metal Products Co., which devoted itself to
      wartime projects. It was this company that GM purchased in 1920, bringing Ketter-
      ing into the fold: General Motors Research and Development Center (1995), pp.
      4–7.
90   "tourists' " lab: Leslie (1983), pp. 120–21.
90   The lab quickly proved its importance: Detroit lab details come chiefly from ibid.,
      pp. 182–88; General Motors Research and Development Center (1995), p. 7.
90   the little guy still had options: Hounshell (1996), p. 28; cf. Vagtborg (1975).
91   first industrial fellowship: the Boston laundry project begat the American Institute of
      Laundering: Vagtborg (1975), pp. 124–25.
91   Mellon Institute: I rely chiefly on details provided by the Mellon Institute Library.
      These include: Anon. (1937); Mellon Institute (1971); "Welcome to Mellon Insti-
      tute" brochure. Cf. Vagtborg (1945), pp. 125–35.
91   The first industrial research fellowship began on March 1, 1911, but the institute it-
      self was founded as The Mellon Institute of Industrial Research and School of Spe-
      cific Industries in 1913. The Mellons gave the institute to the University of
      Pittsburgh in 1921. When it incorporated and broke off from the University in 1927,
      the name was shortened to Mellon Institute of Industrial Research. In 1962, it be-
      came the Mellon Institute.
91   "put cotton shirts on wieners": Anon. (1937), p. 108.
91   Silly Putty: Warrick (1990), p. 28.
91   Battelle: Mees (1950), pp. 107ff; Vagtborg (1975), pp. 135–50.
92   Barely zinged by the Great Depression: Graham (1985A), p. 51, notes that while 300
      of the 1,500 industrial labs that had opened by 1930 closed down, and many others
      scaled back operations, other aspects of business often fared far worse during the
      Great Depression. Moreover, from 1927 to 1933, and from 1933 to 1940, the num-
      ber of corporate research professionals rose impressively. See Hounshell (1996),
      p. 36, cited below.
92   From General Motors came four-wheel brakes and Freon: see Killeffer (1948), p. 43;
      Leslie (1983), pp. 188–89, 223ff.; Bartlett (1941B), pp. 56–57.
92   None of these could match DuPont: Details of interwar years and launch of funda-
      mental polymer chemistry studies are drawn chiefly from Hounshell and Smith
      (1988), esp. pp. 350–60 (a book critical for understanding the full story of DuPont
      research); Hounshell (1996), pp. 39–46; Hughes (1989), pp. 177–80. Cf. Chandler
      (1990), pp. 175ff; Mees (1950), pp. 123–25; Miller (1997); Stine (1941); Stine bio
      supplied by DuPont public relations. Some details and scene-setting from personal
      visit.
92   "a week . . . talking": quoted in Hughes (1989), p. 178.
93   nylon discovery: Hounshell and Smith (1988), pp. 268–71; Hounshell (1996), pp.
      39–40.
93   As the fruits of corporate labs: For Kettering and Revenue Act of 1936, see Kevles
      (1979), pp. 273–74. For National Research Council efforts, Hounshell (1996), p. 36;
      Wise (1985), p. 215.
93   "royal road to riches": Hounshell (1996), p. 36.
93   share of U.S. patents: figures from Jewkes et al. (1958), pp. 104–105; cf. Sanders
      (1962), p. 62.
93   Rising European competition: The assessment is drawn from widespread reading.
      By far the single best source is Chandler (1990). For EMI and ICI stories, see esp.
      pp. 355ff.
94   "R&D" becomes part of the public lexicon: Hughes (1989), p. 139.
94   M.I.T. physics department received so many calls: Kevles (1979), p. 273.
94   number of corporate research professionals: figures cited in Hounshell (1996), p. 36.

94   research featured in annual reports: Graham (1985A), p. 52.

94   winds of coming war whipped the flames: drawn from widespread reading. See Hounshell (1996); Buderi (1996).

94   General Electric's research effort: details from Anon. (Undated).

95   An elbow-to-elbow crowd: account taken from Bell Labs press release, July 1, 1948, supplied by AT&T archives; "The Transistor," a description of the event, with photos, in *Bell Laboratories Record* 26, no. 8 (Aug. 1948): 321–24; Smits (1985), pp. 15–16. Cf. Buderi (1996), pp. 308–309; Riordan and Hoddeson (1997), pp. 163–64.

95   "Scientific research . . . framework": quoted in Riordan and Hoddeson (1997), p. 164.

95   Corporate research primed to enter its golden age: widespread reading. See esp. Hounshell (1996), pp. 41, 49; Kevles (1979), pp. 243–44. Cf. Graham (1985A); Greenberg (1966). For the particularly strong influences of the Manhattan Project and the M.I.T. Radiation Lab, see Buderi (1996) throughout; Vagtborg (1975), pp. 56–57.

95   central tenets laid out by Vannevar Bush: Bush (1945); Hounshell (1996), p. 42; Kevles (1979), esp. p. 347.

96   "After World . . . future": quoted in Vandendorpe (1997), p. 5.

96   the spigot stayed open: For figures, I rely on Birr (1966), p. 69. Graham (1985B), p. 183, puts federal spending in perspective. Cf. Graham (1985A); Hounshell (1996), pp. 47ff. For effect on electronics in particular see Hounshell (1996), pp. 48–49.

96   many U.S. firms established or acquired labs in Europe: ibid., pp. 54–55.

96   more than a hundred applied research institutes were formed: Vagtborg (1975), pp. xiv–xv.

97   In concert with this great expansion: A great source on how the mission and style of research changed is Chandler (1990). See also Mees (1950), pp. 47–48.

97   One factor in this metamorphosis: For antitrust policy and its effects on research, I draw extensively on Mowery and Nelson (1996), pp. 205–206; Mowery and Teece (1996), pp. 113–14; Roe (1996), pp. 107–108.

97   Competition also spurred this evolution: Bartlett (1941B), esp. p. 42; Chandler (1990), pp. 188ff.

97   "In the . . . advance": Bartlett (1941B), p. 42.

97   In light of such pressures: For general post-WWII and Cold War climate of research, I rely on widespread interviews and readings. Specific examples are cited below.

97   "In the . . . company": Davenport interview, Sept. 13, 1997.

98   companies elevated research to divisional status: Graham (1985B), p. 185.

98   GE moved to a hilltop: Birr (1957), pp. 140ff.

98   "I asked . . . doing": quoted in Odlyzko (1995), pp. 1–2.

98   Big Science sired Big Industrial Science: For a good examination of how postwar changes led firms to engage in longer-range, more fundamental research see Chandler (1990), pp. 605–17; Graham (1985A); Hounshell (1996); Hawkins (1950), pp. 98ff; Jewkes et al. (1958), p. 127. Interesting figures are in Worley (1962).

99   "The shibboleths . . . technologies": Hounshell (1996), p. 41.

99   General Electric: Numbers of development labs pieced together from Hawkins (1950), pp. 98ff; Birr (1957), pp. 140ff.

99   The sentiments of scientists: I rely on Kevles (1979), pp. 369, 386–88. Cf. Graham (1985B). For problems industry had in attracting scientists, I draw on Armstrong interview, March 14, 1996; Hounshell (1996), p. 49; Graham (1985A), pp. 54–59.

100   "The new . . . context": Graham (1985A), p. 60.

100   GE and post-WWII basic research: Birr (1957), pp. 140ff.

100   DuPont and post-WWII basic research: chiefly, Hounshell (1996), pp. 43–46. Cf. J. Miller (1997).

100   "retain its . . . research": quoted in Hounshell (1996), p. 44.

101   "new nylons": quoted in ibid., p. 45.
101   Over at AT&T: For general ramping up of basic research, see Hoddeson (1977); Buderi (1996), pp. 311–12, 323–24; W. Brown interview, Dec. 16, 1997.
101   "Sputnik reset . . . clock": W. Brown interview, Dec. 16, 1997.
102   "I am . . . department": quoted in Hounshell (1996), p. 43.
102   "scientific laboratory": quoted in ibid., p. 43.
102   "fundamental research . . . transportation": quoted in ibid.
102   "much that . . . States": quoted in Associated Press (1953). Other details of Ford's Research and Engineering Center: Hounshell (1996), pp. 43, 65 (note 58); Associated Press (1953); Nevins and Hill (1962), p. 373.
102   Car king General Motors: details of GM's basic research expansion come from Hounshell (1996), p. 46; Sloan (1964), p. 250; Stetson (1956). Additional Hafstad details: Who's Who in America 1980–81.
102   "The dramatic . . . era": Sloan (1964), p. 250.
102   "His appointment . . . problems": ibid.
102   GM Technical Center: details drawn from Sloan (1964), pp. 248, 259–63; Leslie (1983), pp. 317–18; Raskin (1956); Stetson (1956).
103   "We must . . . research": quoted in Stetson (1956).
103   "I hope . . . endowed": ibid.
103   "Here in . . . it": quoted in Leslie (1983), p. 318. For Kettering's views on pure science see ibid.; Hounshell (1996), p. 65 (note 56).
103   U.S. Steel launched: USX public relations.
103   Dow and Dow Corning: Warrick interview, March 19, 1998.
103   Merck: Hounshell (1996), p. 46; Buderi (1996), p. 469.
103   Over in the oil sector: for Gulf and Chevron, Anon. (1937); Mellon Institute (1971); Hounshell interview, Aug. 27, 1997. Exxon details from Vandendorpe (1997), p. 8.
103   Few industries pursued: Westinghouse details from Hounshell (1996), p. 64 (n. 51); Westinghouse Research; Westinghouse Research and Development, documents supplied by Westinghouse Science and Technology Center library. Hughes details from HRL Web page; conversation with Dave Weeks, HRL staff. Since Dec. 1997, the old Hughes lab has been jointly owned by Hughes Electronics and Raytheon and known as HRL.
104   Even with accelerators: RCA details from Graham (1985A), pp. 60–70; Graham (1985B), pp. 187ff; Graham (1986), pp. 68–74; Hounshell (1996), p. 46.
104   "building block research": term quoted in Graham (1985A), p. 64.
104   "sustaining research": quoted in ibid., p. 65.
104   Then there was IBM: For early days of IBM research, I rely on Bashe et al. (1986), pp. 522–70; Brennan (1971). Cf. Pugh (1995).
104   "By pure . . . considerations": quoted in Bashe et al. (1986), p. 524.
106   "The welcome . . . industry": Raskin (1956), p. 15.
106   "Twenty years . . . clothes": ibid.
106   "the technical . . . enjoyed": Anon. (1956).
107   "When I . . . worship": Gomory interview, April 8, 1997.

FOUR: OUT OF THE PLUSH-LINED RUT

108   "Scientific leadership . . . success": Armstrong interview, March 21, 1996.
108   "Although it . . . rut": Killeffer (1948), p. 69.
108   The Japanese worked: account is drawn from Gomory and McGroddy interviews; Gomory (1989A, 1989B).
109   "The whole . . . U.S.": Gomory fax of Aug. 26, 1999.
109   "Believe me . . . behind": Gomory interview, July 25, 1996.

110   Research . . . had gotten away from just about everybody: Drawn from extensive interviews and reading. Specific examples are cited below. However, a good overview is provided by the annual *BusinessWeek* R&D Scoreboard issues and related articles. See esp. Buderi (1991A, 1991B); Coy (1993A, 1994, 1995).

111   "The fantastic . . . things": Penzias interview, April 10, 1997.

111   "The model . . . in": ibid.

111   But the argument for another approach: again, drawn from widespread interviews and readings. For good overviews, see Odlyzko (1995); Hounshell (1996).

111   "DuPont had . . . opportunities": Hounshell (1996), p. 50.

112   "I don't . . . years": quoted in Buderi (1992A), p. 104.

112   "It seemed . . . explosions": Preston (1991), p. 80. Other steel industry particulars, pp. 79–81.

112   "the nation's . . . objectives": quoted in Hounshell (1996), p. 68 (n. 73).

113   RCA and the Videodisc: Graham (1985B), pp. 185ff.

113   It was in this gloomy climate: account drawn from Gomory and McGroddy interviews; Gomory (1989A, 1989B).

113   "We had . . . that": McGroddy interview, Feb. 3, 1997.

114   "One of . . . benefit": Gomory interview, Aug. 26, 1999.

114   IBM's research division was not without its problems: Armstrong and McGroddy interviews. For RISC, Cocke interviews; Gomory correspondence; Pugh (1995), pp. 314–16.

114   "By then . . . technology": Pugh (1995), p. 316.

115   Gomory had tried: Gomory interviews; Gomory (1989A, 1989B).

115   "I do . . . credibility": Gomory (1989B), p. 30.

116   Xerox was reeling: Smith and Alexander (1988); Jacobson and Hillkirk (1986); Brown, Myers, Spinrad interviews.

116   "wonderful bribe": quoted in Jacobson and Hillkirk (1986), p. 257.

116   Somewhat similar maladies dogged General Electric's: Robb and Schmitt interviews, correspondence; cf. Hawkins (1950), pp. 99–102.

117   The same sorts of sagas: extensive interviews. Carrubba and Philips details are in Coy (1993A).

117   "so lower-level . . . vine": quoted in ibid., p. 104.

118   The U.S. economic downturn: see Graham (1985B), p. 181.

118   "research bloodbath": term used by Birnbaum, interview of Dec. 12, 1996.

118   R&D spending: see *BusinessWeek* annual R&D Scoreboard issues.

118   "To use . . . have": quoted in Buderi (1992A), p. 105.

118   "I want . . . leave": quoted in Buderi (1993A), p. 78.

118   Not every research organization survived: details of Gulf chiefly from Hounshell interview, Aug. 27, 1997. A few additional points, as well as Westinghouse particulars, stem from Hounshell interview, August 27, 1997. U.S. Steel, Westinghouse, and RCA details drawn largely from company public relations staffs.

119   "Well, that . . . anymore": Linke interview, Oct. 17, 1996. Other NEC details from visits and interviews.

119   "People have . . . together": Armstrong interview, March 21, 1996.

120   "[writers] forgot . . . that": Blanz interview, Nov. 7, 1996.

120   "I have . . . researchers": ibid.

120   "technology push," "market pull": for a good, brief description of these widely discussed principles, as they relate to research, see Weyrich (1997).

120   "Then by . . . who": Horn interview, July 24, 1998.

120   To some degree: German dye industry discussion from Beer (1959), pp. 90ff.

120   For Bell Labs: various interviews.

121   Xerox codevelopment teams: Myers (1996), p. 145.

121   modern day Skunkworks: Gwynne (1997), p. 18ff. The term as it applies to research was apparently coined by Lockheed's Irven Culver.

121   And this was only an inkling: company examples taken from extensive interviews.

121   skyrocketing price of doing science: Mowery and Nelson (1996), pp. 114–15.

121   "largely unknown . . . generation": Graham (1985B), p. 180.

121   Uncle Sam had gotten: details of federal actions to spur more applied R&D come from Graham (1985B), pp. 190ff; Hounshell (1996), pp. 52–54; Mowery and Nelson (1996), pp. 113–14. Bloch background: *Who's Who in America 1988–1989.* For list of government-sponsored industrial R&D programs see Lester (1998), chart, p. 295.

122   research consortia: more specific information is derived from Mowery and Nelson (1996), pp. 115–23.

122   Industry-university alliances: Hounshell (1996), pp. 52–53; Mowery and Nelson (1996), op. cit.

122   HP alliances with universities: Rosenberg interview, Dec. 13, 1996.

122   global competitive alliances: Myers interview, Oct. 29, 1997; Mowery and Nelson (1996), op. cit. They note that international strategic alliances had long been common in mining and oil production, and, since WWII, among manufacturing firms as well. The difference since around 1975 has been the proliferation into a wider range of industries—and their spread into research, product development, and production for world markets. Cf. Hounshell (1996), pp. 54–55.

123   "Well, no . . . returns": Rosenberg interview, Dec. 13, 1996.

123   "ensembles of relationships": Myers interview, Oct. 29, 1997.

123   "These are . . . laboratory": ibid.

124   "When researchers . . . innovated": J. S. Brown interview, Dec. 10, 1997.

124   "didn't have . . . universe": Gomory interview, April 8, 1997.

124   "You can . . . different": ibid.

124   A bright light shone: details of Porter at the National Innovation Summit and the Porter-Stern work from on-scene reporting; Porter and Stern (1998, 1999); Council on Competitiveness (1998).

124   rising share of revenues from new products: see, for example, Coy (1995).

125   Of particular focus: For rise of Japan and other nations, facts and figures are assembled from Hounshell and Smith (1988), pp. 55–56; Buderi (1992A), p. 106; Porter and Stern (1998); Porter at National Innovation Summit. Patent figures: Buderi (1992B); Coy (1993B).

126   For DuPont cycles of basic research, see Miller (1997). A few additional details from DuPont bio of Kwolek; Hounshell (1996), p. 432. Communications cycles: Brinkman and Lang (1999).

127   "practice of . . . research": Brinkman and Lang (1999).

127   "There is . . . research": Spinrad interview, Dec. 16, 1996.

127   "Japan and . . . otherwise": Porter at innovation summit.

128   "Few businesses . . . it": Mees (1950), p. 43.

128   "a milieu . . . grounded": J. S. Brown interview, Feb. 7, 1998.

128   "Creating the . . . stuff": ibid.

FIVE: IBM: TAKING THE ASYLUM

> Details of IBM's organization, budget, plans, and strategy come
> from a series of interviews at the Yorktown-Hawthorne, Almaden,
> Zurich, and Tokyo labs. Only major examples are broken out for
> separate citations.

129   "At IBM . . . glorious": Meyerson interview, Feb. 17, 1999.

129   "It is . . . been": Armstrong interview, April 8, 1999.

130   "man's eternal . . . knowledge": quoted in Bashe et al. (1986), p. 562.

130   The inmates, though: Meyerson and silicon-germanium details from Meyerson interviews; Isaac interview, July 24, 1998. Cf. Bliss (1998); Steinberg (1998).

130   "I could . . . about": Meyerson interview, Feb. 17, 1999.

130   "Things like . . . for": ibid.

131   "It's the . . . faster": Horn interview, July 24, 1998.

131   IBM's roots: Except where noted, story of IBM's early years pieced together primarily from Pugh (1995), pp. 1–60; Watson and Petre (1990), pp. 10–15; Dordick (1995).

132   Top Twelve Achievements: Dordick (1995).

133   "Balance sheets . . . future": quoted in both Pugh (1995), p. 37, and Watson and Petre (1990), pp. 14–15.

133   "greatest sales pitches": Watson and Petre (1990), p. 15.

134   lab of J. Royden Peirce: Peirce's Manhattan lab launched a small research effort when IBM acquired it in 1922, but there seems to be no record of it making any significant contributions. Also in 1922, Watson formed the Future Demands Committee—upgraded seven years later to departmental status—with the idea of making recommendations on "the creation and development of new products to cover new fields": quoted in Pugh (1995), p. 112. The department had been inspired by National Cash Register's Future Demands Committee, on which Watson had served; another attendee was Charles F. Kettering, well before he resigned from NCR in 1909 to found the Dayton Engineering Laboratories Company. The IBM effort was disbanded in 1934—only to be reconstituted in 1940. It then fell dormant a second time in 1943 and would not reemerge until after World War II. See Pugh (1995), pp. 109–116.

134   Only twenty-six of fifty-nine key patents: Pugh (1995), pp. 77–78.

135   "We have . . . work": quoted in ibid., p. 55.

135   It was in this climate: Watson lab at Columbia is taken largely from Bashe et al. (1986), pp. 527–36; Brennan (1971). Cf. Pugh (1995), pp. 127–29.

136   "The rubber . . . fireflies": quoted in Brennan (1971), p. 21.

136   "it is . . . Laboratory": quoted in ibid., p. 34.

136   The lab, though: details of Poughkeepsie lab are drawn chiefly from Bashe et al. (1986), pp. 536–44; Pugh (1995), pp. 149ff. For additional Palmer and IBM electronics background, see Pugh (1995), pp. 81–87, 133–36. Supported by Landauer interview, Oct. 30, 1997.

136   "I understand . . . it?": quoted in Pugh (1995), p. 150.

137   "We were . . . useful": Landauer interview, Oct. 30, 1997.

137   The lab helped IBM's bottom line: Brennan (1971); Pugh (1995), p. 168. Ultimately 5,600 604s were built and installed during the machine's ten-year lifetime: Pugh (1995), p. 152.

137   It became a focal point: R&D work force figures from Dordick (1995), pp. 3–4; Pugh (1995), pp. 237ff.

137   Edison's bust: Bashe et al. (1986), p. 542.

137   Even before people moved in: separation of research from development is best described in Bashe et al. (1986), pp. 544–63. Another factor in the decision to break off research was an internal study of other research organizations conducted in spring 1955. Dubbed the Fox Hill report, it focused on the experience of Westinghouse, which had tried to increase the relevance of its distinguished lab by placing it inside the engineering group. The move had sparked an exodus of top scientists, adding fodder to the argument that IBM should hasten the separation of "R" and "D."

138   "The fashion . . . air": Landauer interview, Oct. 30, 1997.

139   The first research manager: Tucker's background and experiences are drawn chiefly from Tucker interview, Oct. 29, 1997. Details of his actions and Research activities at the time are supported by Cohen et al. (1979), p. 15; Bashe et al. (1986)—for cryotron project see esp. pp. 568ff; Fowler and Landauer interviews.

139 "There was . . . research": Tucker interview, Oct. 29, 1997.

140 "And that . . . lost": ibid.

140 After Tucker left: brief Anderson details from various interviews, bio in *Who's Who in America 1992–1993*.

140 Gomory was born: Gomory background and experiences are drawn chiefly from Gomory interviews; supported by Bashe et al. (1986); Fowler, Landauer, and Armstrong interviews, among others.

141 "You got . . . technology": Gomory interview, Aug. 26, 1999.

141 "When I . . . it": ibid.

142 "it was . . . right": ibid.

142 "It smoked . . . embarrassed": Gomory interview, Aug. 20, 1996.

143 "Alex had . . . to": Armstrong interview, May 2, 1996.

143 "[The] freedom . . . Müller": ibid.

143 As IBM's research boss: Armstrong's attitude and actions from detailed interviews.

144 by the time Gomory retired: Joint Programs status from Gomory (1989B), p. 32; Gomory interviews.

144 In the end, though: account of IBM's troubles is drawn from many interviews, esp. those of Armstrong, Fowler, Gomory, Landauer, McGroddy, and Price. Cf. Fowler (1995).

144 "Research was . . . that": Fowler interview, Oct. 30, 1997.

145 "localized transport . . . semiconductors": ibid.

145 "And I . . . doing": ibid.

145 "it was . . . them": Blanz interview, Nov. 7, 1996.

145 "That's one . . . both": Armstrong interview, April 8, 1999.

145 "It *was* . . . both": Gomory interview, Aug. 26, 1999.

145 "I got . . . budgets": Gomory fax of Aug. 26, 1999.

146 "because IBM . . . internally": Gomory interview, Aug. 26, 1999.

146 "They thought . . . wackos": Kovac interview, July 23, 1998.

146 "We were . . . away": Armstrong interview, April 8, 1999.

146 "He was . . . fell": ibid.

146 McGroddy was a tough: McGroddy's experiences drawn from interviews and correspondence, both with him and others.

147 "As they . . . technology": McGroddy interview, May 9, 1999.

148 "Our motto . . . obvious": McGroddy fax of Oct. 30, 1998.

148 but "people . . . was": McGroddy interview, Feb. 3, 1997.

149 "The only . . . product": McGroddy interview, May 9, 1999.

149 A trial by fire: payroll figures from Anon. (1998B); R&D figures from *BusinessWeek* R&D Scoreboards.

149 "Lou, three . . . well": McGroddy interview, Jan. 22, 1997.

150 Speech Recognition: details drawn primarily from Chen and Nahamoo interviews; supporting information from IBM public relations.

150 "A lot . . . have": McGroddy interview, October 1998.

151 "map the . . . perfectly": Chen interview, July 23, 1998.

152 "Jim McGroddy . . . IBM": Pulleyblank interview, March 5, 1999, amended in e-mail of Sept. 26, 1999.

152 Far fewer innovations: for details on IBM's chip advances, see Steinberg (1998); Bliss (1998); Narisetti (1998A, 1998B); J. Ramo (1997); Zuckerman (1997). For storage advances, a good source is Tristram (1998).

153 IBM Accomplishments: materials provided by IBM Research; Summers interview, Feb. 19, 1999.

153 "They appear . . . Fujitsu": J. Jones interview, March 12, 1999.

154 "What has . . . it": Meyerson interview, Feb. 17, 1999.

154 "We've just . . . world": Theis interview, Feb. 17, 1999.

154 "I'd like . . . heyday": Horn interview, July 24, 1998.

155 "We're entering . . . management": Horn interview, Feb. 19, 1999.
155 "I like . . . again": Stein interview, Feb. 19, 1999.
157 "Hello Mitch . . . you": lab visit of Feb. 19, 1999.
157 "What time . . . on?": Stein interview, Feb. 19, 1999.
157 *The X-Files* . . . usual": lab visit of Feb. 19, 1999.
157 "I came . . . platform": Stein interview, Feb. 19, 1999.
158 One Face of Pervasive Computing: drawn primarily from Almaden lab visit of March 26, 1998.
158 "The culture . . . later": Selker interview, March 26, 1998.
158 *Physical*: PAN from Zimmerman interviews.
158 "the Trojan . . . capabilities": Selker interview, March 26, 1998.
158 The fall 1996: general speech recognition details come primarily from Nahamoo interview, July 23, 1998; supported by information provided by IBM public relations. For MedSpeak: Roukos interview, Jan. 17, 1997. TeamBuilder: Greenwood interview, Feb. 16, 1999; Schecter (1999).
159 *Graphical:* Dryer interviews.
159 *Cognitive:* Dryer, Petkovic interviews.
159 "I see . . . time": Selker interview, March 26, 1998.
159 IBM's first First: Viveros interviews.
160 Machine outthinks man: Deep Blue facts and figures from IBM press materials.
160 "How do . . . champion?": Pulleyblank interview, Feb. 18, 1999.
160 Such is the: Pulleyblank provided a comprehensive overview of deep computing.
160 "Deep computing . . . Blue": Pulleyblank interview, Feb. 18, 1999.
161 One ambitious project: Pulleyblank, Takriti interviews; Hoy et al. (1998). Cf. Spiegler (1999).
162 "We have . . . important": Pulleyblank interview, Feb. 18, 1999.
162 What's more, even as IBM: computational biology overview is drawn primarily from Nunes, Barry Robson, and Colin Harrison interviews; Zipkin (1998). Some Teiresias details from Rigoutsos interview, Feb. 18, 1999.
162 "My passion . . . discipline": B. Robson interview, Feb. 18, 1999.
163 "These guys . . . areas": Horn interview, July 24, 1998.
163 Security: The Virus Wars: drawn from extensive interviews with Steve White, Kephart, Chess, Morar; Kephart et al. (1997); lecture materials furnished by White.
164 Charles and Elaine Palmer: Palmer interviews and IBM supporting materials.
164 "Don't worry . . . since": E. Palmer interview, Feb. 18, 1999.
164 "There is . . . concrete": C. Palmer interview, Feb. 18, 1999.
164 "You realize . . . history": Steve White interview, July 23, 1998.
165 "The key . . . key": E. Palmer interview, Feb. 18, 1999.
165 "It doesn't . . . inside": ibid.
165 "It's a . . . home": C. Palmer interview, Feb. 18, 1999.
166 A computer virus is: Credit for introducing the term computer virus goes to Fred Cohen, now a management consultant and principal member of the technical staff at Sandia National Laboratory in Livermore, California. In fall 1983, Cohen was taking a computer security class at the University of Southern California when he began investigating the possibility of people deliberately fashioning malevolent code—conducting experiments into the infectious nature and ultimate detectability of such creations. On November 12, he demonstrated his work to the class, prompting professor Len Adelman—later the A in the cryptographic security firm RSA Corp.—to say something along the lines of: "It's kind of like . . . a virus." Cohen seized on the analogy, which he unveiled in his landmark paper, "Computer Viruses—Theory and Experiments," presented in Toronto the following spring at a technical committee meeting of the International Federation of Information Processing Societies: Fred Cohen, e-mail correspondence, October 1998. Antivirus sales figures: Dataquest.

166   "The writers . . . world": Kephart interview, Jan. 16, 1997.
167   One selection is: demo of archived viruses courtesy of Dave Chess.
167   "Virus sends . . . corinne": Chess demo.
167   "Well, you . . . treaty": S. White interview, July 23, 1998.
168   "This should . . . concept": Chess demo.
169   "The compression . . . down": S. White interview, July 23, 1998.
169   "Patterns and . . . economies": Kephart interview, Jan. 16, 1997. Modified during subsequent conversations and e-mails.
170   "stop the . . . computing": S. White interview, July 23, 1998.
171   "This is . . . computers": ibid.
171   "He took . . . mean?": Isaac interview, July 24, 1998.
171   "IBM lost . . . lost": Murray talk, December 1998. Talk was co-sponsored by *Technology Review* magazine and the Council on Competitiveness.
171   "We still . . . success": Isaac interview, July 24, 1998.
171   "The real . . . Motorola": Armstrong interview, April 8, 1999. Slightly modified in conversation of July 13, 1999.
172   "We've been . . . us": Horn interview, Feb. 19, 1999.
172   "the high-tech . . . corn": Woodall interview, April 13, 1999 and subsequent e-mail clarification.
173   "Ask Bernie . . . that": anecdote related by Horn and verified by other IBM personnel.

SIX: HOUSE OF SIEMENS

> The Siemens archives provided invaluable help in locating documents and running down answers to my many historical questions.

174   "I do . . . research": Weyrich interview, July 3, 1997.
174   "The future . . . happen": Zorn interview, July 1, 1997.
174   Around 7:30 on the morning: Beckurts's legacy and death are largely derived from Gumbel and O'Boyle (1986), with supporting obituaries and details supplied by Siemens.
175   Although it's hard: for details of WWII losses, see esp. Siemens (1957B), pp. 258ff.
175   Siemens AG: For Siemens pre-1900 history, I draw extensively on Siemens (1957A); Chandler (1990), pp. 392ff, 463–65; Feldenkirchen (1994).
177   In the early 1900s: 1900s growth and Siemensstadt formation: Siemens (1957A) throughout, esp. pp. 166, 192, 263–68, 304ff; Chandler (1990), pp. 465–73; Siemens AG (1997A). Cf. Feldenkirchen (1995), esp. timelines.
177   "A similar . . . D.C.": Chandler (1990), p. 469.
177   Germany's defeat in WWI: for effects of WWI and WWII, and activities, I rely on Siemens (1957B); Chandler (1990), pp. 539ff.
178   By the 1950s: Details of Siemens post-WWII are drawn from Siemens (1957B), pp. 235, 260ff; Siemens AG (1997A); plus various interviews, corporate reports, and documents.
179   "Everywhere you . . . Siemens": widely repeated saying conveyed by Zorn, interview of July 1, 1997.
179   The Path of Research: except where noted, history of Siemens R&D and its first centralized research organization is drawn from Siemens (1957A) throughout, esp. pp. 35ff, 60–62, 91, 134–56, 178ff, 204–207; Siemens (1957B), esp. pp. 35ff, 252–55; Kaske (Undated), document provided by Siemens archives, translation thanks to Dr. Giuseppe Oppo.

179 These men and a handful of others: A few details of Frölich's work are taken from Siemens (1957B), p. 35. Some X-ray work specifics from ibid., pp. 79ff.

180 Siemens Science and Technology Firsts: Siemens material, with some semiconductor details drawn from Braun (1992).

180 "almost unsurveyable . . . laboratories": Siemens (1957B), p. 253. For details of the mergers and their effects on research, see Siemens (1957B), pp. 82–85, 252–55; Chandler (1990), pp. 463–73.

181 The first lab: For Bolton and tantalum lamp, see Siemens (1957A), pp. 15, 286ff; Siemens (1957B), pp. 35ff.

181 "amongst the . . . laboratories": Siemens (1957B), p. 36.

182 Gustav Hertz: some details from bio in Feldtkeller and Goetzeler (1994), pp. 78–83, translation assistance from Vivien Marx.

182 Erwin Müller: some details from bio in Feldtkeller and Goetzeler (1994), pp. 170–75, translation assistance from Vivien Marx; Anon. (1995), *Science* article.

182 A last area: For Schottky's work, here and throughout, I rely chiefly on Riordan and Hoddeson (1997), pp. 62–68, 84–85; bio in Feldtkeller and Goetzeler (1994), pp. 70–75, translation with help from Vivien Marx; Braun (1992), pp. 448ff; *New Penguin Dictionary of Electronics*, 1981 paperback edition.

182 The most fundamental: Description of the lab's location, organization, and focus comes largely from Schindler interview, July 2, 1997, supplemented by documents, plans, and fact checking from the Siemens archives, esp. Kaske (undated), op. cit.; Trendelenburg (1975), translation by Dr. Giuseppe Oppo. Finkelnburg background: Feldtkeller and Goetzeler (1994), pp. 139–46. Welker background: Feldtkeller and Goetzeler (1994), pp. 176–81.

182 "The way . . . together": Schindler interview, July 2, 1997.

183 The biggest payoff: floating zone details from Siemens AG (1997A), p. 68; Braun (1992), p. 477. Welker's work is drawn from Schindler interview; Braun (1992), pp. 481–82. Cf. Pearson and Brattain (1973), p. 1804.

183 In the early 1960s: story of the new Erlangen facility and the effects of the formation of Siemens AG on research stem chiefly from Siemens AG (1997A), pp. 64–65, 74; Feldenkirchen (1997), translation by Dr. Giuseppe Oppo; Kaske (undated). Note: The original Corporate Technology Division, or *Zentralbereich Technik*—a different ZT from today's—was directed initially by Helmut Wilhelms.

184 The emphasis on applied: Schindler interview and subsequent correspondence; Runge interview, July 1, 1997, and subsequent conversations; Beckurts material, Gumbel and Boyle (1986), with supporting details provided by Siemens. For chip catch-up programs, see esp. Friedrich and Becker (1991).

185 The new research boss: Weyrich background and views from Weyrich interviews; Weyrich (1997); Weyrich c.v.; numerous supporting interviews, esp. Schmitter and Zorn; supporting materials on ZT's current structure, funding, and practices from Siemens public relations. His concept of Fourth Generation R&D represents a unique extension and modification of the concept of Third Generation R&D, as advanced by a group at Arthur D. Little. See Roussel et al. (1991).

185 "But we . . . Siemens": Schmitter interview, Nov. 18, 1998.

186 "Our aim . . . market": Weyrich interview, July 3, 1997.

186 "Some of . . . forever": background interview with Siemens source.

186 With business unit complaints: Many Weyrich actions were related to a company-wide initiative launched in 1993. Called TOP—for Time-Optimized Processes—it was derived from techniques such as Total Quality Management, benchmarking, and reengineering. The goal was to increase productivity not just through traditional efficiency and cost-cutting measures, but by spurring innovation and the growth of new businesses. Its 1998 extension, TOP-Plus, was an effort to extend the initiative by adopting the best practices of the first few years—and by reaching more parts of

the company. TOP details were related in several Siemens interviews, esp. Schmitter; Teresko (1996).

187 Claus Weyrich: Fourth Generation R&D: Weyrich interviews; Weyrich (1997).
187 "When we . . . before": Weyrich interview, July 3, 1997.
187 "small fraction . . . market": Weyrich (1997).
187 "People found . . . pull": Weyrich interview, July 3, 1997.
187 "We want . . . company": ibid.
188 "That was . . . market-oriented": Weyrich interview, Dec. 15, 1998.
188 "You can . . . them": Weyrich interview, July 3, 1997.
188 "Some things . . . myself": Weyrich interview, Dec. 15, 1998.
188 "I said . . . thesis": ibid.
189 "But they . . . future": Dietmar Theis interview, Nov. 19, 1998.
189 "We don't . . . units": Zorn interview, July 1, 1997.
189 "There's a . . . successful": Weyrich interview, Dec. 15, 1998.
189 The first few: details of early Perlach days from Runge interview, July 1, 1997.
190 Much of this activity: Information and Communications activities from visit of July 1997 and interviews with Kämmerer, Harke, Wirtz, and others; supporting organizational diagrams.
190 The epicenter of: fingerprint and other biometric material drawn from Wirtz interview, July 1, 1997; Hierold interview, Jan. 15, 1999; Eberl (1998A); demo of November 1998; Kämmerer and e-mail correspondence; Siemens supporting material.
191 Bathtub Inspiration: July 1997 demo; Kämmerer interview, July 1, 1997; Eberl (1998C); supporting company documents.
191 "visitors intuitively . . . do": quoted in Eberl (1998C), p. 34.
192 "This makes . . . air": quoted in Eberl (1998A), p. 16.
194 "This is . . . directly": quoted in ibid. (1998A), p. 18.
194 "With such . . . Internet": Hierold interview, Jan. 15, 1999.
194 Two and a half: Erlangen details from visits of July 1997, November 1998; company background materials.
195 "Seven years . . . applications": Vetter interview, July 2, 1997.
195 Nowhere is this focus: high-temperature superconductivity work is drawn from Vetter, Schindler, and Krämer interviews; e-mail correspondence with Neumüller; Tsakiridou (1998B).
196 "All the . . . normal": Krämer interview, Nov. 20, 1998.
197 "Generally, the . . . on": Neumüller e-mail of Jan. 18, 1999.
198 Set in a woodsy: Princeton lab background and mission, plus overview to projects, primarily taken from Grandke interviews, company documents, demos, and visits.
198 "I'm convinced . . . have": Grandke interview, Nov. 7, 1996.
198 "We are . . . resort": ibid.
199 The guru of virtual: general details from Chiu and Bani-Hashemi interviews; Eberl (1998B).
199 "We're trying . . . them": Bani-Hashemi interview, Dec. 21, 1998.
200 Innovation Fields: Hejazi interview and correspondence; Schmitter, Weyrich, Zawadzki, and Zorn interviews; Siemens AG (1997B).
201 "You try . . . outlook": Hejazi interview, Dec. 18, 1998.
201 "It's visionary . . . us": ibid.
202 "If the . . . spot": quoted in Eberl (1998B), p. 27.
202 "If you're . . . movie": Bani-Hashemi interview, Dec. 21, 1998.
202 "so what . . . data": ibid.
202 "That would . . . ultimate": ibid.
203 Even as the colonoscope: other imaging and visualization details from Alok Gupta and Chiu interviews; lab demos; Tsakiridou (1998A).
203 "bread and . . . 3-D": Alok Gupta interview, Dec. 21, 1998.

203   Everything is also going: Multimedia/Video research overview from Hsu and Pizano interviews; Tsakiridou (1998A).
203   "Our job . . . advantage": Hsu interview, Nov. 7, 1996.
203   The stir-fry: WIRE, DICE, and LIAISON details from Hsu, Pizano, Goose interviews and correspondence.
204   "You have . . . messages": demo of April 9, 1998.
204   "Message three . . . again": ibid.
204   "Our idea . . . passenger": Hsu interview, April 9, 1998.
205   "I think . . . me": Schindler interview, July 2, 1997.
206   "Well, what . . . done": Weyrich interview, July 3, 1997.

SEVEN: NEC: BALANCING EAST AND WEST

207   "The big . . . world": Goto interview, Oct. 14, 1998.
207   "The how . . . thing": Ishiguro interview, March 6, 1997.
207   "*I am . . . us*": Khang speech, draft dated April 23, 1990, transcript provided by NEC.
208   Dawon Khang's words: description of the Institute's opening provided by interviews with attendees. Cf. MacPherson (1990); "New Basic Research Institute Is Established in the United States," NEC press release of June 30, 1988.
209   This perseverance has: patent and citation figures from Anon. (1997F).
209   "It should . . . roots": Khang speech, op. cit.
209   NEC sprang to life: NEC's history is pieced together from NEC Corp. (1984); Kobayashi (1991). Note that the company's nominal shareholder was International Standard Electric Co., an ITT subsidiary. For general Japanese history see Braudel (1993), pp. 276–300.
209   "the spark . . . powder": quoted in Braudel (1993), p. 292.
211   The architect of NEC's: Kobayashi background in Kobayashi (1991), pp. 1ff. In 1974, the changes he put in place won him the prestigious Deming Individual Prize for Operation Quality; before his tenure ended, NEC itself would win three other Deming awards.
211   "I want . . . world": quoted in Kobayashi (1991), p. 13.
211   Kobayashi also oversaw: Partly as a result of pressure to show quarterly profit, ITT had begun selling off its stake in Nippon Electric in the 1930s. 1998 ranks taken from NEC press materials and annual report.
211   Although NEC's research history: For Japan's R&D evolution, I rely on Kodama interview, Oct. 15, 1998; Kodama (1998). Although Japanese industry has always drawn on academic consultants, the country seems to have largely bypassed a period of heavy reliance on academic consultants before setting up their own research organizations. That stands in contrast to American and European firms.
212   Just behind this startling: NEC's expansion comes from company material. For the general Japanese picture, I draw on Hounshell (1996), p. 55.
213   "That, we hate": Kodama interview, Oct. 15, 1998.
213   NEC opened its first: The prime source for NEC's research history is NEC Corp. (1984), throughout. Niwa biography furnished courtesy of NEC Corp. Niwa resigned from NEC in 1947 as executive vice president, and the next year took over as president of Tokyo Denki University. Four years before his death in 1975, he was awarded Japan's Medal of Culture.
215   "Because the . . . needs": Kobayashi (1991), p. 196.
215   The first manager: This phase of the evolution is described in NEC Corp. (1984), pp. 55ff; Kobayashi (1991), pp. 13–19, 38–39, 196. Degawa bio furnished by NEC Corp.
215   Mickey Uenohara: background, philosophy, and experiences come from Uenohara interview and extensive e-mail follow-ups; Klamann (1992).

215 "The Central . . . do": Uenohara interview, March 3, 1997.
216 "Mickey" Uenohara: Except where noted, all quotes from Uenohara interview, March 3, 1997, corrected and clarified via e-mail correspondence.
216 "If I . . . objective": Uenohara e-mail of March 3, 1999.
217 "I intended . . . units": Uenohara e-mail of Jan. 15, 1999.
217 "So I . . . divisions": Uenohara interview, March 3, 1997.
217 "For such . . . it": ibid.
217 "helped me . . . philosophy": Uenohara e-mail of Jan. 15, 1999.
218 "The key . . . motivated": Uenohara interview, March 3, 1997.
218 "And if . . . committed": ibid.
218 "I took . . . units": Uenohara e-mail of Jan. 15, 1999.
218 "To do . . . strategy": Uenohara e-mail of Jan. 16, 1999.
219 "From the . . . small": Uenohara interview, March 3, 1997.
219 "But if . . . technology": ibid.
219 "Research managers . . . money": ibid.
220 The challenge of: Ishiguro views from interview of March 6, 1997. Description of NEC research budget, scope, and mission from numerous interviews, supported by company press and background materials.
221 "As a . . . changing": Ishiguro interview, March 6, 1997.
221 "It takes . . . future": ibid.
221 Despite its lack: details of CRL taken from two visits, company documents and organizational charts, public relations assistance. The four CRL labs and their predecessors are the result of a 1980 reorganization that saw the formal adoption of the C&C mission—and simultaneously broke the previously all-in-one central lab into a series of smaller, largely autonomous groups. See NEC Corp. (1984), pp. 72–73, 83; Kobayashi (1991), p. 128.
222 *Senmon Shoku Sedo*: details from Endo interview, Oct. 14, 1998, and subsequent e-mail follow-ups with him and public relations officials; Kobayashi interview, Oct. 20, 1998.
223 "the role . . . managers": Endo interview, Oct. 14, 1998.
223 "That is . . . experience": ibid.
224 One of CRL's biggest: Display Device Research Lab details from Hayama and Murai interviews and subsequent e-mails; related company press releases and background material.
225 "We complain . . . display": Hayama interview, Oct. 14, 1998.
225 "I still . . . development": ibid.
225 "I think . . . now": Goto interview, Oct. 14, 1998.
225 A concrete example: details of Iji's work come from interview and subsequent e-mails; supporting company materials.
226 "In general . . . so": Iji interview, Oct. 14, 1998.
228 Basic Research: Tsukuba: Tsukuba lab details taken from extensive interviews, especially Kobayashi and Lang, made during two visits; follow-up e-mail notes; supporting company materials.
228 "The Fundamental . . . practical": Kobayashi interview, Oct. 20, 1998.
228 "For near . . . smaller": ibid.
229 "At the . . . smallest": ibid.
229 "New Information": details of Miwa's work from interview of Oct. 20, 1998.
229 "I think . . . right": ibid.
230 "We can . . . world": ibid.
230 While many of the fundamental: For Iijima's work, I rely on numerous interviews and e-mail correspondence; citation details from Anon. (1997F), with breakdown of papers provided courtesy of *ScienceWatch*.
230 "didn't study hard": Iijima interview, Oct. 20, 1998.
230 "much, much better": ibid.
231 "The people . . . now": ibid.

231 "Maybe it . . . dream": ibid.
231 Basic Research: Princeton: Princeton lab overview details from visits and interviews, esp. with Gear; supporting company documents; Weber (1992); Corcoran (1992).
232 "Although they . . . lines": Gear interview, Oct. 17, 1996.
232 "This is paradise": Altshuler interview, Oct. 17, 1996.
232 On the hardware end: Linke's work is drawn primarily from Linke, Redmond, Chadi interviews; e-mail correspondence; supporting materials describing their work. Cf. Linke et al. (1998); Ryskin et al. (1998).
233 "I see . . . PC": Linke interview, April 10, 1998, modified in subsequent e-mails.
233 "What I . . . researcher": Linke interview, Feb. 3, 1999.
234 Watermarking: story pieced together from Gear, Philbin, Tarjan interviews and correspondence; supporting materials.
235 "Everything was . . . meeting": Philbin interview, Sept. 24, 1997.
235 "the litany . . . episodes": copy of remarks provided by Tarjan via e-mail.
235 "I see . . . flexible": ibid.
236 "Other storage . . . volume": Linke interview, Feb. 3, 1999.
237 "That we . . . that": ibid.
237 "In that . . . material": ibid.
238 "It's not . . . storage": ibid.
238 The story Andrew Goldberg: for the archival intermemory, I rely on A. Goldberg and Yianilos interviews; Goldberg and Yianilos (1997).
239 "Your ideas . . . wars": A. Goldberg interview, April 10, 1998.
239 "Even if . . . memory": Yianilos interview, June 5, 1998.
239 "We don't . . . works": ibid.
240 "When a . . . smile": ibid.
240 For the corporation: for details of the defense contract scandal, see Strom (1998).
240 "The how . . . thing": Ishiguro interview, March 6, 1997.
240 "How well . . . waiting": Khang speech, draft dated April 23, 1990, transcript provided by NEC.

EIGHT: THE PIONEERS: GENERAL ELECTRIC AND BELL LABS

242 "The truth . . . years": Berninger interview, Nov. 12, 1996.
242 "My boss . . . day": Anthony interview, June 19, 1998.
242 One of Thomas Edison's: details of The Knolls primarily from personal visits; company documents; Berninger interview, Nov. 12, 1996.
243 "I would . . . know": Anthony interview, Nov. 12, 1996.
243 Despite its storied past: overview of history taken from former company historian and public relations manager George Wise; company documents; Robb and Schmitt interviews and correspondence.
244 Tom Anthony and Harvey Cline: drawn from Anthony interviews, e-mail correspondence with Anthony and Cline; GE press releases.
244 "Good lawyers": Anthony interview, Nov. 12, 1996.
244 "When I . . . it": quoted in "GE Researcher Reaches 150-Patent Milestone," press release of May 26, 1997.
245 "I have . . . invention": quoted in "GE Researcher Harvey Cline Hits 150-Patent Milestone," press release of Oct. 2, 1998.
245 "Both beekeeping . . . dissimilar": Anthony e-mail of July 6, 1999.
246 "We received . . . century": Schmitt e-mail of Sept. 21, 1999.
246 Robb was the: Robb experiences and philosophy taken from Robb interviews and correspondence; Robb (undated).
246 "I found . . . down": Robb interview, June 23, 1999.
247 "They all . . . between": ibid.

247 "All the . . . orders": quote assembled, with Robb's permission, from interviews of June 23 and 30, 1999.

247 "and we . . . okay": Robb interview, June 30, 1999.

247 "And so . . . shops": Robb interview, June 23, 1999.

247 Lonnie Edelheit: Edelheit background, views, and philosophy taken from Edelheit interview, June 19, 1998; Edelheit (1995); Hammond, Jr. (1994); cf. Robb interview, June 30, 1999.

248 Design for Six Sigma: Sneeringer, Edelheit, Berryman interviews and follow-up e-mails; some Performix and LightSpeed details from company materials.

248 "He [Lonnie] . . . kid": Sneeringer interview, June 19, 1998.

248 "It's really . . . businesses": Berryman interview, June 28, 1999.

249 "So the . . . snowball": Sneeringer interview, June 24, 1999.

249 "For the . . . another": ibid., with subsequent e-mail clarification.

250 "And that's . . . that": Edelheit interview, June 19, 1998.

250 "Now we . . . cheap": ibid.

250 That being said: lab organization and funding are derived largely from Edelheit, Goldberg, Berninger, and Pfoh interviews; supporting company materials.

251 Lonnie Edelheit: Edelheit interview, June 19, 1998; Edelheit (1995), with e-mail follow-up; Hammond, Jr. (1994).

251 "GE is . . . culture": Edelheit interview, June 19, 1998.

251 "It's a . . . practices": ibid.

251 "Being vital . . . marketplace": Edelheit (1995), p. 17.

251 "To the . . . vulnerable": Edelheit interview, June 19, 1998.

252 "We don't . . . revenue": Edelheit e-mail of Sept. 17, 1999.

252 People know GE: for applied statistics description, I rely on Hahn, Doganoksoy, and Martin interviews.

253 "one, two, three, and four": quoted by Martin, interview of June 30, 1999. Confirmed by company.

253 "All of . . . company": Hahn interview, Nov. 12, 1996.

254 "Our job . . . opinions": ibid.

254 "Not only . . . roaches": Hahn interview, June 24, 1999.

254 The trauma patient: services description is drawn from D. Smith, Hahn, and Martin interviews and follow-up e-mails; supporting company material.

255 "expensive wrench turners": quoted by GE vice president Tom Dunham in "GEMS—R&D Center Service Teamwork Leads Company-Wide Services Initiative," document provided by GE.

255 "By adding . . . like": GE 1998 annual report, p. 3.

255 "The same . . . business": D. Smith interview, June 23, 1999.

256 "Now, that's . . . flying": Edelheit interview, June 19, 1998.

257 "You as . . . late": Hahn interview, June 24, 1999.

257 Ever since William Coolidge: For digital X-ray story, I rely on Griffing and Pfoh interviews and e-mail follow-ups; "GE Digital X-ray Detector Technology," August 19, 1997 draft of Lonnie Edelheit's speech for an upcoming press conference.

257 "That's going . . . thing": Edelheit interview, June 19, 1998.

258 "We were . . . deal": Griffing interview, July 9, 1999.

258 "We went . . . program": ibid.

259 "The idea . . . quickly": ibid.

259 "You need . . . done": ibid.

260 "Not much . . . years": Berninger interview, Nov. 12, 1996.

260 "The intensity . . . week": ibid., modified in e-mail of Sept. 17, 1999.

260 "is that . . . out": Anthony interview, June 19, 1998.

260 "In the . . . used": Anthony interview, Nov. 12, 1996.

261 "There's a . . . interaction": Gelperin interview, July 27, 1999.

262 "There's been . . . way": Penzias interview, April 19, 1997.

262   Bell Labs opened: lab history, evolution, and overview to current size and status pieced together from many readings and interviews, esp. with Penzias and Brinkman. See also Penzias (1989, 1995), esp. pp. 70–75; Draper (1990). For early details of Penzias's life and how the Bell breakup affected AT&T: Bernstein (1984), pp. 213–14.
262   "Years ago . . . live": Penzias interview, April 10, 1997.
263   "Washington was . . . jewel": Brinkman interview, Nov. 15, 1996.
263   "When I . . . miniature": ibid.
264   "They figured . . . pleased": ibid.
264   "This is . . . for": ibid.
264   "Everybody worked . . . money": ibid.
264   "I finally . . . anymore": Penzias interview, June 5, 1998.
264   "While colleagues . . . methods": Penzias (1995), p. 72.
265   " 'Eighty-nine . . . paradigm": Penzias interview, April 10, 1997.
265   "too researchy": Brinkman interview, Dec. 16, 1997.
265   "His support . . . change": Penzias interview, June 5, 1998.
265   "In my . . . heads": ibid.
266   "I didn't . . . it": ibid.
266   Penzias revealed his changes: For summary of Penzias's July 1990 address to employees, see Draper (1990).
266   "reverse the . . . research": quoted in ibid.
266   "communicate relentlessly": Penzias interview, June 5, 1998.
266   "And there . . . points": ibid.
267   "Since the . . . standards": Penzias (1995), p. 72.
267   "Not just . . . research": Penzias interview, April 10, 1997.
267   Lucent Technologies, insiders like: overview to contemporary organization is drawn from extensive visits and interviews, esp. with Netravali, Brinkman, Murray, and Tyson. See also Gwynne (1997B).
268   "I had . . . 13": Netravali interview, Dec. 16, 1997.
269   "R&D talent . . . world": ibid.
269   "In the . . . pull": ibid.
269   Since the early 1990s: details of joint and breakthrough projects come from Mel Cohen interviews and various follow-up communications with company; Gwynne (1997B).
270   "Some of . . . us": Mel Cohen interview, Dec. 15, 1997.
270   "It was . . . kind": interview with confidential source.
270   "It could . . . painful": Ritchie interview, Nov. 15, 1996.
270   "They wanted . . . adapted": Penzias interview, April 10, 1997.
271   "Lucent has . . . fate": Stormer interview, Nov. 15, 1996.
271   "This was . . . had": ibid.
271   Despite that passion: for an overview to how Lucent stacks up with competitors see Narisetti (1998B); Hill and Clark (1998).
271   "If we . . . interests": Mel Cohen interview, June 15, 1998.
271   "overdo it . . . science": Brinkman interview, Nov. 15, 1996.
272   "At the . . . middle": ibid.
272   "The opposite . . . unapplied": Penzias interview, April 10, 1997.
272   Penzias models of research: ibid.
273   "The one . . . academic": ibid.
273   "So there . . . equipment?": ibid.
274   Staffed by about 140: overview to Physical Research Laboratory from Murray and Brinkman interviews.
274   David J. Bishop labors: MEMS work is drawn primarily from multiple interviews with Bishop, supplemented with e-mail correspondence and various other discussions about the field throughout the course of writing this book.
274   "has a . . . wireless": Bishop interview, May 29, 1998.

275 Cherry Murray: Murray interviews.
275 "I was . . . Science": Murray interview, Dec. 16, 1997.
275 "was that . . . that": ibid.
275 "the director . . . edge": Murray talk of Oct. 29, 1999, Cambridge, MA.
275 "Now we're . . . so": ibid.
276 "Our horse . . . MEMS": Bishop interview, July 27, 1997.
276 "they very . . . issue": ibid.
277 "that amount . . . question": ibid.
277 "cheap, fast, small, and robust": ibid.
277 "This is . . . technology": ibid.
277 "We definitely . . . out": ibid.
277 While it's easy: description of Gelperin's work drawn from Gelperin interviews; fol-
    low-up e-mail correspondence.
277 "rats, pigeons, and undergraduates": Gelperin interview, Feb. 24, 1997.
278 "Where the . . . happen": Gelperin interview, April 9, 1998.
278 "Some folks . . . bothered": ibid.
279 If Gelperin's research: For Tyson's work, I rely on Tyson interviews.
279 "He's discovered . . . say?": Murray interview, Dec. 16, 1997.
279 "At the . . . years": Tyson interview, June 1, 1998.
279 "I'm a . . . pickax": Tyson interview, April 8, 1998.
280 "What I . . . ocean": ibid.
281 "Bell Labs . . . country": Townes interview, June 1, 1998. Quote, based on an infor-
    mal conversation of March 1997, was modified at this time.
281 "There is . . . be": Penzias interview, April 10, 1997.
281 "Isolated . . . it": ibid.

NINE: CHILDREN OF THE SIXTIES: XEROX AND HEWLETT-PACKARD

283 "It's not . . . fit": Myers interview, Oct. 29, 1997.
283 It's a classic video: details and quotes related by Suchman, interviews of Dec. 5,
    1997, Jan. 10, 1998. The film apparently debuted in 1983 before designers and engi-
    neers attending the annual Computer-Human Interaction conference in Boston.
284 This group—this approach to innovation: J. S. Brown philosophy and experiences
    are drawn from extensive interviews and readings. See Brown (1991, 1992, 1995,
    1997A, 1997B, 1997C); Brown and Gray (1995).
284 "Our goal . . . produced": J. S. Brown interview, Dec. 10, 1997.
284 "cyclone of research": PARC promotional material.
285 The whole sorry matter: For details of early Xerox history and especially PARC's tri-
    umphs and fumbles, here and throughout, I rely on Smith and Alexander (1988); Ja-
    cobson and Hillkirk (1986)—with more specific citations below—supplemented
    primarily by Brown, Myers, and Spinrad interviews.
285 "invention implemented": Brown (1997B), p. xxvi.
285 The first product: for early Xerox growth figures, see Smith and Alexander (1988),
    pp. 117ff.
286 It was in the early: For PARC's early days and accomplishments, see esp. Smith and
    Alexander (1988), pp. 97–98, 117ff, 151–52, 195–96.
287 "It was . . . do": quoted in Jacobson and Hillkirk (1986), p. 257.
287 Then the bottom: For antitrust issues see Jacobson and Hillkirk (1986), pp. 70ff;
    Smith and Alexander (1988), pp. 117ff. For changing market share information and
    Xerox's cultures of inefficiency: Jacobson and Hillkirk (1986), pp. 3, 170ff, 215ff;
    Smith and Alexander (1988) throughout.
287 "Xerox was . . . foundations": Spinrad interview, Dec. 16, 1996.
287 "Research in . . . opposite": quoted in Smith and Alexander (1988), p. 157.

287   By the late 1970s: For PARC's darkest days, I rely chiefly on Smith and Alexander (1988), esp. pp. 215–54.

287   "This is . . . it!": quoted in ibid., p. 216.

288   "What I . . . past": quoted in ibid., p. 250.

288   Legend has it: laser printing and Ethernet payoffs came up in numerous interviews. See Port (1997).

288   "probably justifies . . . PARC": quoted in Port (1997), p. 99.

288   Mark Myers: Myers background and philosophy and CRT practices from Myers interviews; Myers (1996); Myers and Rosenbloom (1996), pp. 217ff. Cf. John Seely Brown interviews. Some details on how Spencer affected research from Rosenbloom interview, Jan. 27, 1998.

289   "profoundly roughly right": Myers interview, Oct. 29, 1997.

289   At the modern Xerox: details of Xerox 2000 and its successors from Spinrad interviews.

289   "The network . . . internally": excerpt provided by Spinrad.

290   "This type . . . timing": Myers (1996), p. 144.

290   *Emerging markets and technologies:* protobusiness interviews provided by Xerox public relations.

291   Xerox's BMWs: Myers (Feb. 2, 1998), Rosenbloom (Jan. 27, 1998), D. Robson (Feb. 9, 1998), and J. S. Brown (Feb. 7, 1998) interviews.

291   "I just . . . discourse": Brown interview, Feb. 7, 1998.

291   "If you . . . opportunity": Myers interview, Feb. 2, 1998.

292   "So we . . . charge?": Myers interview, Feb. 2, 1998.

292   "What PARC . . . Xerox": J. S. Brown interview, Dec. 10, 1997.

292   A variety of projects: Workscapes of the Future details from Beverly Harrison, Fishkin, Frederick, and Want interviews; supporting documents; Workscapes tour by Lois Wong.

293   "Innovation means . . . implemented": Brown (1997B), p. xxvi.

293   "It was . . . itself": J. S. Brown interview, Dec. 10, 1997.

293   "We're looking . . . flows": ibid.

293   "The researchers . . . conversation": ibid.

293   "It is . . . worlds": Brown (1997B), p. xxiv.

294   "We believe . . . ago": statement provided by Harrison.

294   Wandering around PARC: details of MEMS work taken primarily from Berlin interview, Dec. 4, 1997, and follow-up communications. For re:reading, see Xerox PARC (1999).

295   Another far-out form: description of smart dust and Aspect Oriented Programming from Kiczales interview and communications.

295   So far, anyway: story of the PARC anthropology and workplace studies effort is derived from interviews with virtually all participants named—as well as supporting documents and follow-up communications; Orr (1995); Blomberg, Suchman, et al. (1996).

296   Mark Yim: Details and quotes from interview of Jan. 26, 1999, and follow-up communications.

296   "And then . . . person": Yim interview, Jan. 26, 1999.

296   "For an . . . identification": Orr (1995), p. 3.

297   "These are . . . true": ibid.

297   "the first . . . locomotion": ibid.

297   "Since you . . . useful": ibid.

298   "In the . . . business": Suchman interview, Jan. 10, 1998.

298   "The point . . . solve": ibid.

298   "That was . . . per se": J. S. Brown interview, Jan. 9, 1998.

299   "The idea . . . widgets": Suchman interview, Nov. 14, 1997, with subsequent e-mail clarification.

301   A second thrust: For details of Xerox service research, I rely mainly on Suchman, Orr, and Bobrow interviews.

302   "It's easy . . . there": Bobrow interview, Dec. 5, 1997.

302   "One of . . . about": Suchman interview, Nov. 14, 1997.

303   "radical research . . . impact": J. S. Brown interview, May 17, 1999.

303   "It's almost . . . going": J. S. Brown interview, April 30, 1999.

303   "Most companies . . . indigestion": quoted in Packard (1995), p. 52.

303   "I actually . . . technology": Friedrich interview, March 18, 1999.

304   Not long after: Birnbaum experiences and philosophy from Birnbaum interviews; widespread HP interviews; supporting company materials. For WBIRL, see also Mieszkowski (1998).

304   "We're not . . . checks": Birnbaum interview, Dec. 12, 1996, modified slightly in follow-up communications.

305   The main HP Labs: for early company history, I rely mainly on Packard (1995), with supporting company materials.

305   Hewlett-Packard—a coin flip: For details of first few years, see esp. Packard (1995), pp. 35–46.

305   Sales exploded during WWII: growth figures here and throughout taken from Packard (1995), pp. 77, and timeline in back of the book. Current size and plans taken from HP documents, including 1998 annual report; Burrows (1998); and widespread interviews.

305   The first hard step: For Barney Oliver background, see Packard (1995), pp. 21–22; Packard (c. 1995). The evolution of HP's R&D organization, leading to the creation of HP Labs, is pieced together from Packard (1995), pp. 140–42; Packard (c. 1995); Hewlett (1983); Cutler and Marconi interviews.

306   "it was . . . fields": Hewlett (1983), p. viii.

306   Ultimately, the quest: details of HP Labs' creation are found in March 3, 1966, press release announcing the event, supplied by Karen Lewis, HP archives.

306   "Barney was . . . things": Cutler interview, May 26, 1999.

307   Under Oliver's tutelage: For an overview to HP's most significant technical achievements in this period, see "Historical Highlights of Hewlett-Packard Company," company document; Packard (1995), esp. pp. 105–21. For overview of its computer strategy and struggles, I rely on pp. 101ff; Birnbaum and Marconi interviews.

307   "A number . . . computation": Hewlett (1983), p. viii.

308   "The growth . . . ourselves": ibid.

309   MC^2 got off to: for recent innovations, I draw on many interviews and supporting company documents.

309   "It was . . . Labs": Birnbaum interview, July 23, 1999.

310   Joel Birnbaum: Birnbaum interview, Dec. 12, 1996.

310   "It's not . . . breathing": Birnbaum interview, Dec. 12, 1996.

310   Early in 1998: details of Birnbaum's departure and subsequent HP (especially the split) and HP Labs changes are derived from widespread interviews, including those with Birnbaum and Lampman. For company's woes at the time, see Hamilton (1999A).

312   "From my . . . sense": Birnbaum interview, July 23, 1999.

312   "I said . . . it": ibid.

313   "With the . . . strategies": Lampman interview, May 27, 1999.

314   New Business Development: I rely chiefly on Corben, Willis, and Barry Bronson interviews, with follow-up e-mails. Assessment of HP's success in digital cameras is drawn from Gerard interview.

314   "incredibly arrogant mission": Corben interview, Dec. 12, 1996.

314   "For a . . . fit": ibid.

314   "That is . . . strategy": Gerard interview, June 4, 1999.

314   It's been a war: For overview of HP's information appliances effort and general vi-

sion of computing, I drawn on interviews with Birnbaum, Frid, Lampman, Norman, and Mark T. Smith; Lampman (1998).

315   "This is . . . skills": Corben interview, July 8, 1998.
316   "we're going . . . world": Lampman interview, May 27, 1999.
317   "MC² is . . . appliances": M. T. Smith interview, March 18, 1999.
317   "There's absolutely . . . works": with his permission, quote is a combination of Smith's statements of March 18 and June 15, 1999.
317   "negative value": M. T. Smith interview, March 18, 1999.
317   "What if . . . you?": ibid.
317   "The phone . . . that": ibid.
318   "It means . . . them": ibid.
318   "I have . . . happen": ibid.
318   Pony-tail swishing: Stan Williams's story is derived from Williams interviews and follow-up communications; Collier et al. (1999); Heath et al. (1998); Williams (1998).
319   "That's the . . . after": R. S. Williams interview, March 19, 1999.
319   "What we're . . . transistor": ibid.
319   "Conservatively, I'll . . . now": R. S. Williams interview, June 16, 1999.
321   "We'd be . . . noun": quoted by R. S. Williams, interview of March 19, 1999.
321   "set off a firestorm": ibid.
321   "We figured . . . what": ibid.
322   "we actually . . . need": ibid.
322   "We haven't . . . yet": ibid.
322   "I quit . . . frustration": Norman interview, March 18, 1999.
322   "friendly critic": Norman interview, July 8, 1999.
322   "HP can't . . . now": Norman interview, March 18, 1999.
323   "So whether . . . organization": Birnbaum interview, Dec. 12, 1996.
323   "The future . . . Crazy": Norman interview, July 8, 1999, with amendments made in e-mail of July 10, 1999.
323   But by that fall: For HP turnaround details and hiring of Fiorina, see Hamilton (1999A, 1999B); Hamilton and Blumenstein (1999).
323   "So maybe . . . transition": Norman e-mail of July 11, 1999.
323   "really disappointing": Birnbaum interview, July 23, 1999.
323   "I had . . . truth": ibid.
324   "I always . . . organization": ibid.
324   "What I've . . . do": ibid.
324   "There's a . . . ending": ibid.

TEN: THE NEW PIONEERS: INTEL AND MICROSOFT

General description of Intel lab and its mission come from visits of December 3, 1997; March 5, 1998; as well as a series of interviews with Moore, Pollack, Vadasz, Wirt, Yu. Cf. Takahashi (1996). Project descriptions derived from interviews and demonstrations from the individual researchers quoted below.

326   "I gather . . . interesting": Moore interview, March 4, 1998.
326   "I went . . . throttle": Rutman interview, March 5, 1998.
327   "My kid . . . years": ibid.
327   So goes the future: For overview of the antitrust actions against Intel, see Wilke and Takahashi (1998); Takahashi (1999A, 1999B).
328   "We're laying . . . along": Wirt interview, March 5, 1998.
328   "The chance . . . future": Yu interview, Dec. 3, 1997.
329   "In my . . . operating": Moore interview, March 4, 1998.

329   The key to Intel's: Fairchild background from Vadasz, Moore, and Yu interviews; Moore (1994, 1996, 1998). Cf. Lenzner (1995).

329   "The processing . . . research": Moore interview, March 4, 1998.

329   "After we . . . that": ibid.

329   "Essentially, as . . . transfer": ibid.

330   "So when . . . way": ibid.

331   "Both Andy . . . lie": ibid.

331   "We were . . . yourself": ibid.

332   "Frankly, I . . . topics": Vadasz interview, March 6, 1998.

332   "aggressive new . . . rise": Wirt interview, March 11, 1998.

333   "We can't . . . software": Pollack e-mail of Sept. 19, 1999.

335   "The difference . . . interesting": Rutman interview, Dec. 3, 1997.

335   "Beyond evolving . . . here": ibid.

335   "That's regular . . . right": Rutman interview, March 5, 1998.

335   "You can start now": ibid.

335   "To broadcasters . . . shows": ibid.

336   Take Monica Lewinsky: In the summer of 1999, too late for incorporation into the profile, Boon-Lock Yeo left Intel for a start-up company. His wife, Minerva, took over as head of MRL's video technology effort.

336   "We ought . . . material": Yeo interview, March 5, 1998.

337   "Maybe there's . . . works": ibid.

337   "Computers right . . . us": Bradski interview, March 5, 1998.

338   "There is . . . this": Bradski e-mail of March 20, 1998.

338   "The pieces . . . together": ibid.

338   It's been a battle: Smart Computing, Anytime, Anywhere discussion from Pollack interview. For Intel's view on information appliances, see Takahashi (1999C); Lohr and Markoff (1998). For overview of its communications networking strategy: Takahashi (1999D, 1999E).

339   "Our objective . . . computing": Pollack interview, Aug. 18, 1999.

340   "The real . . . technologies": ibid.

340   "Smart Computing . . . them": Pollack e-mail of Sept. 19, 1999.

341   "I'm not . . . relevant": Rutman interview, March 10, 1998.

341   "When you . . . product": Wirt interview, March 11, 1998.

341   "In products . . . interested": Yu interview, Dec. 3, 1997.

341   "is the . . . has": ibid.

342   "We learned . . . this": Vadasz interview, March 6, 1998.

342   "What we . . . interesting": Moore interview, March 4, 1998.

342   "They've done . . . future": ibid.

342   "If quantum . . . here": ibid.

343   Microsoft—No Software Ivory Tower: The Microsoft Research organization profile is derived primarily from personal visits and interviews with all the principals except Myhrvold, supplemented by company-provided background materials, press releases, and individual biographies. Much is also to be gleaned from Stross (1997); Markoff (1995, 1996); Lohr (1997); Myhrvold (1997, 1998); Edstrom and Eller (1998); Knobel (1998); Downey (1996); Anon. (1997E).

343   "Even in . . . that": quoted in Stross (1997).

343   The boxes had: descriptions of Cedar Court comes from visit of March 1999. Details of the various MSR moves from company public relations representatives.

344   Knowing the genesis: early lab history comes primarily from Rashid and Ling interviews; Stross (1997).

345   "The only . . . yourself": quoted in Stross (1997), p. 4 of reprint.

346   "the power . . . involved": Knobel (1998), p. 4 of reprint.

347   "Basically what . . . company": Rashid interview, June 30, 1998.

347   "We wanted . . . twosies": ibid.

347 "Last one . . . lights": quoted in Stross (1997), p. 6 of reprint.
348 Daniel Ling: All material from Ling interview, with a few clarifications and amendments through follow-up e-mails. Regarding his comment that Bill Gates wanted to see more progress on developing better ways to get bugs out of software, in early 1999 MSR formed its Programming Productivity Research Center, an effort to do just that.
349 "religious statement": Stone interview, Sept. 9, 1998.
349 "I thought . . . people": ibid.
349 "What's happened . . . size": Rashid interview, June 30, 1998.
349 And the potential payoff: For a look at Microsoft's evolving vision of computing see Bank and Clark (1999).
350 HutchWorld: Stone, Cheng, Clark, Drucker, Kollock interviews; supporting materials.
350 "But then . . . it": Stone interview, June 29, 1998.
350 "But we . . . technology": Clark interview, July 10, 1998.
351 "We initially . . . community": Stone interview, Sept. 13, 1999, modified in e-mail message of Sept. 16, 1999.
351 "continuous partial attention": Stone interview, Sept. 13, 1999.
351 "they can . . . enough": ibid.
351 "Our intent . . . them": Clark interview, July 10, 1998.
352 "And that's . . . so": Ling interview, June 30, 1998.
352 "My impression . . . fire": IBM Research interview.
352 And then there's Myhrvold: details of option cash-in from Stross (1997); other details from company materials.
353 "Talking to . . . sense": Edstrom and Eller (1998), p. 13.
353 "a second . . . Internet": quoted in Markoff (1995), p. D5.
353 "They already . . . did?": J. S. Brown interview, Sept. 10, 1998.
353 "That was . . . funnels": ibid., with subsequent e-mail adjustment.
353 "If their . . . today": Wohl interview, Oct. 9, 1998.
354 Yet another ominous: Talisman details from D. Clark (1996); Kajiya (Sept. 16, 1998) and Michael Cohen (June 30, 1998) interviews. Cf. Helm (1997).
354 "The market . . . games": Kajiya e-mail of Sept. 21, 1999.
354 "You can . . . past": Draves interview, July 1, 1998.
355 "At the . . . technology": Rosen interview, July 1, 1998.
355 "He knows . . . topics": ibid.
356 "We are . . . that": quoted in Downey (1996), p. 14.
356 "so pervasive . . . them": Rashid interview, March 22, 1999.
356 "You get . . . small": Myhrvold (1997).
357 If you ever: For natural language processing work I rely on Vanderwende and Dolan interviews.
358 "The grammar . . . time": Ling interview, June 30, 1998.
359 "We don't . . . crud": Vanderwende interview Sept. 16, 1998, e-mail of Oct. 8, 1998.
359 With his Scottish burr: MS eBooks story comes from Hill and Brass interviews, with follow-up e-mail corrections and clarifications. Details of the controversy surrounding Microsoft's claimed invention of ClearType come thanks to reader comments after my related column in *Upside* magazine, August 1999, p. 132.
359 "A hundred . . . book": Brass interview, March 23, 1999.
360 "I really . . . really": Brass interview, May 11, 1999.
360 "The book . . . attention": Hill interview, March 23, 1999.
360 "My whole . . . read": ibid.
361 "jaggies . . . more like . . . recognition": ibid.
362 "So while . . . displays": Gibson Web site: http://www.grc.com/cleartype.htm.
362 "If somebody . . . decades?": Brass interview, July 22, 1999.

362 "The patent . . . available": ibid.

362 "all the . . . screen": ibid.

362 Only months after: For Microsoft Reader details, see "Microsoft Announces New Software for Reading on Screen," press release of Aug. 30, 1999.

363 "Wood has . . . step": Brass interview, March 23, 1999.

363 Around the lab: Flow discussion comes primarily from Kajiya interview, Sept. 16, 1998, supplemented by Michael Cohen, Guenter, Toyama interviews and related demos.

364 "The PC . . . flow": Kajiya interview, Sept. 16, 1998.

364 "The question . . . device?": ibid.

365 "The kind . . . engineering": Armstrong interview, June 8, 1998.

366 "So is . . . raised": Rashid interview, June 30, 1998.

366 "When you're . . . theoretical": ibid.

366 "We can do that": ibid.

366 "Nathan is . . . operation": Brown interview, Sept. 10, 1998.

366 "We're seeing . . . labs": Garey interview, June 6, 1998, with e-mail follow-up.

CONCLUSION: The Innovation Marathon

368 The scene was: NAE meeting from Schmitt interviews; Gomory interview, Aug. 26, 1999.

368 "We began . . . science?": Schmitt interview, July 13, 1999.

368 "So I . . . Fine": ibid., modified slightly in interview of Aug. 25, 1999.

370 All of which: Porter and Stern (1998, 1999); Council on Competitiveness (1998); Stern interview, July 29, 1999.

370 "While it . . . index": Rosenbloom interview, July 15, 1999.

371 "We are . . . innovations": Penzias e-mail of Aug. 11, 1999.

371 "What we . . . economy": Lerner interview, July 30, 1999. Modified in e-mail of Aug. 23, 1999.

INTERVIEWS

No site indicates telephone interview. Correspondence and short conversations are not included.

Dan Abramowicz—November 12, 1996, Schenectady, NY
Dennis Adler—June 30, 1998, Redmond, WA
Rod Alferness—November 15, 1996, Murray Hill, NJ
Ross Allen—March 19, 1999, Palo Alto, CA
Boris Altshuler—October 17, 1996, Princeton, NJ
Tom Anthony—November 12, 1996, June 19, 1998, both Schenectady, NY
John Armstrong—March 14, 1996, Cambridge, MA; March 21, May 2, and September 4, 1996, Cambridge, MA; April 25, 1997, Cambridge, MA; June 8, 1998; April 8, 1999; July 13, 1999
Shojiro Asai—March 4, 1997, Hatoyama, Japan
Victor Bahl—June 30, 1998, Redmond, WA
Gary Baldwin—March 19, 1999, Palo Alto, CA
Gene Ball—July 1, 1998, Redmond, WA; September 17, 1998
Ali Bani-Hashemi—December 21, 1998
Roger Barga—July 1, 1998, Redmond, WA
Robert C. Barrett—March 26, 1998, San Jose, CA
Alden S. Bean—November 21, 1996
Gordon Bell—March 16, 1999, San Francisco, CA
Andy Berlin—December 4, 1997, Palo Alto, CA
Walter H. Berninger—November 12, 1996, Schenectady, NY
Steve Bernstein—December 4, 1997, Palo Alto, CA
Maurice Berryman—June 28, 1999
John Best—March 10, 1997, San Jose, CA
Inderpal S. Bhandari—January 17, 1997, Hawthorne, NY
Joel Birnbaum—December 12, 1996, Palo Alto, CA; July 23, 1999
David Bishop—December 15, 1997, Murray Hill, NJ; May 29, 1998; July 27, 1999; August 26, 1999
W. Ekkehard Blanz—November 7, 1996, Princeton, NJ
Jeanette Blomberg—December 5, 1997, Palo Alto, CA; January 12, 1998
Daniel G. Bobrow—December 5, 1997, Palo Alto, CA; January 2, 1998
Christian Borgs—March 23, 1999, Redmond, WA
J. Michael Bowman—October 16, 1996, Newark, DE; September 17, 1997

Gary R. Bradski—March 5, 1998, Santa Clara, CA; March 11, 1998
Dick Brass—March 23, 1999, Redmond, WA; May 11, 1999; July 22, 1999
J. W. Bray—November 12, 1996, Schenectady, NY
Jack Breese—July 1, 1998, Redmond, WA
Bill Brinkman—November 15, 1996, December 16, 1997, both Murray Hill, NJ
David Allan Bromley—April 3, 1997; April 23, 1997
Barry Bronson—March 18, 1999, Palo Alto, CA; June 15, 1999
John Seely Brown—December 10, 1997; January 9, 1998; February 7, 1998; September 10, 1998; April 30, 1999; May 17, 1999
Walter Brown—December 16, 1997, Murray Hill, NJ
Pieter Buhler—June 30, 1997, Rüschlikon, Switzerland
Roger A. Burges—June 27, 1997, Sandwich, Great Britain
Scott Callon—November 12, 1996; February 27, 1997
Federico Capasso—December 15, 1997, Murray Hill, NJ
Stuart K. Card—December 4, 1997, Palo Alto, CA
Kent Carey—October 16, 1998, Kawasaki, Japan
James Chadi—February 1999
Hsuan Chang—November 12, 1996, Schenectady, NY
Suresh Chari—February 18, 1999, Hawthorne, NY
Surajit Chaudhuri—July 1, 1998, Redmond, WA
Jennifer Chayes—March 23, 1999, Redmond, WA
C. Julian Chen—July 23, 1998, Yorktown Heights, NY; August 3, 1998
Lili Cheng—March 22, 1999, Redmond, WA
Dave Chess—July 23, 1998, Hawthorne, NY
Ming-Yee Chiu—November 7, 1996, Princeton, NJ
Ann Marie Clark—July 10, 1998
John Cocke—November 27, 1996
Mel Cohen—December 15, 1997, Murray Hill, NJ; June 15, 1998
Michael F. Cohen—June 30, 1998, March 23, 1999, both Redmond, WA
Rob Colterbank—September 25, 1998
Sharon M. Connor—December 13, 1996, Palo Alto, CA
Rick Corben—December 12, 1996, Palo Alto, CA; July 8, 1998
Len Cutler—May 26, 1999
David Dack—July 14, 1997, Bristol, Great Britain
Amitava Datta—November 7, 1996, Princeton, NJ
Lee Davenport—September 13, 1997, Cambridge, MA; October 29, 1997, Greenwich, CT; November 25, 1997
Alex de Angelis—December 10, 1996
Christos Dimitrakopoulos—February 17, 1999, Yorktown Heights, NY
Ananth Dodabalapur—March 2, 1998
Necip Doganaksoy—November 12, 1996, Schenectady, NY
Bill Dolan—July 1, 1998, Redmond, WA
Richard Draves—July 1, 1998, Redmond, WA
Steven M. Drucker—June 29, 1998, Seattle, WA; March 22, 1999, Redmond, WA
D. Christopher Dryer—March 26, 1998, March 17, 1999, both San Jose, CA
Thomas W. Ebbesen—March 20, 1997
Edward J. Eckert—November 15, 1996, Murray Hill, NJ
Lewis "Lonnie" Edelheit—June 19, 1998, Schenectady, NY
William A. Edelstein—November 12, 1996, Schenectady, NY
Don Eigler—March 10, 1997, San Jose, CA
Nobuhiro Endo—October 14, 1998, Kawasaki, Japan
Peter R. Farrow—June 27, 1997, Sandwich, Great Britain
Usama Fayyad—July 1, 1998, Redmond, WA
Ken Fishkin—March 16, 1999, Palo Alto, CA

Myron D. Flickner—March 17, 1999, San Jose, CA
Jerry Foschini—June 5, 1998
Alan Fowler—October 30, 1997, Yorktown Heights, NY
Ron Frederick—March 16, 1999, Palo Alto, CA
Marcos Frid—December 13, 1996, Palo Alto, CA
Rich Friedrich—March 18, 1999, Palo Alto, CA
David Frohlich—July 14, 1997, Bristol, Great Britain
Robert Frosch—December 2, 1996, Cambridge, MA
Jack Fuhrer—November 25, 1996, Princeton, NJ
John Funge—March 5, 1998, Santa Clara, CA
Armando Garcia—January 17, 1997, Hawthorne, NY
Michael Garey—June 6, 1998
C. William Gear—early 1996; October 17, 1996, Princeton, NJ; February 11, 1999
Alan Gelperin—February 24, 1997; April 9, 1998, Murray Hill, NJ; May 27, 1998; July 27, 1999, Murray Hill, NJ
Jim Gemmell—March 16, 1999, San Francisco, CA
Alexis Gerard—June 4, 1999
James K. Gimzewski—June 30, 1997, Rüschlikon, Switzerland
Frederique Giraud—February 18, 1999, Hawthorne, NY
Andrew V. Goldberg—April 10, 1998, Princeton, NJ
Howard S. Goldberg—November 12, 1996, Schenectady, NY
Ralph Gomory—July 25, August 20, 1996, both New York, NY; April 8, 1997; August 26, 1999
Stuart Goose—April 9, 1998, Princeton, NJ
Satoshi Goto—October 14, 1998, Kawasaki, Japan
Ambuj Goyal—February 17, 1999, Yorktown Heights, NY
Marcel Graf—June 30, 1997, Rüschlikon, Switzerland
Thomas Grandke—November 7, 1996, April 9, 1998, both Princeton, NJ
Jim Gray—March 16, 1999, San Francisco, CA
Paul Green, Jr.—March 9, 1999
Michael Greenwood – February 16, 1999, Tarrytown, NY
Bruce Griffing—July 9, 1999
John Grinham—July 14, 1997, Bristol, Great Britain
Radek Grzeszczuk—March 5, 1998, Santa Clara, CA
Brian Guenter—June 30, 1998, Redmond, WA; September 17, 1998
Pierre Guéret—June 30, 1997, Rüschlikon, Switzerland
Jeremy Gunawardena—March 21, 1997; July 14, 1997, Bristol, Great Britain
Alok Gupta—December 21, 1998
Anoop Gupta—March 23, 1999, Redmond, WA
Gerald J. Hahn—November 12, 1996, Schenectady, NY; June 24, 1999
Ken Harada—March 4, 1997, Hatoyama, Japan
Ulrike Harke—July 1, 1997, Munich, Germany
Beverly Harrison—February 12, 1999
Colin G. Harrison—February 18, 1999, Hawthorne, NY
Tomihiro Hashizume—March 4, 1997, Hatoyama, Japan
Hiroshi Hayama—October 14, 1998, Kawasaki, Japan
David Heckerman—March 22, 1999, Redmond, WA
Shahram Hejazi—December 18, 1998
Erika Hibino—July 24, 1997
Christofer Hierold—January 15, 1999
Bill Hill—March 23, 1999, Redmond, WA
Jan Hoffmeister—July 26, 1999
Mark Holler—March 5, 1998, Santa Clara, CA
Mark W. Hopkins—October 16, 1996, Newark, DE

Paul Horn—July 24, 1998, February, 19, 1999, both Yorktown Heights, NY
David Hounshell—August 27, 1997
Richard E. Howard—November 15, 1996, Murray Hill, NJ
Arding Hsu—November 7, 1996, April 9, 1998, both Princeton, NJ
Xuedong (X.D.) Huang—June 30, 1998, Redmond, WA
Stephen H. Hunt—March 5, 1998, Santa Clara, CA
Sumio Iijima—February 26, 1997, October 20, 1998, both Tsukuba, Japan
Masatoshi Iji—October 14, 1998, Kawasaki, Japan
Randy Isaac—July 24, 1998, Yorktown Heights, NY
Waguih Ishak—March 19, 1999, Palo Alto, CA
Tatsuo Ishiguro—March 6, 1997, Tokyo, Japan
Toshiaki Iwanaga—February 28, 1997, Kawasaki, Japan
Kazuo Iwano—February 26, 1997, Yamato, Japan
Adam Jaffe—July 28, 1999
Phil Janson—June 30, 1997, Rüschlikon, Switzerland
Bill Janssen—January 27, 1998
David E. Johnson—July 24, 1998, Yorktown Heights, NY
John B. Jones, Jr.—March 12, 1999
Lauretta Jones—January 16, 1997, Yorktown Heights, NY
Kohichi Kajitani—February 26, 1997, Yamato, Japan
James Kajiya—September 16, 1998
Tak Kamae—March 3, 1997, October 16, 1998, both Kawasaki, Japan
Bernhard Kämmerer—July 1, 1997, Munich, Germany
Yawara Kaneko—October 16, 1998, Kawasaki, Japan
H. Edward Karrer—December 16, 1996, Palo Alto, CA
Jeffrey O. Kephart—January 16, 1997, July 23, 1998, both Hawthorne, NY
Roger Kerry—June 27, 1997, Sandwich, Great Britain
Gregor Kiczales—January 23, 1998
Alison Kidd—July 14, 1997, Bristol, Great Britain
Akio Kinoshita—December 2, 1996, November 24, 1997, both Cambridge, MA
Toshiro Kita—October 19, 1998, Tokyo, Japan
Kohroh Kobayashi—October 20, 1998, Tsukuba, Japan
Fumio Kodama—October 15, 1998, Tokyo, Japan
Teno Kohashi—March 4, 1997, Hatoyama, Japan
Peter Kollock—June 29, 1998, Seattle, WA
Carol Kovac—July 23, 1998, Yorktown Heights, NY
Hans-Peter Krämer—November 20, 1998, Erlangen, Germany
Krishna Kumar—January 17, 1997, Hawthorne, NY
Karl Kümmerle—June 30, 1997, Rüschlikon, Switzerland
Kazuhiro Kuwabara—October 19, 1998, Tokyo, Japan
Lynn Labun—April 4, 1997
Thomas Lackner—November 19, 1998, Munich, Germany
John LaMattina—October 2, 1996, Groton, CT
Dick Lampman—December 13, 1996, Palo Alto, CA; May 27, 1999
Rolf Landauer—October 30, 1997, Yorktown Heights, NY
Chuck Larson—August 29, 1996
Minyoung Lee—November 12, 1996, Schenectady, NY
Wilfried Lenth—March 11, 1997, San Jose, CA
Josh Lerner—on or about July 30, 1999
Baruch Lev—July 19, 1999
Bob C. Liang—March 5, 1998, Santa Clara, CA
J. Alexander Liddle—November 15, 1996, Murray Hill, NJ
Dan Ling—June 30, 1998, Redmond, WA
Richard A. Linke—October 17, 1996, April 10, 1998, both Princeton, NJ; February 3, 1999

Albert Liu—December 11, 1996, Palo Alto, CA
Andrew Liu—November 24, 1997, Cambridge, MA; January 8, 1998
Dave Lomay—July 1, 1998, Redmond, WA
Joseph Lombardino—October 2, 1996, Groton, CT
Steve Loughran—July 14, 1997, Bristol, Great Britain
John M. Lucassen—July 23, 1998, Yorktown Heights, NY
Mark Lucente—February 19, 1999, Yorktown Heights, NY
Paul Maglio—March 17, 1999, San Jose, CA
Juan-Carlos Maier—November 20, 1998, Erlangen, Germany
Henrique Malvar—July 1, 1998, Redmond, WA
John Manferdelli—March 23, 1999, Redmond, WA
Paul Mankiewich—December 15, 1997, Murray Hill, NJ
Richard Marconi—December 11 and 16, 1996, Palo Alto, CA; May 24, 1999
William S. Mark—March 18, 1999, Menlo Park, CA
Nancy R. Martin—June 30, 1999
Frank Mayadas—August 20, 1996, New York, NY
Michael E. McGrath—June 10, 1997, Weston, MA
James McGroddy—January 22, 1997; February 3, 1997; October 1998; May 9, 1999
David F. McQueeney—January 17, 1997, Yorktown Heights, NY; April 5, 1999
Bob Melcher—July 24, 1998, Yorktown Heights, NY
Jim Menendez—December 22, 1998
Thomas Mertelmeier—November 20, 1998, Erlangen, Germany
Bernard S. Meyerson—February 17, 1999, Yorktown Heights, NY; April 15, 1999
Fred Mintzer—January 16, 1997, Yorktown Heights, NY
Laurie S. Mittelstadt—December 12, 1996, Palo Alto, CA
Johji Miwa—October 20, 1998, Tsukuba, Japan
Takashi Mizutani—October 20, 1998, Tsukuba, Japan
Gordon E. Moore—March 4, 1998, Santa Clara, CA
Thomas P. Moran—December 4, 1997, Palo Alto, CA
John F. Morar—July 23, 1998, Hawthorne, NY
Joe Morone—September 12 and 19, 1996
Hiroshi Motoyama—March 4, 1997, Hatoyama, Japan
A. Currie Munce—March 10, 1997, San Jose, CA; February 17, 1999
Konrad Mund—July 2, 1997, Erlangen, Germany
Craig Mundie—March 23, 1999, Redmond, WA
Hideya Murai—February 28, 1997, Kawasaki, Japan
Cherry Murray—December 16, 1997, April 8, 1998, July 30, 1999, all Murray Hill, NJ
Mark B. Myers—October 29, 1997, Stamford, CT; February 2, 1998
David Nahamoo—July 23, 1998, Yorktown Heights, NY
Shinya Nakagawa—March 3, 1997, Kawasaki, Japan
Vivek Narasayya—July 1, 1998, Redmond, WA
Arun Netravali—December 16, 1997, Murray Hill, NJ
Katsumi Nihei—October 14, 1998, Kawasaki, Japan
Parry M. Norling—October 16, 1996, Wilmington, DE
Donald A. Norman—March 18, 1999, Atherton, CA; July 8, 1999
Sharon Nunes—April 7, 1999
Yukinori Ochiai—February 26, 1997, Tsukuba, Japan
Neil O'Connell—July 14, 1997, Bristol, Great Britain
Stephen M. Omohundro—October 17, 1996, Princeton, NJ
Yasushi Ooi—February 28, 1997, Kawasaki, Japan
Steven B. Oppen—October 17, 1996, Princeton, NJ
Julian E. Orr—December 5, 1997, Palo Alto, CA
Charles Palmer—February 18, 1999, Hawthorne, NY; April 16 and 19, 1999
Elaine Palmer—February 18, 1999, Hawthorne, NY
Alex Pentland—November 24, 1997, Cambridge, MA

Arno Penzias—April 10, 1997; June 5, 1998
Dragutin Petkovic—March 26, 1998, San Jose, CA
Armin Pfoh—June 25, 1999
James Philbin—September 24, 1997; October 2, 1997; October 14, 1997
J. C. Phillips—February 22, 1997
Mark Pinto—November 15, 1996, Murray Hill, NJ
Arturo Pizano—November 7, 1996, Princeton, NJ; December 16, 1998
Fred Pollack—August 18, 1999
Erich Port—June 30, 1997, Rüschlikon, Switzerland
Balaji Prabhakar—July 14, 1997, Bristol, Great Britain
Peter Price—October 30, 1997, Yorktown Heights, NY
William R. Pulleyblank—February 18, 1999, Hawthorne, NY; March 5, April 1 and 2, 1999
Hiroyoshi Rangu (Roy Lang)—February 26, 1997, Tsukuba, Japan
J. R. Rao—February 18, 1999, Hawthorne, NY
Rick Rashid—June 30, 1998, March 22, 1999, both Redmond, WA
D. Raychaudhuri—October 17, 1996, Princeton, NJ
Ian R. Redmond—April 10, 1998, Princeton, NJ; February 5, 1999
Dave Reynolds—July 14, 1997, Bristol, Great Britain
Whitman Richards—December 2, 1996, November 24, 1997, both Cambridge, MA
Isidore Rigoutsos—February 18, 1999, Hawthorne, NY
Dennis Ritchie—November 15, 1996, Murray Hill, NJ
Mark Ritter—February 17, 1999, Yorktown Heights, NY
Walter Robb—June 23, 1999; June 30, 1999
George Robertson—March 23, 1999, Redmond, WA
Barry Robson—February 18, 1999, Hawthorne, NY
Dave Robson—February 9, 1998
Paul M. Romer—April 11, 1997
Daniel Rosen—July 1, 1998, Redmond, WA
Steven Rosenberg—December 13, 1996, Palo Alto, CA
Richard Rosenbloom—January 27, 1998; May 26, 1999; July 15, 1999
Salim Roukos—January 17, 1997, Yorktown Heights, NY
Hartmut Runge—July 1, 1997, Munich, Germany
Serge Rutman—December 3, 1997, March 5, 1998, both Santa Clara, CA; March 10, 1998
Howard E. Sachar—January 16, 1997, Yorktown Heights, NY
David Safford—February 18, 1999, Hawthorne, NY
Peter Saraga—July 10, 1997, Redhill, Great Britain
Heinrich Schindler—July 2, 1997, Erlangen, Germany
Roland Schmitt—July 13, 1999; August 25, 1999
Ernst Schmitter—November 18, 1998, Munich, Germany; December 11, 1998
Kevin M. Schofield—June 30, 1998, Redmond, WA
Warren Seering—December 2, 1996, November 24, 1997, both Cambridge, MA
Ted Selker—March 26, 1998, San Jose, CA; April 3, 1998
Steven Shafer—March 23, 1999, Redmond, WA
Talal Shamoon—October 15, 1997
Bill Sherwood—September 9, 1996
Peter Shor—September 25, 1998
Polly Siegel—December 13, 1996, Palo Alto, CA
Charles Simonyi—March 22, 1999, Redmond, WA
Barton Smith—March 11, 1997, San Jose, CA
Dan Smith—June 23, 1999
Marc Smith—March 22, 1999, Redmond, WA
Mark T. Smith—March 18, 1999, Palo Alto, CA; June 15, 1999

Mark Sneeringer—June 19, 1998, Schenectady, NY; June 24, 1999
Robert Solow—September 19, 1996
Jun'ichi Sone—October 20, 1998, Tsukuba, Japan
Kohki Sone—March 7, 1997, Yokosuka, Japan
William J. Spencer—December 27, 1996, Austin, TX
Robert Spinrad—December 16, 1996, Palo Alto, CA; January 26, 1998
Mark Stefik—December 4, 1997, Palo Alto, CA
Mitch Stein—February 19, 1999, Yorktown Heights, NY, March 30, 1999
Phil Stenton—July 14, 1997, Bristol, Great Britain
Scott Stern—July 29, 1999
Linda Stone—June 29, 1998, Seattle, WA; September 9, 1998; September 13, 1999
Horst L. Stormer—November 15, 1996, Murray Hill, NJ
Ramesh Subramonian—March 5, 1998, Santa Clara, CA
Lucy Suchman—November 14, 1997; December 5, 1997, Palo Alto, CA; January 10, 1998; February 21, 1999
Srinivas Sukumar—October 10, 1996; December 12, 1996, Palo Alto, CA
Phillip D. Summers—July 23, 1996, January 17, 1997, October 30, 1997, February 19, 1999, all Yorktown Heights, NY
George T. G. Swaye—June 27, 1997, Sandwich, Great Britain
Yo Tabuse—February 26, 1997, Tsukuba, Japan
Nobuaki Takanashi—February 28, 1997, October 14, 1998, both Kawasaki, Japan
Akihiko Takano—March 4, 1997, Hatoyama, Japan
Samer Takriti—April 9, 1999
Tetsuya Tamura—February 28, 1997, Kawasaki, Japan
Chao Tang—October 17, 1996, Princeton, NJ
Robert E. Tarjan—October 17, 1996, Princeton, NJ; October 15, 1997
John Taylor—July 23, 1997
Dietmar Theis—November 19, 1998, Munich, Germany
Tom Theis – February 17, 1999, Yorktown Heights, NY
David Thompson—February 17, 1999
Charles Townes—June 1, 1998
Kentaro Toyama—July 1, 1998, Redmond, WA
Dave Trecker—September 10, 1996; September 20, 1996, Groton, CT
James B. Treece—March 5, 1997, Tokyo, Japan
Randall H. Trigg—December 5, 1997, Palo Alto, CA
Helmut Trischler—July 3, 1997, Munich, Germany
Xiaoyuan Tu—March 5, 1998, Santa Clara, CA
Gardiner Tucker—October 29, 1997, Westport, CT
Tony Tyson—April 8, 1998, Holmdel, NJ; June 1, 1998
Makoto Uchiyama—March 7, 1997, Yokosuka, Japan
Michiyuki Uenohara—March 3, 1997, Tokyo, Japan
Robert Uhl—October 1997
Tom Uhlman—December 16, 1997, Murray Hill, NJ
Leslie Vadasz—March 6, 1998
Sophie Vandebroek—February 10, 1998
Lucy Vanderwende—July 1, 1998, Redmond, WA; September 16 and 17, 1998
Jürgen Vetter—July 2, 1997, Erlangen, Germany
Mahesh Viswanathan—July 23, 1998, Yorktown Heights, NY
Marisa Viveros – February 17, 1999, Yorktown Heights, NY; March 29, 1999
Helmut Volkmann—November 18, 1998, Munich, Germany
Kirby G. Vosburgh—November 12, 1996, June 19, 1998, both Schenectady, NY
Bob Waites—March 18, 1999, Palo Alto, CA
Roy Want—March 16, 1999, Palo Alto, CA
Todd Ward—July 23, 1998, Yorktown Heights, NY

Earl Warrick—March 19, 1998
Warren K. Waskiewicz—November 15, 1996, Murray Hill, NJ
Richard Webb—July 12, 1999
Elisabeth Weinberger—July 2, 1997, Munich, Germany
Claus Weyrich—July 3, 1997, Munich, Germany; December 15, 1998
Alice White—December 15, 1997, Murray Hill, NJ
Steve R. White—July 23, 1998, Hawthorne, NY
Lou Whitken—December 13, 1996, Palo Alto, CA
Turner Whitted—March 23, 1999, Redmond, WA
Elizabeth A. Williams—November 12, 1996, Schenectady, NY
R. Stanley Williams—March 19, 1999, Palo Alto, CA; June 16, 1999
Barry Willis—July 28, 1998
Robert Wilson—June 17, 1998
Richard Wirt—March 5 and 6, 1998, Santa Clara, CA; March 11, 1998
Brigitte Wirtz—July 1, 1997, Munich, Germany
Amy Wohl—October 9, 1998
Jerry Woodall—April 13, 1999
Hirohito Yamada—April 10, 1998, Princeton, NJ
Mihalis Yannakakis—June 5, 1998
Boon-Lock Yeo—March 5, 1998, Santa Clara, CA
Peter N. Yianilos—April 10, 1998; June 5, 1998; February 22, 1999
Mark Yim—January 26, 1999
John N. Yochelson—July 29, 1999
Sadahiko Yokoyama—February 28, 1997, Kawasaki, Japan
Tadashiro Yoshida—October 19, 1998, Tokyo, Japan
Albert Yu—January 17, 1997; December 3, 1997, Santa Clara, CA
Andrzej (Andy) Zawadzki—December 16, 1998
Shumin Zhai—March 17, 1999, San Jose, CA
Thomas G. Zimmerman—March 26, 1998, March 17, 1999, both San Jose, CA
Gerhard Zorn—July 1, 1997, Munich, Germany

BIBLIOGRAPHY

Adams, Walter, Ed. 1982. *The Structure of American Industry.* New York, Macmillan.

Adams, Walter, and Hans Mueller. 1982. "The Steel Industry." In *The Structure of American Industry*, W. Adams, Ed. New York, Macmillan: 73–135.

Anon. 1915A. "Government Co-operation with Our Industries—I." *Scientific American*, January 30: 99, 108.

Anon. 1915B. "Doing Without Europe—II." *Scientific American*, February 6: 128.

Anon. 1915C. "Doing Without Europe—III." *Scientific American*, February 13: 157, 164–66.

Anon. 1915D. "Doing Without Europe—IV." *Scientific American*, February 20: 176.

Anon. 1915E. "Doing Without Europe—V." *Scientific American*, February 27: 196.

Anon. 1915F. "Doing Without Europe—VI." *Scientific American*, March 6: 223.

Anon. 1915G. "Formal Opening of the Transcontinental Telephone Line." *The Telephone Review*: 3–9.

Anon. 1937. "History of the Institute and the Industrial Fellowship System." *The Crucible*, May: 103–111.

Anon. 1956. "Molding the Future." *New York Times*, May 17: 30.

Anon. 1995. "Private Philanthropy Establishes Erwin Mueller Professorship in Physics." *Science Journal*, Spring: 13.

Anon. 1996A. "Prospects Mixed for U.S. R&D Spending." *Research Technology Management*, September-October: 6–7.

Anon. 1996B. "IRI Urges Better Economy for Technology Innovation." *Research Technology Management*, September-October: 7–8.

Anon. 1997A. "Industrial Research Institute's R&D Trends Forecast for 1997." *Research Technology Management*, January-February: 11–15.

Anon. 1997B. "U.S. Industry Spending More on R&D." *Research Technology Management*, January-February: 3.

Anon. 1997C. "Key Sectors Lead Surge in R&D Intensity." *Research Technology Management*, January-February: 3–4.

Anon. 1997D. "Public Science Crucial to U.S. Technology, Patent Study Finds." *Research Technology Management*, July-August: 4.

Anon. 1997E. "Restless in Seattle." *New Scientist*, August 2: 22–25.

Anon. 1997F. "The Finest in Physical Sciences." *ScienceWatch*, November/December: 1–2.

Anon. 1998A. "Leveraging IBM's Technical Strengths: A Conversation with Nick Donofrio." *IBM Research*, Nos. 1-2: 30–32.

Anon. 1998B. "The Rebirth of IBM." *The Economist*, June 6: reprint edition.

Anon. Undated. *Out of the Research Laboratory*, Publication of General Electric, from Hall of Electrical History, Schenectady, NY.

Armstrong, John A. 1994A. "Is Basic Research a Luxury Our Society Can No Longer Afford?" *The Bridge* 24(2).

——. 1994B. "What Is a Science or Engineering Ph.D. For?" Lecture paper provided by author.

——. 1996. "University-Industry Relations: What Do We Know?" SPIE International Symposium on Microlithography, Santa Clara, CA.

Arrow, Kenneth J. 1962A. "Comment." In *The Rate and Direction of Inventive Activity*. Princeton, Princeton University Press: 353–358.

——. 1962B. "Economic Welfare and the Allocation of Resources for Invention." In *The Rate and Direction of Inventive Activity*. Princeton, Princeton University Press: 609–625.

Associated Press. 1953. "President Recalls Souping Up Ford 'T.'" *New York Times*, May 21: 46.

Avishai, Bernard, and William Taylor. 1989. "Customers Drive a Technology-driven Company: An Interview with George Fisher." *Harvard Business Review*, November-December: 107–114.

Bank, David. 1997. "Microsoft Takes Big Step in Strategy of PC Programs for Server Terminals." *Wall Street Journal*, May 12: B4.

——. 1998A. "In the Microsoft Endgame, a Puzzle." *Wall Street Journal*, April 24: B1.

——. 1998B. "Dumb Machines, Smart Networks." *Wall Street Journal*, November 16: R8.

Bank, David, and Don Clark. 1999. "Microsoft Broadens Vision Statement Beyond PCs." *Wall Street Journal*, July 23: A3.

Barnholt, Edward W. 1997. "Fostering Business Growth with Breakthrough Innovation." *Research Technology Management*, March-April: 12–16.

Bartlett, Howard R. 1941A. "The Development of Industrial Research in the United States." *Research—A National Resource, II—Industrial Research*, Report of the National Research Council to the National Resources Planning Board. Washington, DC, U.S. Government Printing Office: 19–42.

——. 1941B. "Development of Organized Research Within Individual Companies." *Research—A National Resource, II—Industrial Research*, Report of the National Research Council to the National Resources Planning Board. Washington, DC, U.S. Government Printing Office: 42–77.

Basalla, George. 1988. *Evolution of Technology*. Cambridge, England, Cambridge University Press.

Bashe, Charles J., Lyle R. Johnson, et al. 1986. *IBM's Early Computers*. Cambridge, MA, MIT Press.

Baxter, James Phinney, III. 1947. *Scientists Against Time*. Boston, Little, Brown.

Beer, John J. 1958. "Coal Tar Dye Manufacture and the Origins of the Modern Industrial Research Laboratory." In *The Development of Western Technology Since 1500*, T. P. Hughes, Ed. New York, Macmillan: 129–138.

——. 1959. *The Emergence of the German Dye Industry*. Urbana, IL, University of Illinois Press.

Benedek, Emily. 1998. "The Human-centered Interface." *IBM Research*, Nos. 1-2: 16–20.

Bernal, J. D. 1953. *Science and Industry in the Nineteenth Century*. London, Routledge & Kegan Paul.

Bernstein, Jeremy. 1984. *Three Degrees Above Zero: Bell Labs in the Information Age*. New York, Scribner's.

Birr, Kendall. 1957. *Pioneering in Industrial Research: The Story of the General Electric Research Laboratory*. Washington, DC, Public Affairs Press.

——. 1966. "Science in American Industry." In *Science and Society in the United States*. In D. D. Van Tassel and M. G. Hall, Eds. Homewood, IL, Dorsey Press: 35–80.

Blau, John. 1997A. "Global Networking Poses Management Challenge/Risk." *Research Technology Management,* January-February: 4–5.

——. 1997B. "Siemens Nixdorf Moves to Knowledge Culture." *Research Technology Management,* May-June: 3–4.

Bliss, Jeff. 1998. "IBM Aims Chip Plan at Cellular Market." *The Journal News,* Oct. 12. Reprint.

Blomberg, Jeanette, Lucy Suchman, et al. 1996. "Reflections on a Work-oriented Design Project." *Human-Computer Interaction* 11: 237–265.

Boyd, T. A., Ed. 1961. *Prophet of Progress: Selections from the Speeches of Charles F. Kettering.* New York, E. P. Dutton.

Brandt, Richard L. 1997. "Andy Grove on Intel: Magic or Monopoly?" *Upside,* September: 82–93.

Braudel, Fernand. 1993. *A History of Civilizations.* New York, Penguin Books.

Braun, Ernest. 1992. "Selected Topics from the History of Semiconductor Physics and Its Applications." In *Out of the Crystal Maze,* L. H. Hoddeson, Ed. New York, Oxford University Press: 443–488.

Brennan, Jean Ford. 1971. *The IBM Watson Laboratory at Columbia University: A History.* Armonk, NY, IBM.

Brinkman, W. F., and D. V. Lang. 1999. "Physics and the Communications Industry." Article written for a special edition of *Reviews of Modern Physics* to celebrate the centennial of the American Physical Society.

Brooks, John. 1975. *Telephone: The First Hundred Years.* New York, Harper & Row.

Brown, John Seely. 1991. "Research that Reinvents the Corporation." *Harvard Business Review,* January-February: 102–111.

——. 1992. "Reflections of/on the Document." *XPloration Magazine.*

——. 1995. "Assessing Corporate Research Restructuring at Xerox." National Research Council Workshop on Assessing Corporate Research Restructuring, Washington, DC.

——, Ed. 1997A. *Seeing Differently: Insights on Innovation.* Boston, Harvard Business Review.

——. 1997B. "Introduction: Rethinking Innovation in a Changing World." In *Seeing Differently: Insights on Innovation,* J. S. Brown, Ed. Boston, Harvard Business Review: ix-xxviii.

——. 1997C. "Changing the Game of Corporate Research: Learning to Thrive in the Fog of Reality." In *Technological Innovation: Oversights and Foresights,* Garud, Nayyar, and Shapira, Eds. Cambridge, England, Cambridge University Press: 95–110.

Brown, John Seely, and Estee Solomon Gray. 1995. "The People Are the Company." *Fast Company:* 78–82.

Brozen, Yale. 1962. "The Future of Industrial Research and Development." In *The Rate and Direction of Inventive Activity.* Princeton, Princeton University Press: 273–276.

Buderi, Robert. 1988. "Universities Buy into the Patent Chase." *The Scientist,* December 12: 1, 4–5.

——. 1989A. "Inside H-P's High-powered Think Tank." *The Scientist,* April 3: 1, 6–7.

——. 1989B. "U.S. Companies Hike Investment in Foreign R&D." *The Scientist,* May 29: 1.

——. 1991A. "The Brakes Go On in R&D." *BusinessWeek,* July 1: 24–26.

——. 1991B. "A Tighter Focus for R&D." *BusinessWeek,* October 25: 170–172.

——. 1992A. "On a Clear Day You Can See Progress." *BusinessWeek,* June 29: 104–106.

——. 1992B. "Global Innovation: Who's in the Lead?" *BusinessWeek,* August 3.

——. 1993A. "American Inventors Are Reinventing Themselves." *BusinessWeek,* January 18: 78–82.

——. 1993B. "The Case of the Catalytic Chemist." *BusinessWeek,* January 18: 80.

——. 1996. *The Invention That Changed the World.* New York, Simon & Schuster.

——. 1997. "Corporate Research: It's Alive." *Upside*, September: 118–132.

Burke, Connie. 1993. "A Fond Farewell to Bob Frosch." *Outlook*, May: 1–6.

Burrows, Peter. 1998. "Lew Platt's Fix-It Plan for Hewlett-Packard." *BusinessWeek*, July 13: 128–131.

Bush, Vannevar. 1945. *Science—The Endless Frontier*. Washington, DC, U.S. Government Printing Office.

BusinessWeek. 1994. "R&D Scoreboard." *BusinessWeek*, June 27: 81–103.

——. 1995. "R&D Scoreboard." *BusinessWeek*, July 3: 1–18.

Callon, Scott. 1995. *Divided Sun: MITI and the Breakdown of Japanese High-tech Industrial Policy, 1975–1993*. Stanford, California, Stanford University Press.

Campbell-Kelly, Martin, and William Aspray. 1996. *Computer*. New York, Basic Books.

Carey, John. 1997A. "The New Kings of the Lab." *BusinessWeek*, May 26: 170.

——. 1997B. "What Price Science?" *BusinessWeek*, May 26: 166–170.

Carnegie, Andrew. 1948. *The Autobiography of Andrew Carnegie*. Boston, Houghton Mifflin.

Chandler, Alfred D., Jr. 1977. *The Visible Hand*. Cambridge, MA, Belknap Press/Harvard University Press.

——. 1990. *Scale and Scope*. Cambridge, MA, Belknap Press of Harvard University Press.

Chase, Victor D. 1996. "On the Horizon: Plastic Semiconductors." *Bell Labs News*, November 15.

Chesbrough, Henry W., and David J. Teece. 1996. "When Is Virtual Virtuous? Organizing for Innovation." *Harvard Business Review*, January-February: 65–73.

Chiesa, Vittorio. 1996. "Strategies for Global R&D." *Research Technology Management*, September-October: 19–25.

Clark, Kim B. 1989. "What Strategy Can Do for Technology." *Harvard Business Review*, November-December: 94–98.

Clark, Don. 1996. "Microsoft Will Describe Chip Project to Upgrade PCs." *Wall Street Journal*, August 6: B4.

Cohen, Hirsh, Seymour Keller, et al. 1979. "The Transfer of Technology from Research to Development." *Research Management*, May: 11–17.

Cohen, I. Bernard. 1948. *Science, Servant of Man*. Boston, Little, Brown.

Collier, C. P., et. al. (1999). "Electronically Configurable Molecular-Based Logic Gates." *Science*, July 7: unnumbered reprint.

Comello, Vic. 1998. "Magnetic Storage Research Aiming at High Areal Densities." *R&D Magazine*, December: 14–19.

Conant, James B. 1951. *Science and Common Sense*. New Haven, CT: Yale University Press.

Cook, William J. 1998. "Men in Blue." *U.S. News & World Report*, February 16: 44–49.

Cooper, Franklin S. 1941. "Location and Extent of Industrial Research Activity in the United States." *Research—A National Resource, II—Industrial Research*, Report of the National Research Council to the National Resources Planning Board. Washington, DC, U.S. Government Printing Office: 173–187.

Corcoran, Elizabeth. 1992. "New Head of NEC Research Institute in US Hopes to Change Our View of Computers." *Nature* 360, 12 November: 97.

——. 1997. "Intel to Enlist US Scientists at Weapons Labs." *Boston Globe*, September 11: D1.

Coughlin, Kevin. 1998. "Quantum Leap." *The Star Ledger* (Newark), July 12: Section 3, p. 1.

Council on Competitiveness. 1998. *Going Global: The New Shape of American Innovation*. Washington, DC, Council on Competitiveness.

Coy, Peter. 1993A. "In the Labs, the Fight to Spend Less, Get More." *BusinessWeek*, June 28: 102–104.

——. 1993B. "The Global Patent Race Picks Up Speed." *BusinessWeek*, August 9: 57–58.

——. 1994. "What's the Word in the Lab? Collaborate." *BusinessWeek*, June 27: 78–80.

——. 1995. "Blue-Sky Research Comes Down to Earth." *BusinessWeek*, July 3: 78–80.

Daniels, George H., Ed. 1972. *Nineteenth-Century American Science: A Reappraisal.* Evanston, IL, Northwestern University Press.

David, Edward E., Jr. 1994. "Science in the Post–Cold War Era." *The Bridge,* Spring: 3–8.

Dordick, Rowan L. 1995. "50 Years of Research." *IBM Research,* 1–12 (special reprint).

——. 1998. "The Convenience of Small Devices." *IBM Research,* No. 3: 33–36.

Downey, Roger. 1996. "The prophet margin." *Eastsideweek,* July 31: 12–18.

Doyle, T. C. 1996. "What Makes HP Tick." *VARBusiness,* May 1: unpaginated reprint.

Draper, Joyce. 1990. "Area 11 Redefines Roles." *Bell Labs News,* July 31: 1.

Ebbesen, Thomas W. 1996. "Carbon Nanotubes." *Physics Today,* June: 26–32.

Eberl, Ulrich. 1998A. "The Key You Can't Lose." *Research and Innovation I.* 14–19.

——. 1998B. "Fantastic Journey." *Research and Innovation I.* 26–27.

——. 1998C. "Talking to Computers—With Your Hands." *Research and Innovation I.* 32–34.

Edelheit, Lewis S. 1995. "Renewing the Corporate R&D Laboratory." *Research Technology Management,* November-December: 14–18.

Edstrom, Jennifer, and Marlin Eller. 1998. *Barbarians Led by Bill Gates.* New York, Henry Holt.

Eidt, Clarence M., Jr., and Roger W. Cohen. 1997. " 'Reinventing' Industrial Basic Research." *Research Technology Management,* January-February: 29–36.

Eldred, Emmett W., and Michael E. McGrath. 1997. "Commercializing New Technology—I." *Research Technology Management,* January-February: 41–47.

Enos, John L. 1962. "Invention and Innovation in the Petroleum Refining Industry." In *The Rate and Direction of Inventive Activity.* Princeton, Princeton University Press: 299–321.

Erickson, Jim. 1996. "Deep Thinkers Still Digging for Friendly Computer." *Seattle Post-Intelligencer,* August 28.

Erker, Paul. Undated. "Forschung und Entwicklung in der Transistortechnologie. Entscheidungszwänge und Handlungsspielräume am Beispiel Siemens und Philips, 1947–1960." Freie Universität Berlin.

Farre, Tom. 1996. "Hewlett-Packard: A Giant You Can Count On." *VARBusiness,* July 15: 54.

Feldenkirchen, Wilfried. 1994. *Werner von Siemens: Inventor and International Entrepreneur.* Columbus, Ohio State University Press.

——. 1995. *Siemens: 1918–1945.* Munich/Zurich, Piper.

——. 1997. *Siemens: Von der Werkstaff zum Weltunternehmen.* Munich/Zurich, Piper.

Feldtkeller, Ernst, and Herbert Goetzeler. 1994. *Pionere der Wissenschaft bei Siemens.* Munich, Publicis MCD Verlag.

Florida, Richard. 1998. "Other Countries' Money." *Technology Review,* March-April: 29–36.

Fowler, Alan B. 1995. "What Has Happened to Research at Industrial Laboratories?" Draft of article for *APS News,* December.

Freundlich, Naomi. 1991. "Getting Everybody into the Act." *BusinessWeek,* October 25: 149–152.

Friedrich, H., and F. S. Becker. 1991. "The European Microelectronics Industry and Its Challenge for the Nineties." *International Symposium on VLSI Technology, Systems and Applications, Proceedings,* Taipei, Taiwan.

Garud, Raghu, Praveen Rattan Nayyar, et al., Eds. 1997. *Technological Innovation: Oversights and Foresights.* Cambridge, England, Cambridge University Press.

General Motors Research and Development Center. 1995. *The General Motors Research and Development Center: 75 Years of Inspiration, Imagination and Innovation.* General Motors Corporation.

Goldberg, Andrew V., and Peter N. Yianilos. 1997. *Towards an Archival Intermemory.* Technical report #97-149. Princeton, NJ, NEC Research Institute.

Gomes, Les. 1998. "Microsoft Finds Windows CE Is No Juggernaut." *Wall Street Journal*, July 1: B1.

——. 1998. "Bigger and Smaller." *Wall Street Journal*, November 16: R6.

Gomory, Ralph E. 1983. "Technology Development." In *The Race for the New Frontier: International Competition in Advanced Technology*. New York, National Academy of Sciences/Simon & Schuster: 81–94.

——. 1989A. "From the 'Ladder of Science' to the Product Development Cycle." *Harvard Business Review*, November-December: 99–105.

——. 1989B. "Moving IBM's Technology from Research to Development." *Research Technology Management*, November-December: 27–32.

Goodstein, David. 1993. "Scientific Ph.D. Problems." *The American Scholar* 62(2), Spring: 215–220.

——. 1995. "Peer Review after the Big Crunch." *American Scientist*, September-October: 401–402.

Gorowitz, Bernard, Ed. 1981. *A Century of Progress: The General Electric Story*. Schenectady, NY, Hall of History Foundation.

Graham, Margaret B. W. 1985A. "Industrial Research in the Age of Big Science." In *Research on Technological Innovation, Management and Policy*, Vol. 2, R. S. Rosenbloom, Ed. Greenwich, CT, JAI Press: 47–79.

——. 1985B. "Corporate Research and Development: The Latest Transformation." *Technology in Society* 7: 179–195.

——. 1986. *RCA and the VideoDisc: The Business of Research*. Cambridge, England, Cambridge University Press.

Greenberg, D. S. 1966. "Basic Research: The Political Tides Are Shifting." *Science* 152, 24 June: 1724–1726.

Griliches, Zvi. 1962. "Comment." In *The Rate and Direction of Inventive Activity*. Princeton, Princeton University Press: 346–353.

Gross, Neil. 1992A. "Inside Hitachi." *BusinessWeek*, September 28: 92–100.

——. 1992B. "Where Pure Science Calls the Shots." *BusinessWeek*, September 28: 99.

Guerlac, Henry E. 1987. *Radar in World War II*. Los Angeles, Tomash Publishers; New York, American Institute of Physics.

Gumbel, Peter, and Thomas F. O'Boyle. 1986. "Siemens's Beckurts Guided Firm into Crucial New Areas." *Wall Street Journal—Europe*, July 21: 8.

Gwynne, Peter. 1997A. "Skunk Works, 1990s-style." *Research Technology Management*, July-August: 18–23.

——. 1997B. "Fresh Light Shines on Lucent." *R&D Magazine*, May: Unpaginated reprint.

——. 1998. "Resurrecting Big Blue." *R&D Magazine*, April: 50–54.

Haber, L. F. 1958. *The Chemical Industry During the Nineteenth Century*. Oxford, Oxford at the Clarendon Press.

——. 1971. *The Chemical Industry 1900–1930*. Oxford, Clarendon Press.

Hamilton, David P. 1999A. "H-P Net Soars 34%, Eclipsing Forecasts." *Wall Street Journal*, May 18: A3.

——. 1999B. "H-P's Net Rises 37%, Exceeds Expectations." *Wall Street Journal*, August 17: A3.

Hamilton, David P., and Rebecca Blumenstein. 1999. "H-P Names Carly Fiorina, a Lucent Star, to Be CEO." *Wall Street Journal*, July 20: B1.

Hammond, William F., Jr. 1994. "GE Scientists Take Practical Approach." *Schenectady Sunday Gazette*, October 2.

Hardy, Quentin. 1999. "Motorola's New Research Efforts Look Far Afield." *Wall Street Journal*, June 17: B6.

Hauser, John R., and Florian Zettelmeyer. 1997. "Metrics to Evaluate R,D&E." *Research Technology Management*, July-August: 32–38.

Hawkins, Laurence A. 1950. *Adventure into the Unknown: The First Fifty Years of the General Electric Research Laboratory*. New York, William Morrow.

Hayes, Brian. 1995. "The Square Root of NOT." *American Scientist*, July-August.

Heath, James R., et al. 1998. "A Defect-tolerant Computer Architecture: Opportunities for Nanotechnology." *Science* 280, 12 June: 1716–1721.

Helm, Leslie. 1997. "Chasing the Holy Grail of Graphics." *Los Angeles Times*, August 4.

Henderson, Rebecca. 1994. "Managing Innovation in the Information Age." *Harvard Business Review*, January-February: 100–105.

Heppenheimer, T. A. 1996. "What Made Bell Labs Great." *Invention & Technology*, Summer: 46–56.

Hewlett, William. 1983. "Introduction." In *Inventions of Opportunity: Matching Technology with Market Needs. Selections from the Pages of the Hewlett-Packard Journal.* Palo Alto, CA, Hewlett-Packard Company.

Hewlett-Packard Co. 1983. *Inventions of Opportunity: Matching Technology with Market Needs. Selections from the Pages of the Hewlett-Packard Journal.* Palo Alto, CA, Hewlett-Packard Company.

Hill, G. Christian, and Don Clark. 1998. "Motorola to Slash Staff, Take Big Charge." *Wall Street Journal*, June 5: A3.

Hoddeson, Lillian Hartmann. 1977. "The Roots of Solid-State Research at Bell Labs." *Physics Today*, March: 23–30.

———. 1980. "The Entry of the Quantum Theory of Solids into the Bell Telephone Laboratories, 1925–40: A Case-Study of the Industrial Application of Fundamental Science." *Minerva* 18(3): 422–447.

———. 1981. "The Discovery of the Point-Contact Transistor." *Historical Studies in the Physical Sciences* 12(1): 41–76.

Hoddeson, Lillian Hartmann, et al., Eds. 1992. *Out of the Crystal Maze.* New York, Oxford University Press.

Horn, Paul. 1997. "Creativity and the Bottom Line." *Financial Times*, November 17.

———. 1999. "Information Technology Will Change Everything." *Research Technology Management*, January-February: 42–47.

Hounshell, David A., and John Kenly Smith, Jr. 1988. *Science and Corporate Strategy.* Cambridge, England, Cambridge University Press.

Hounshell, David A. 1996. "The Evolution of Industrial Research in the United States." In *Engines of Innovation: U.S. Industrial Research at the End of an Era*, R. S. Rosenbloom and W. J. Spencer, Eds. Boston, Harvard Business School Press: 13–85.

———. 2000. "The Medium Is the Message, or How Context Matters: The RAND Corporation Builds an Economics of Innovation, 1946–1962." In *Systems, Experts, and Computers*, Thomas P. Hughes and Agatha Hughes, Eds. Cambridge, MA, MIT Press.

Hoy, Emma, Samer Takriti, et al. 1998. "Divide and Conquer—A Threshold Model." *Energy & Power Risk Management* 2(10): reprint.

Hughes, Thomas Park. 1964. *The Development of Western Technology Since 1500.* New York: Macmillan.

———. 1989. *American Genesis.* New York, Viking.

Ihde, A. J. 1957. "Chemical Industry, 1780–1900." *Journal of World History* IV(1): 957–984.

Industrial Research Institute. 1995. *IRI Strategic Planning Process for 1996–2000.* Washington, DC, Industrial Research Institute.

———. 1995. *R&D Trends Forecast for 1996.* Washington, DC, Industrial Research Institute.

———. 1996A. *IRI Position Statement on U.S. Economic and Technology Policy.* Washington, DC, Industrial Research Institute.

———. 1996B. *Industrial Research and Development Facts.* Washington, DC, Industrial Research Institute.

———. 1997. "Industrial Research Institute's R&D Trends Forecast for 1997." *Research Technology Management*, January-February: 11–15.

Jacobson, Gary, and John Hillkirk. 1986. *Xerox: American Samurai*. New York, Macmillan.

Jaffe, Adam. 1998. "The U.S. Patent System in Transition: Policy Innovation and the Innovation Process." Paper prepared for the Research Policy Symposium on Technology Policy. July.

Jenkins, George. 1989. "Venture Capital Is Cautious: Who Will Seed Startups?" *Harvard Business Review*, November-December: 117.

Jewkes, John, David Sawers, et al. 1958. *The Sources of Invention*. London, Macmillan.

Johnson, Marvin M. 1996. "Finding Creativity in a Technical Organization." *Research Technology Management*, September-October: 9–11.

Jordan, Brigitte. 1992. "New Research Methods for Looking at Productivity in Knowledge-intensive Organizations." Working paper.

Kaske, Karlheinz. Undated. *Forschung und Entwicklung im Weltunternehmen Siemens*. Publication undetermined.

Kaysen, Carl, Ed. 1996. *The American Corporation Today*. New York, Oxford University Press.

Kephart, Jeffrey O., Gregory B. Sorkin, et al. 1997. "Fighting Computer Viruses." *Scientific American*, November: 88–93.

Kettering, Charles F. 1961A. "Head Lamp of Industry." In *Prophet of Progress: Selections from the Speeches of Charles F. Kettering*, T. A. Boyd, Ed. New York, E. P. Dutton.

———. 1961B. " 'Research' Is a High-Hat Word." In *Prophet of Progress: Selections from the Speeches of Charles F. Kettering*, T. A. Boyd, Ed. New York, E. P. Dutton.

———. 1961C. "Invention and Inventors." In *Prophet of Progress: Selections from the Speeches of Charles F. Kettering*, T. A. Boyd, Ed. New York, E. P. Dutton.

Kevles, Daniel J. 1972. "On the Flaws of American Physics: A Social and Institutional Analysis." In *Nineteenth-Century American Science: A Reappraisal*, G. H. Daniels, Ed. Evanston, IL, Northwestern University Press: 133–151.

———. 1975. "Scientists, the Military, and the Control of Postwar Defense Research: The Case of the Research Board for National Security, 1944–46." *Technology and Culture* 16, January: 20–47.

———. 1979. *The Physicists: The History of a Scientific Community in Modern America*. New York, Vintage.

Killeffer, D. H. 1948. *The Genius of Industrial Research*. New York, Reinhold Publishing.

King, R. W., and G. C. Southworth. 1924. "The Meaning of Research to the Telephone Investor." *Bell Telephone Quarterly* III(2): 65–75.

Klamann, Edmund. 1992. "NEC's Corporate 'Ambassador' Spreads Gospel of Globalism." *Nikkei Weekly*, October 5: 4.

Klein, Burton H. 1962. "The Decision Making Problem in Development." In *The Rate and Direction of Inventive Activity*. Princeton, Princeton University Press: 477–497.

Kline, Ronald R. 1992. *Steinmetz, Engineer and Socialist*. Baltimore, The Johns Hopkins University Press.

Knobel, Lance. 1998. "Nathan's Law." *worldlink*, May/June.

Kobayashi, Koji. 1991. *The Rise of NEC*. Cambridge, MA, Blackwell Business.

Kodama, Fumio. 1998. "Industry Creation and Technical Evolution." International Scientific Panel Meeting, Manchester, England.

Kolata, Gina. 1995. "High-tech Labs Say Times Justify Narrowing Focus." *New York Times*, September 26: C1, C7.

Kostoff, Ronald N. 1997. "Identifying Research Program Technical Risk." *Research Technology Management*, May-June: 10–12.

Kuhn, Thomas S. 1962. "Comment." In *The Rate and Direction of Inventive Activity*. Princeton, Princeton University Press: 450–457.

Kuznets, Simon. 1962. "Inventive Activity: Problems of Definition and Measurement." In *The Rate and Direction of Inventive Activity*. Princeton, Princeton University Press: 19–42.

Lampman, Dick. 1998. "The Future of Computing." IDC Asia/Pacific IT Forum, Sydney, Australia, HP Labs Web page.

Landauer, Rolf. 1997. "Fashion in Science and Technology." Draft of Opinion article supplied to author.

Larson, Charles F. 1995. *R&D in Industry. Research and Development FY 1996*. Washington, DC, American Association for the Advancement of Science. AAAS Report 20: 33–38.

———. 1996. *R&D in Industry. Research and Development FY 1997*. Washington, DC, American Association for the Advancement of Science. AAAS Report 21: 31–36.

Layton, Edwin. 1972. "Mirror-Image Twins: The Communities of Science and Technology." In *Nineteenth-Century American Science: A Reappraisal*, G. H. Daniels, Ed. Evanston, IL, Northwestern University Press: 210–230.

Lengyel, Jed. 1998. "The Convergence of Graphics and Vision." *Computer*, July: 46–53.

Lenzner, Robert. 1995. "The Reluctant Entrepreneur." *Forbes*, September 11: 162.

Leslie, Stuart W. 1983. *Boss Kettering*. New York, Columbia University Press.

Lester, Richard K. 1998. *The Productive Edge*. New York, W. W. Norton.

Levine, Daniel S. 1996. "Famous Lab Redesigns Itself to Give Tech a Human Face." *San Francisco Business Times*, March 15–21: 2A.

Liebling, Herman I. 1962. "Comment on Sanders and Kuznets." In *The Rate and Direction of Inventive Activity*. Princeton, Princeton University Press: 85–90.

Linke, R. A., et al. 1998. "Holographic Storage Media Based on Optically Active Bistable Defects." *Journal of Applied Physics* 83(2), January 15: 661–673.

Lohr, Steve. 1997. "Microsoft Plans to Establish a Research Lab in England." *New York Times*, June 18: D1.

———. 1998. "I.B.M. Opens the Doors of Its Research Labs to Surprising Results." *New York Times*, July 13: D1.

Lohr, Steve, and John Markoff. 1998. "Computing's Next Wave Is Nearly at Hand." *New York Times*, December 28: D1.

MacPherson, Kitta. 1990. " 'Hothouse' for Science." *The Star-Ledger* (Newark), May 3: 19.

Malone, Michael S. 1995. "Can Silicon Graphics Hold Off Hewlett-Packard?" *Fortune*, October 30: 119–126.

Maney, Kevin. 1998. "IBM Labs Develop New Mark of Distinction." *USA Today*, January 15: reprint.

Markoff, John. 1995. "Microsoft Quietly Puts Together Computer Research Laboratory." *New York Times*, December 11: D1.

———. 1996. "Microsoft Plans 300% Increase in Spending for Basic Research in 1997." *New York Times*, December 9.

———. 1997. "A Race to Catch a Japanese Star on Blue Lasers." *New York Times*, October 27: D1.

———. 1998A. "Disk-drive Madness: How Far Can It Go?" *New York Times*, February 26: 10.

———. 1998B. "I.B.M. to Introduce Disk Drive of Tiny Size and Big Capacity." *New York Times*, September 9: C2.

Marschak, Thomas A. 1962. "Strategy and Organization in a System Development Project." In *The Rate and Direction of Inventive Activity*. Princeton, Princeton University Press: 509–548.

Marshall, A. W., and W. H. Meckling. 1962. *Predictability of the Costs, Time, and Success of Development*." In *The Rate and Direction of Inventive Activity*. Princeton, Princeton University Press: 461–475.

McAuliffe, Kathleen. 1995. "The Undiscovered World of Thomas Edison." *Atlantic Monthly*, December: 80–93.

McHugh, Josh. 1998. "For the Love of Hacking." *Forbes*, August 10: 94–100.

Mees, C. E. Kenneth, and John A. Leermakers. 1950. *The Organization of Industrial Research.* New York, McGraw-Hill.

Mellon Institute. 1971. "Mellon Institute Through the Years," *Mellon Institute News,* June.

Meyer-Thurow, Georg. 1982. "The Industrialization of Invention: A Case Study from the German Chemical Industry." *Isis* 73: 363–381.

Mieszkowski, Katharine. 1998. "Barbara Waugh." *Fast Company,* December: 146.

Miller, Howard S. 1972. "The Political Economy of Science." In *Nineteenth-Century American Science,* G. H. Daniels, Ed. Evanston, IL, Northwestern University Press: 95–112.

Miller, Joseph A. 1997. "Discovery Research Re-emerges in DuPont." *Research Technology Management,* January-February: 24–28.

Minasian, Jora R. 1962. *The Economics of Research and Development.* Princeton, Princeton University Press: 93–141.

Mitchell, Russ. 1998. "Don't Cry for Me, Janet Reno." *U.S. News & World Report,* June 1: 40–45.

Moore, Gordon E. 1994. "The Accidental Entrepreneur." *Engineering & Science,* Summer: 23–30.

———. 1996. "Some Personal Perspectives on Research in the Semiconductor Industry." In *Engines of Innovation: U.S. Industrial Research at the End of an Era,* R. S. Rosenbloom and W. J. Spencer, Eds. Boston, Harvard Business School Press: 165–174.

———. 1998. "The Role of Fairchild in Silicon Technology in the Early Days of 'Silicon Valley.' " *Proceedings of the IEEE* 86(1): 53–62.

Morris, Charles R., and Charles H. Ferguson. 1993. "How Architecture Wins Technology Wars." *Harvard Business Review,* March-April: 86–96.

Mowery, David C., and Richard R. Nelson. 1996. "The U.S. Corporation and Technical Progress." In *The American Corporation Today.* C. Kaysen, Ed. New York, Oxford University Press: 187–241.

Mowery, David C., and David J. Teece. 1996. "Strategic Alliances and Industrial Research." In *Engines of Innovation: U.S. Industrial Research at the End of an Era,* R. S. Rosenbloom and W. J. Spencer, Eds. Boston, Harvard Business School Press.

Mueller, Willard F. 1962. "The Origins of Basic Inventions Underlying Du Pont's Major Product and Process Innovations, 1920 to 1950." In *The Rate and Direction of Inventive Activity.* Princeton, Princeton University Press: 323–353.

Myers, Mark B. 1996. "Research and Change Management in Xerox." In *Engines of Innovation: U.S. Industrial Research at the End of an Era,* R. S. Rosenbloom and W. J. Spencer, Eds. Boston, Harvard Business School Press: 133–149.

Myers, Mark B., and Richard S. Rosenbloom. 1996. "Rethinking the Role of Industrial Research." In *Engines of Innovation: U.S. Industrial Research at the End of an Era,* R. S. Rosenbloom and W. J. Spencer, Eds. Boston, Harvard Business School Press: 209–228.

Myhrvold, Nathan. 1997. "What's the Return on Research?" *Fortune,* December 7.

———. 1998. "Supporting Science." *Science,* on-line version, October 23.

Narin, Francis. 1993. "Patent Citation Analysis: The Strategic Application of Technology Indicators." *Patent World,* April: 25–30.

Narin, Francis, Michael B. Albert, et al. 1992. "Technology Indicators in Strategic Planning." *Science and Public Policy* 19(6), December: 368–381.

———. 1993. "What Patents Tell You about Your Competition." *Chemtech,* February: 52–59.

Narin, Francis, and Anthony Breitzman. 1995. "Inventive Productivity." *Research Policy* 24: 507–519.

Narin, Francis, and Dominic Livastro. 1992. "Status Report: Linkage between Technology and Science." *Research Policy* 21: 237–249.

Narisetti, Raju. 1997. "IBM's Gerstner to Stay Five More Years." *Wall Street Journal*, November 21: A3.

———. 1998A. "IBM Wins 1,724 Patents for No. 1 Spot on '97 List, but Fruits of R&D Fall 8%." *Wall Street Journal*, January 12: B16.

———. 1998B. "IBM, Looking to Hot Wireless Market, Will Sell Digital Chip Mimicking TI's. *Wall Street Journal*, June 10: B6.

National Bureau of Economic Research. 1962. In *The Rate and Direction of Inventive Activity: Economic and Social Factors*. Princeton, NJ: Princeton University Press.

National Inventors Hall of Fame. 1995. *The National Inventors Hall of Fame*. Akron, OH, National Inventors Hall of Fame Foundation.

National Research Council. 1941. *Research—A National Resource*. Washington, DC, U.S. Government Printing Office.

———. 1983. *The Race for the New Frontier: International Competition in Advanced Technology*. New York, National Academy of Sciences/Simon & Schuster.

National Science Board. 1998. *Science and Engineering Indicators*. Arlington, VA, National Science Foundation.

NEC Corp. 1984. *NEC Corporation: The First 80 Years*. Tokyo, NEC Corporation.

Nee, Eric. 1993. "John Seely Brown: An Interview with Eric Nee." *Upside*, December: 24–34.

———. 1997A. "Rich McGinn." *Upside*, February: 82–86, 122–129.

———. 1997B. "What Have You Invented for Me Lately?" *Forbes*, July 28: 76.

Nelson, Richard R. 1962A. "Introduction." In *The Rate and Direction of Inventive Activity*. National Bureau of Economic Research. Princeton, Princeton University Press: 3–16.

———. 1962B. "The Link Between Science and Invention: The Case of the Transistor." In *The Rate and Direction of Inventive Activity*. National Bureau of Economic Research. Princeton, Princeton University Press.

Nevins, Allan, and Frank Ernest Hill. 1962. *Ford: Decline and Rebirth 1933–1962*. New York, Charles Scribner's Sons.

Nichols, Nancy A. 1994. "Scientific Management at Merck: An Interview with CFO Judy Lewent." *Harvard Business Review*, January-February: 89–99.

Norman, Donald A. 1998. *The Invisible Computer*. Cambridge, MA, MIT Press.

Odlyzko, Andrew. 1995. "The Decline of Unfettered Research." Unpublished paper supplied to author.

Ody, Penelope. 1999. "Customers Scan Their Palms Before Buying." *Financial Times*, on-line version, March 24.

Orr, Julian E. 1995. "Ethnography and Organizational Learning: In Pursuit of Learning at Work." In *Organizational Learning and Technological Change*, S. Bagnara, C. Zucchermaglio, and S. Strucky, Eds. New York and Berlin, Springer-Verlag.

Packard, David. 1995. *The HP Way*. New York, HarperBusiness.

———. c. 1995. *Bernard M. Oliver*. Palo Alto, CA, Hewlett-Packard.

Pawlak, Andrzej M. 1996. "Developing Technology with R&D Customers." *Research Technology Management*, September-October: 44–47.

Payson, Steven. 1998. "R&D as a Percent of GDP Is Highest in Six Years." National Science Foundation, Division of Science Resources Studies Data Brief. October 16.

Pearson, G. L., and Walter H. Brattain. 1973. "History of Semiconductor Research." *Proceedings of the IRE* 43(12): 1794–1806.

Pease, Arthur F. 1998A. "The Transparent Turbine." *Research and Innovation* I. 49–52.

———. 1998B. "Telemedicine: The Digital Transformation." *Research and Innovation* II. 7–8.

Peck, Merton J. 1962. "Inventions in the Postwar American Aluminum Industry." In *The Rate and Direction of Inventive Activity*. Princeton, Princeton University Press: 279–298.

Penzias, Arno. 1989. *Ideas and Information*. New York, W. W. Norton.

———. 1995. *Harmony: Business, Technology & Life After Paperwork*. New York, Harper-Business.

Phillips, J. C. 1996. "Past and Future of Academic and Industrial Research in Solid State Physics." Paper prepared for European Physical Society address.

Pisano, Gary P., and Steven C. Wheelwright. 1995. "High-tech R&D." *Harvard Business Review*, September-October: 93–105.

Plafker, Ted. 1997. "Western Companies Go Slow on China R&D Operations: Quality, Intellectual Property Are Concerns." *Research Technology Management*, May-June: 2–3.

Port, Otis. 1996A. "The Silicon Age? It's Just Dawning." *BusinessWeek*, December 9: 148–152.

———. 1996B. "Whose Brainchild Was the Brain Chip?" *BusinessWeek*, December 9: 152.

———. 1997. "Xerox Won't Duplicate Past Errors." *BusinessWeek*, September 29: 98–99.

Porter, Michael E., and Scott Stern. 1998. "Evaluating U.S. Innovative Capacity: The Innovation Index." Presentation given at the National Innovation Summit, Cambridge, MA. March 13.

———. 1999. *The New Challenge to America's Prosperity: Findings from the Innovative Index*. Washington, DC, Council on Competitiveness.

Preston, Richard. 1991. *American Steel*. New York, Prentice Hall.

Pruitt, Bettye H., and George David Smith. 1986. "The Corporate Management of Innovation: Alcoa Research, Aircraft Alloys, and the Problem of Stress-corrosion Cracking. In *Research on Technological Innovation, Management and Policy*, Vol. 3, R. S. Rosenbloom, Ed. Greenwich, CT, JAI Press: 33–81.

Pugh, Emerson. 1995. *Building IBM: Shaping an Industry and Its Technology*. Cambridge, MA, MIT Press.

Pursell, Carroll. 1972. "Science and Industry." In *Nineteenth-Century American Science: A Reappraisal*, G. H. Daniels, Ed. Evanston, IL, Northwestern University Press: 231–248.

Rae-Dupree, Janet. 1999. "The Future of Technology." *San Jose Mercury News*, January 11: E1.

Ramo, Joshua Cooper. 1997. "Chips Ahoy." *Time*, October 6: 72

Ramo, Simon. 1989. "National Security and Our Technology Edge." *Harvard Business Review*, November-December, 115–120.

Rapaport, Richard. 1995. "What Does a Nobel Prize for Radio Astronomy Have to Do with Your Telephone?" *Wired*, April: 124–130, 174–178.

Raskin, A. H. 1956. "Key Men of Business—Scientists." *New York Times Magazine*, May 13: 15.

Reich, Leonard S. 1977. "Research, Patents, and the Struggle to Control Radio: A Study of Big Business and the Uses of Industrial Research." *Business History Review* 51: 208–235.

———. 1980. "Industrial Research and the Pursuit of Corporate Security: The Early Years at Bell Labs." *Business History Review* 54(4), Winter: 504–529.

———. 1983. "Irving Langmuir and the Pursuit of Science and Technology in the Corporate Environment." *Technology and Culture* 24: 199–221.

Reingold, Nathan. 1972. "American Indifference to Basic Research: A Reappraisal." In *Nineteenth-Century American Science: A Reappraisal*, G. H. Daniels, Ed. Evanston, IL, Northwestern University Press: 38–62.

Riordan, Michael, and Lillian Hoddeson. 1997. *Crystal Fire*. New York, W. W. Norton.

Robb, Walter L. Undated. Untitled manuscript detailing the history of cross-sectional imaging.

Roe, Mark J. 1996. "From Antitrust to Corporation Governance? The Corporation and the Law: 1959–1994." In *The American Corporation Today*, C. Kaysen, Ed. New York, Oxford University Press: 102–127.

Rosenbloom, Richard S., Ed. 1985. *Research on Technological Innovation, Management and Policy*, Vol. 2. Greenwich, CT, JAI Press.

——, Ed. 1986. *Research on Technological Innovation, Management and Policy*, Vol. 3. Greenwich, CT, JAI Press.

Rosenbloom, Richard S., and William J. Spencer. 1996A. "The Transformation of Industrial Research." *Issues in Science and Technology* 12(3): 68–74.

——, Eds. 1996B. *Engines of Innovation: U.S. Industrial Research at the End of an Era.* Boston, Harvard Business School Press.

——. 1996C. "Introduction: Technology's Vanishing Wellspring." In *Engines of Innovation: U.S. Industrial Research at the End of an Era*, R. S. Rosenbloom and W. J. Spencer, Eds. Boston, Harvard Business School Press: 1–9.

Roussel, Philip A., Kamal N. Saad, et al. 1991. *Third Generation R&D*. Boston, Harvard Business School Press.

Ryskin, A. I., et al. 1998. "Mechanisms of Writing and Decay of Holographic Gratings in Semiconducting CdF_2:Ga." *Journal of Applied Physics* 83(4), February 15: 2215–21.

Sanders, Barkev S. 1962. "Some Difficulties in Measuring Inventive Activity." In *The Rate and Direction of Inventive Activity*. Princeton, Princeton University Press: 53–77.

Schecter, Bruce. 1999. "Seeing the Light." *THINK RESEARCH*, No. 1: 10.

Schmookler, Jacob. 1957. "Inventors Past and Present." *Review of Economics and Statistics* 39(3), August: 321ff.

——. 1962A. "Comment to Kuznets." In *The Rate and Direction of Inventive Activity*. Princeton, Princeton University Press: 43–51.

——, 1962B. "Changes in Industry and in the State of Knowledge as Determinants of Industrial Invention." In *The Rate and Direction of Inventive Activity*. Princeton, Princeton University Press: 195–232.

Scigliano, Eric. 1999. "The Tide of Prints." *Technology Review*, January-February: 63–67.

Service, Robert F. 1997. "Patterning Electronics on the Cheap." *Science* 278(17), October: 383–384.

Shockley, William. 1957. "On the Statistics of Individual Variations of Productivity in Research Laboratories." *Proceedings of the IRE*, March: 279–290.

——. 1973. "The Invention of the Transistor—An Example of Creative-Failure Methodology." *Proceedings of Conference on the Public Need and the Role of the Inventor*, National Bureau of Standards.

Siegel, Irving H. 1962. "Scientific Discovery and the Rate of Invention." In *The Rate and Direction of Inventive Activity*. Princeton, Princeton University Press: 441–450.

Siemens AG. 1971. *Meilensteine—10 Jahre Forschungszentrum: Zentrale Entwicklung und Forschung.*

——. 1997A. *150 Years of Siemens*. Munich, Siemens AG.

——. 1997B. "Technology Trends in the 21st Century." *News and Views*, special newsletter 2, May 2.

Siemens, Georg. 1957A. *History of the House of Siemens*, Vol. 1. Freiburg/Munich: Karl Alber.

——. 1957B. *History of the House of Siemens*, Vol. 2. Freiburg/Munich: Karl Alber.

Sloan, Alfred P., Jr. 1964. *My Years with General Motors*. Garden City, NY, Doubleday.

Smith, Douglas K., and Robert C. Alexander. 1988. *Fumbling the Future: How Xerox Invented, then Ignored, the First Personal Computer*. New York, William Morrow.

Smith, Geoffrey. 1991. "A Warm Feeling Inside." *BusinessWeek*, October 25: 158.

Smith, George David, and Davis Dyer. 1996. "The Rise and Transformation of the American Corporation." In *The American Corporation Today*. C. Kaysen, Ed. New York, Oxford University Press: 28–73.

Smits, F. M., Ed. 1985. *A History of Engineering and Science in the Bell System: Electronics Technology (1925–1975)*. Indianapolis, AT&T Bell Laboratories.

Solow, Robert M. 1956. "A Contribution to the Theory of Economic Growth." *Quarterly Journal of Economics* 70(1), February: 65–94.

——. 1957. "Technical Change and the Aggregate Production Function." *Review of Economics and Statistics* 39(3), August: 312–320.

——. 1960. "Investment and Technical Progress." *Mathematical Methods in the Social Science*. Stanford, CA, Stanford University Press.

Spiegler, Marc. 1999. "Eye of the Storm." *Wired*, February: 105.

Steinberg, Steve G. 1998. "Germanium in Bloom." *Wired*, September: 85.

Stetson, Damon. 1956. "Curtice Bids U.S. Spur Technology." *New York Times*, May 17: 1.

Stevens, Greg A., and James Burley. 1997. "3,000 Raw Ideas = 1 Commercial Success." *Research Technology Management*, May–June: 16–27.

Stimson, Dorothy. 1948. *Scientists and Amateurs*. New York, Henry Schuman.

Stine, Charles M. A. 1941. "Fundamental Research in Industry." *Research—A National Resource, II—Industrial Research*, Report of the National Research Council to the National Resources Planning Board. Washington, DC, U.S. Government Printing Office: 98–107.

Strom, Stephanie. 1998. "NEC's Elder Statesman Resigns in Scandal." *New York Times*, October 24.

Stross, Randall E. 1997. "Mr. Gates Builds His Brain Trust." *Fortune*, December 8.

Swinbanks, David. 1996. "Japan to Hike Spending on S&T 50% by 2000." *Research Technology Management*, September-October: 5–6.

Takahashi, Dean. 1996. "Intel Shifts Its Focus to Long-term, Original Research." *Wall Street Journal*, August 26: B4.

——. 1998. "Showdown Looms over a Chip Giant's Power." *Wall Street Journal*, June 8: B1.

——. 1999A. "Intel Will Keep It Simple in FTC Case." *Wall Street Journal*, March 2: B6.

——. 1999B. "Intel Still Faces a Broader FTC Inquiry." *Wall Street Journal*, March 10: B4.

——. 1999C. "Intel Chiefs Shrug Off Information Appliances Threat." *Wall Street Journal*, August 12: B6.

——. 1999D. "With New Chip, Intel Launches Ambitious Bid for Web Market." *Wall Street Journal*, September 1: B1.

——. 1999E. "Intel Unveils a Prototype of Merced Chip." *Wall Street Journal*, September 1: B6.

Tanner, Ogden. 1996. *25 Years of Innovation: The Story of Pfizer Central Research*. Lyme, CT, Greenwich Publishing Group.

Teresko, John. 1996. "Rethinking the Business of Technology." *Industry Week*, July 1.

Trager, Louis. 1997. "Falling into Place." *San Francisco Examiner*, March 23: B-1.

Trendelenburg, Ferdinand. 1975. *Aus der Geschichte der Forschung im Hause Siemens*. Dusseldorf, Verein Deutsches Ingenieure.

Tristram, Claire. 1998. "The Big, Bad Bit Stuffers of IBM." *Technology Review*, July-August: 45–51.

Tsakiridou, Evdoxia. 1998A. "Shortcut to Reality." *Research and Innovation* I. 44–48.

——. 1998B. "Power Without Impedance." *Research and Innovation* I. 53–57.

——. 1998C. "Six Stars for Great Software." *Research and Innovation* II. 52–56.

Uchitelle, Louis. 1989. "U.S. Companies Lift R&D Abroad." *New York Times*, February 22: D2.

——. 1996. "Basic Research Is Losing Out as Companies Stress Results." *The New York Times*, October 8: A1.

Vagtborg, Harold. 1945. "Industrial Research Pattern of the United States." *Chemical and Engineering News*, November 10: 1943–1945, 2056.

——. 1975. *Research and American Industrial Development: A Bicentennial Look at the Contributions of Applied R&D*. New York, Pergamon.

Vandendorpe, Laura, Ed. 1997. *Basic Research White Paper* (special publication of *R&D Magazine*): 1–32.

Van der Meer, Wim, et al. 1996. "Collaborative R&D and European Industry." *Research Technology Management*, September-October: 15–18.

Van Tassel, David D., and Michael G. Hall, Eds. 1966A. *Science and Society in the United States*. Homewood, IL, Dorsey Press.

———. 1966B. "Introduction." In *Science and Society in the United States*, David D. Van Tassel and Michael G. Hall, Eds. Homewood, IL, Dorsey Press: 1–34.

Warrick, Earl L. 1990. *Forty Years of Firsts*. New York, McGraw-Hill.

Watson, Thomas J., Jr., and Peter Petre. 1990. *Father and Son & Co.: My Life at IBM and Beyond*. New York, Bantam.

Weber, Joseph. 1992. "Pure Research, Compliments of Japan." *BusinessWeek*, July 13: 136–137.

Weiser, Mark, and John Seely Brown. 1996. "The Coming Age of Calm Technology," Paper given to author.

Welker, Heinrich. 1976. "Discovery and Development of III–V Compounds." *IEEE Transactions on Electron Devices* ED-23(7): 664–673.

Werner, Bjorn M., and William E. Souder. 1997. "Measuring R&D Performance—U.S. and German Practices." *Research Technology Management*, May-June: 28–32.

Weyrich, Claus. 1997. "The Role of R&D in a Technology Enterprise." Address to the European Conference on Innovation and Technology: Growth Through Innovation. March 21.

White, Daniel. 1996. "Stimulating Innovative Thinking." *Research Technology Management*, September-October: 31–35.

Whiteley, Roger L., Alden S. Bean, et al. 1996. "Meet Your Competition: Results from the 1994 IRI/CIMS Annual R&D Survey." *Research Technology Management*, January-February: 1–8 (IRI reprint).

———. 1997. "Meet Your Competition: Results from IRI/CIMS Annual R&D Survey for FY '95." *Research Technology Management*, January-February: 16–23.

Wilke, John R., and Dean Takahashi. 1998. "Intel Is Hit with FTC Antitrust Charges." *Wall Street Journal*, June 9: A3.

Williams, R. Stanley. 1998. "Computing in the 21st Century: Nanocircuitry, Defect Tolerance and Quantum Logic." *Phil. Trans. Royal Society of London* A(356): 1783–1791.

Wise, George. 1980. "A New Role for Professional Scientists in Industry: Industrial Research at General Electric, 1900–1916." *Technology and Culture* 21: 408–429.

———. 1984. "Science at General Electric." *Physics Today*, December.

———. 1985. *Willis R. Whitney, General Electric, and the Origins of U.S. Industrial Research*. New York, Columbia University Press.

Worley, James S. 1962. "The Changing Direction of Research and Development Employment among Firms." In *The Rate and Direction of Inventive Activity*. Princeton, Princeton University Press: 233–251.

Xerox Palo Alto Research Center. 1999. *re:reading: Dispatches from the Gutenberg Galaxy*. Pamphlet produced by the Research in Experimental Documents Group, Xerox PARC.

Ziegler, Bart. 1997A. "Gerstner Slashed R&D by $1 Billion; for IBM, It May Be a Good Thing." *Wall Street Journal*, October 6: A1.

———. 1997B. "IBM Shuffles Jobs, Elevating PC Executive." *Wall Street Journal*, July 23: A3.

Zipkin, Ilan. 1998. "Gulliver in Lilliput." *BioCentury* 6(55), June 22: A1.

Zuckerman, Laurence. 1995. "From Mainframes to Global Networking." *New York Times*, December 18: D1–2.

———. 1997. "I.B.M. to Make Smaller and Faster Chips." *New York Times*, Sept. 22: D1.

INDEX

ABOUT THE AUTHOR

ROBERT BUDERI is the author of *The Invention That Changed the World* (Simon & Schuster, 1996). The former technology editor at *BusinessWeek,* he writes the popular "Lab Watch" column for *Upside* magazine and is a contributing editor at *Technology Review.* His articles have appeared in numerous national and international publications, including *Newsweek, Time, Sports Illustrated, The Economist, Science, Nature,* and *The Atlantic Monthly.* He served as an adviser to the BBC's recent television documentary series *Science at War* and is a past Vannevar Bush Fellow at M.I.T. He lives in Cambridge, Massachusetts.